TEMPEH

NATTO

& Other
Tasty Ferments

A Step-by-Step Guide to
Fermenting Grains and Beans

Storey Publishing

Kirsten K. Shockey & Christopher Shockey

Photography by Dina Avila

Foreword by David Zilber

The mission of Storey Publishing is to serve our customers by publishing practical information that encourages personal independence in harmony with the environment.

EDITED BY Deanna F. Cook and Sarah Guare
ART DIRECTION AND BOOK DESIGN BY Michaela Jebb
TEXT PRODUCTION BY Jennifer Jepson Smith
INDEXED BY Christine R. Lindemer, Boston Road Communications

COVER PHOTOGRAPHY BY © Dina Avila, except © Christopher Shockey, back (authors); © Dan Goldberg/Getty Images, front (top left); © Topic Images/Getty Images, front (top right)

INTERIOR PHOTOGRAPHY BY © Dina Avila

ADDITIONAL PHOTOGRAPHY BY © Brent Hofacker/Alamy Stock Photo, 182; © Christopher Shockey, 15, 96, 99; © Ed Wray/Getty Images, 139; © Kirsten Shockey, 373–374, 377, 379; Mars Vilaubi, 50, 111, 125, 155, 175, 253, 275

PHOTO STYLING BY Anne Parker

ILLUSTRATIONS BY © Olga Matskevich/stock.adobe.com, 8 and throughout; © tiff20/stock.adobe.com, 70–71, 302–303

TEXT © 2019 by Kirsten Shockey and Christopher Shockey

The information in this book is true and complete to the best of our knowledge. All recommendations are made without guarantee on the part of the author or Storey Publishing. The author and publisher disclaim any liability in connection with the use of this information.

Be sure to read all instructions thoroughly before using any of the techniques or recipes in this book and follow all safety guidelines.

Storey books are available at special discounts when purchased in bulk for premiums and sales promotions as well as for fund-raising or educational use. Special editions or book excerpts can also be created to specification. For details, please call 800-827-8673, or send an email to sales@storey.com.

Storey Publishing
210 MASS MoCA Way
North Adams, MA 01247
storey.com

Printed in China by Printplus Ltd.
10 9 8 7 6 5 4 3 2 1

Library of Congress Cataloging-in-Publication Data on file

Dedicated to Lyra and Finn, who wholeheartedly embraced sticky beans and fuzzy rice. And to all the children who dive into fermented foods with abandon because they understand instinctually that it is good food. The little humans and their parents of this time face tougher environmental challenges than our generation, who inherited a more polluted and played-out world than our parents. What we all worried about, you have to adapt to and hopefully begin to heal. The planet's future rests on your continued brave food choices. We hope this book plays a small part in that work.

CONTENTS

FOREWORD

BY DAVID ZILBER, COAUTHOR OF *THE NOMA GUIDE TO FERMENTATION*

"Fermentation and civilization are inseparable." These words, penned by the American poet John Ciardi, will stay with me forever.

I was 18 and still in high school when I first stepped into a professional kitchen, a forward-thinking and inventive restaurant whose style was self-described as *pan-Asian fusion* (it was the early 2000s, after all). I found myself surrounded by unusual and unfamiliar ingredients — fermented tofu, shiro shoyu, hatcho miso, natto, nam pla, and rice wine vinegar, to name but a few. These products were far from the mash-up of Caribbean and Jewish food culture I'd known at home. I learned of whole new worlds of flavor and toured many aisles inside dubious downstairs grocery stores in Chinatown, shopping for strange jars whose labels I couldn't read. It was a fascinating headfirst dive into the world of fermented food, and while I appreciated these ingredients for their novelty, functionally, I thought of them as just that: ingredients.

A decade later, after cooking my way through top restaurants, I landed at Noma and took yet another deep dive into the world of fermented food, but the approach this time was different. I had to leave my preconceptions at the door and relearn how to cook. At Noma, everything is considered . . . and then reconsidered. The very notion of edibility is challenged, and through that pursuit, new techniques and flavors are discovered and put to use. In the fermentation lab, ferments aren't just a set of novel ingredients; they're tools. Instead of using traditional fermented foods to remix recipes and create new and exciting dishes, as I had at the beginning of my career, I was remixing traditional fermentation *techniques* to create exciting ferments that would then be used to make even more progressive and exciting dishes. We do this to improve flavor, but other benefits are improved digestion and microbiotic health, and even a philosophically grander view of life. We are creating food on a deeper level than what has come before, and that deeper level can only happen through education and knowledge.

The Shockeys have put together a thorough and masterful work that builds on that same theme — that education and knowledge are essential. Though our daily lives are very different, in many ways, we work in parallel track. But if there's one thing I know about fermenters and their world, it's that they only ever want it to grow. They seek to share their microbial cultures, the bits and bobs, the know-how; and in doing so, they hope to enlarge the community of it all. While cooking might bring people together, fermentation weaves those people into a common tapestry. All life comes from life, all cells from cells, and all culture from culture. As the authors will explain, fermentation has persisted for so long because it has helped civilizations the world over to sustain themselves and thrive.

There's beauty in the idea that you have to share ferments to make them. Because you have to share knowledge, too, to create it. And within these pages, you'll find a wealth of it.

PREFACE

Food is a tool for exchange of culture.
DAVID ZILBER

We each grew up with the standard Western plate — meat, starch, and side of veggies. It varied for Kirsten when her family lived in Southeast Asia and the time her mom ventured off to hummus sandwiches before anyone else had heard of hummus, but those were fleeting. Christopher would eat a total of three vegetables when he met Kirsten, and one of those was iceberg lettuce, but only if it was drenched in blue cheese dressing. The other two — corn on the cob and potatoes — were closer to starches than vegetables, Kirsten argued. When we were in college, the one meal Christopher cooked was a Frito taco casserole; Kirsten made a lot of stroganoff or other meals that softened cheap round steak, but really, we ate mostly quesadillas.

Then a couple about five years older than us (*old* in our 20-year-old eyes) moved in next door. They were vegetarians. We started experimenting with our core meals, and honestly, we had no place to go but up. Our budget was limited, and at first, saving money by not buying meat — combined with enjoying more interesting flavors — was fun. Okay, we also enjoyed watching people squirm when we told them that the chocolate pie in their mouth was really tofu (it was the 1980s and tofu was still unconventional). As we went deeper into our vegetarian diet, we loved that we were eating better food that also had a smaller ecological footprint.

We don't remember when we transitioned from having a few vegetarian meals a week to being subscribers to *Vegetarian Times*, nor do we remember whether this shift was a gradual progression or a lofty declaration. We wore our vegetarian identity like young adults rebelling from the life they were raised with. Sometimes that made traveling difficult. Kirsten, who'd been raised as the daughter of an anthropologist (which meant accepting all food that is offered, for food is the host's greatest gift of self), felt conflicted: what was more important, self-imposed ideals or building relationships through meals? As Christopher worked worldwide in developing rural areas, he too was faced with this collision of culture. When he was invited to a meal where the household had fried their one piece of bologna for him, he ate it with gratitude. We began to see that most of the world does not have the luxury of grinding up 10 pounds of carrots to drink. Food, and protein especially, must be eaten as it is available.

These thoughts were cracks in our armor but by no means the end of our journey toward high-quality, ecologically sustainable food.

Ten years and four children later, pregnancy and nursing had depleted Kirsten; most notably, she was anemic. Wheat and soy took the edge off Kirsten's constant hunger, but they made her feel chronically achy and fatigued; she was gluten intolerant and suspected a soy intolerance as well. Because Kirsten needed protein, we added meat back into our diet. It was at this time that we moved to our homestead, where we could ethically and sustainably raise our own dairy and meat animals. Living as stewards of the land, our understanding of what was sustainable and ecological became more complex and nuanced (like our understanding of fermentation). We found that we could feed a family of six over the course of two years with the gift of life from one steer. We grew fruit and vegetables, milked cows and goats, made jam and cheese and cider, and began to ferment vegetables.

In deference to Kirsten's situation, we avoided soy at our table, except for naturally fermented soy sauce and the occasional miso or tempeh meal (fermented soy did not seem to have any negative effects for her). And then life had a beautiful (cosmically comical) way of making us grow: Storey Publishing asked us to write a book on fermenting legumes and grains — including soybeans. Our first thought was that we — who generally avoid soy — were not the people to write this, and then we realized that maybe our own relationship with soy was perfect.

A few things have happened while we wrote this book — we turned 50, our last child moved away, we ate more soy (fermented) than we had in the last 20 years, and we started growing rows and rows of heirloom beans. And we feel great. Our focus on good food and fermentation — even with a boost in our soy consumption — is treating us well. (We both took blood tests and microbiome samples, because you can now, and found no evidence of sensitivities or worrisome factors.)

We share our food story to show that it is a progression — a continuing fluid journey of trying to navigate feeling good and doing the best we can do. We would be lying if we said we didn't have moments of way too much dogma, but that rigidity has given way to an appreciation for balance. With this book, we hope to share a well-researched, unbiased perspective on fermented foods and how they can contribute to the health and wellness of both humans and the planet. We don't advocate any one diet, and we understand how confusing food choices are at this time. We don't want to add to the confusion but instead to help you understand, appreciate, and enjoy these superfoods just as much as we do.

INTRODUCTION

Food systems have the potential to nurture human health and support environmental sustainability; however, they are currently threatening both. Providing a growing global population with healthy diets from sustainable food systems is an immediate challenge.

THE EAT-LANCET COMMISSION ON FOOD, PLANET, AND HEALTH

We wrote this book to introduce you to foods, both ancient and ultramodern, that come from fermenting legumes and grains. Historically, grains are the most important food to humans, followed closely by legumes. These foods are inexpensive, easy to grow, and easy on the planet, and most of humanity still relies on them for the majority of their nutrition.

Here is the thing — when fermented, these humble ingredients are transformed, becoming more delicious and nutritionally charged. In fact, their benefits can go beyond basic nutrition to positive effects on our health in other ways, making them functional foods. The first step to appreciating what they have to offer is understanding how the microbes responsible for fermentation alter the foods right down to the molecules they are made of. Fermentation is all about breaking down the big molecules of proteins, starches, and fats. Why? Because as Harold McGee, author of *On Food and Cooking*, pointed out, big molecules don't have much flavor, but the smaller ones do. We see this when we turn grain into bread or beer, when we make wine out of fruit juices, when we turn cabbage into sauerkraut, and when we transform soy beans into miso. It is really that simple — fermentation makes flavor.

In the pages ahead, you will get to know and appreciate foods produced by the fermentation of legumes and grains, like miso, tempeh, natto, amazake, koji, and more. There are many reasons to get to know these foods. For some it is that flavor we just talked about, for many it is about eating more sustainable foods, and for all of us it should be about health: fermentation makes food a whole lot better for us. Even if you come to these foods only because you've heard that they're good for you, the flavors will surprise you; you will fall in love, and so will your gut — you may even begin to crave them.

We want to help you enjoy these foods regularly. Our hope is that you will be inspired to make or buy some miso, tempeh, natto, or amazake and get to know these foods. If you are lucky enough to have grown up with these fermented foods as part of your food heritage, we hope our recipes will provide a new twist to some of the standards and perhaps give you a reason to appreciate some long-forgotten flavors from your childhood. For everyone else, we hope you get curious about these foods and learn more. Pick a recipe that looks interesting and give it a try.

You can generally buy traditional ferments like soy sauce, miso, and tempeh at the grocery store, and you'll find recipes for using these foods to make delicious dishes in part III. For those of you who want to ferment these foods yourselves, we've got you covered. Whether you're an experienced fermenter or this is your first foray into the field, this book will give you comprehensive step-by-step instructions and troubleshooting sections. We not only include traditional recipes but also invite you to choose your own adventure by taking the concepts and using nontraditional ingredients to create unique and modern flavors. In short, it's time to play with your food.

It's About the Planet

"Eat food. Not too much. Mostly plants." With these words, food author Michael Pollan famously summed up his eating advice in his book *In Defense of Food: An Eater's Manifesto*, published in 2008. The next year chef Dan Barber, one of the originators of the farm-to-table culinary movement, was anointed one of *Time* magazine's 100 most influential people in the world. In the decade that followed, green smoothies, farm-to-table salads, and other trendy plant-based dishes featuring the super-food veggie du jour reigned supreme in the American food scene. In the middle of all this, however, Barber fired a shot over the revolution that he had championed by essentially exposing the mounting environmental costs to the planet of our A-list-vegetable desires. His radical solution? We all need to eat plants that are restorative to the farmer's soil, so that our appetites are good for the farms and the environment.

If you have ever planted a row of greens and nursed them through summer's heat, when all they really want to do is lie down under the scorching sun and die, you understand the intense labor required for growing many vegetables. To produce them again and again from the same land, farmers need to continually replenish their soil with nutrients because nearly all vegetables are takers — not givers — of nutrients from soil. When there was more land and time available, farmers could plant restorative nitrogen-fixing legumes in a field and give that field a year off from production. Today, when time and costs are paramount, feeding the soil is accomplished by bringing in fertilizers between each planting, and sometimes during the planting when fields are completely played out.

In 1967 there were 3.5 billion people on the planet. In 2018 there were well over 7.6 billion — that is, twice as many humans needing to be fed. During the same period, scientists estimate we lost one-third of our arable land to erosion or pollution. Much of the arable land in agricultural production today is being used to support the demands of our growing global meat consumption. Taken in composite, the need to feed rising populations with decreasing arable lands (which are increasingly used for animal production) is hardly sustainable.

There are things we can do. One of the best is to bring into our diets the economical, nutritious, soil-replenishing plants in the legume family. Like most plants, legumes pull nutrients from the soil as they grow, but they also pull nitrogen from the air and put it back into the soil, leaving it a bit more fertile for the next crop. They also contain high levels of vitamins, minerals, and protein. They are the nutritional workhorses of the plant world. In some cases, like with soybeans, they also have constituents that are not good for us when we digest them.

However, the process of fermentation reduces or entirely removes these antinutrients and toxins while also unlocking the nutrients for us.

Sounds great, right? It is! But many of us did not grow up eating these foods. We don't know how to make them, use them, or enjoy them. We don't crave them. That can change. And we can make a big difference for ourselves and our planet beginning with that first bite.

It's About Culture

What foods do you crave? Why do you crave them and not something else, perhaps something healthier? One answer is that we learn to love the foods we are given from a young age. Think about it. From your first bites of solid food through your rebellious toddler food fights and on to grade-school lunches, it's a journey that starts pretty much out of our control. The culture we were raised in, and the socialization we received from our families, told us what we should eat. Then we grow up and gain our freedom over this important decision we make multiple times a day. Or do we?

Leading science now suggests that our cravings come not from our heads but from our guts.[1] More specifically, from the microbes that live in our guts. It turns out these bacteria have their own cravings and use our 100 million nerve cells between our digestive tract and our brain to place their orders, all while we — their hosts — think it's us making that decision. Bacteria prefer to eat a wide range of foods, from the worst junk food to raw veggies not yet out of the ground.

Why do the bacteria in our gut matter? Because the food choices we make affect our gut, our health, and the health of our environment. If you have ever tried to change your diet only to revert back to your old habits in a few weeks or months, you know how daunting dietary changes can be. The good news is that if the science is true, our cravings mostly come

How Does Blood Sugar Work?

When we eat food, our bodies convert it into sugar (glucose), which goes to our cells to provide them with energy. Ideally, the food-to-energy coming in matches our body's need for fuel, but that's usually not the case. When we eat more food than we can use, producing more blood sugar than our cells need at that time, it gets stored for later use. The storage and the delivery to our cells are managed by the dynamic duo of our liver and pancreas. Our liver both stores and makes glucose. Our pancreas secretes the hormone insulin into our bloodstream, which helps ferry the sugars to the awaiting cells. The rate at which food is converted to glucose differs from food to food; some foods are quickly converted into glucose, spiking glucose levels in our blood, and others have a slower, more sustained rate of conversion.

Good fats, proteins, and fiber all help reduce the rate of glucose conversion. High-sugar and refined carbohydrate foods are quickly converted into sugar, demanding comparable high levels of insulin. High insulin levels over prolonged periods of time can cause our insulin to begin to lose its ability to convince our cells to accept the glucose. They become insulin resistant. Type 2 diabetes may result.

from our gut microbes. Gut microbes crave all kinds of things, good and bad. If you want to change what you are eating, change your gut. How can you do this? One way is by bringing more fermented foods into your diet.

It's About You

A number of major chronic diseases like obesity, type 2 diabetes, cardiovascular disease, and cancer share physiological aberrations and stressors — things like inflammation, oxidative stress, and alterations in the body's metabolism. Fermented legumes and grains have high levels of several bioactive compounds that appear to combat these conditions. Their actions are multifaceted and not easily reduced to simple if-you-eat-this-it-will-cure-that formulas. Our individual diets, microbiome composition, genes, and lifestyle choices all play a part as well. Still, there is mounting evidence from recent studies[2] that the consumption of fermented legumes is a contributing factor in the reduction of obesity, reduced risk of cardiovascular disease, lower cholesterol levels, improved calcium uptake in postmenopausal women, and improved glucose control and reduced insulin resistance in prediabetic and diabetic populations.

Fermentation techniques have been around for as long as humans have had too much food to eat all at once and needed to preserve it for another day. For some cultures, the day's diet is made up of barely 5 percent fermented foods, but for many cultures, as much as 40 percent of the daily diet comes from fermentation. That's just shy of 1 pound of cheeses, breads, pickles, hot sauces, tempeh, charcuterie, miso, beer, wine, cider, sake, salads, soups, fermented legumes, curries, stews, and yes, even fermented desserts consumed every day by over 7 billion people on this planet.

As we've discussed, we find these foods tasty because microbes in us think the same thing and signal us to eat those foods. With probiotics and prebiotics, we can "cultivate" our gut microbial population to support healthy dietary preferences. Studies show that we can make measurable microbial changes in just 24 hours.[3]

The live microbes we ingest, whether in fermented foods or in a probiotic pill, go on one of three possible journeys within us. Some pass right through us, like tourists checking off sites as they pass by them while speeding down the highway. Some slow down, stop, and stick around in our digestive system for a while, like those tourists who actually want to experience a place. The last group tests our tourist metaphor a bit because they become food for the microbes in our gut — their tour ends when they are consumed. This last group is referred to as prebiotics — food for probiotic microbes in us. If we think of probiotics like the schoolkids in the *Magic School Bus* series, magically shrinking in order to adventure through our digestive system, then prebiotics would be the lunches they pack for themselves for the journey.

It's About Flavor

It is an exciting time for food. The pendulum is swinging in favor of diverse flavors, small-batch production, local and wild sourcing, employing microbes and pairing ancient wisdom with science to unlock myriad tastes and techniques, and with them, nutrients. People all over the world are freely exchanging recipes and methods and layering on local traditions and ingredients to create entirely new foodways. (In this book, you will meet just a few of these pioneers.)

With this global exchange of information comes a new vernacular. What do we call these new foods? The dairy industry's insistence that the term *cheese* be used only on dairy cheeses and not on dairy-free cheese like "cheezes" is a contentious but very real example of this struggle over what to call "new" food. We experienced a similar struggle when trying to name our own recipes in this book and our two previous books, *Fermented Vegetables* and *Fiery Ferments*. At some point it made no sense to call everything we made using traditional sauerkraut techniques (dry brining) a sauerkraut (soured cabbage). For a while we used the term *kraut* to describe these foods, but that wasn't enough when we needed to describe the complex, rich, thick condiments that were fermented with a dry brine but in no way resembled a sauerkraut. The word *ferments* is taking the forefront now; it is perhaps way too broad and far from perfect, but it's a generally accepted name for fermented foods in all forms. In this book, we were stumped several times. What, for instance, is a sauce that is flavorful and full of character but not a soy-based sauce? What is a tasty paste that is made like miso but is in no way traditional miso?

While it may be hard to find the right names for these new ferments, one thing they all have in common is this: they are darn tasty. Deliciousness, of course, has driven humankind's food choices since the beginning. We describe the taste sensation of savory deliciousness with the Japanese word *umami*. It is the concept of the fifth taste, or fifth flavor (the others being sweet, sour, bitter, and salty). It denotes savory — think of rich broths, mushrooms, sautéed onions, aged cheeses, and so on — and it fulfills our deep, primal desire for satisfying food.

There is so much we could say about the neurology of taste (and we encourage you to explore the many fascinating books and articles on the subject), but for this book the important piece is that proteins and amino acids are the primary source of umami. When we ferment legumes and grain proteins, we are breaking down these proteins and freeing the amino acids to release flavors. One amino acid in particular, called glutamic acid, is umami's superpower. Along with taste receptors for sweet, sour, salty,

Core-Legume-Fringe

The concept of core-legume-fringe eating was developed by the late anthropologist Sidney Mintz. Mintz believed that the meals of people living in agrarian societies around the world were once basically the same, being constructed of three components: core carbohydrates, accompanying legumes, and fringe condiments to help them get through the blandness of the first two. The carbohydrates provided most of their calories and came from wheat, rice, corn, and millet — half the cast of chapter 3. The legumes provided the majority of their protein and fat and came from soybeans, chickpeas, lentils, black-eyed peas, and common beans — also

in chapter 3. The fringe were powerful flavor-enhancing condiments and sides. These tasty, small-but-robust sides were often fermented for higher nutrition, flavor, and preservation. In Asia, these included many of the legume-based ferments in this book: misos, soy sauces, tempeh, and natto.

Mintz went on to say that our diets become socially reinforced through habit; we come to expect our diet to always be the way it always was. When things around us change, our diets lag. We can be reluctant to try new foods to fill these core-legume-fringe categories. We hope to help change that pattern with this book.

and bitter, we have one for umami. It lights up when in the presence of glutamate, a salt formed from glutamic acid. When glutamic acid is bound up in a protein, we cannot detect it, so it has no flavor to us. To be tasted, it needs to be liberated from protein through a process like fermentation, which allows the acid to combine with other compounds that can be dissolved in our mouths and enjoyed as umami flavor. Enter the ferments in this book . . .

All of the ferments (in a very generic sense) have amazing, albeit very different, umami qualities. The toasted nuttiness of natto has big umami and very little in common with the velvety smooth, sweet, textural taste of white miso, which doesn't offer the same sensory pleasures as its longer-aged cousin red miso. Tempeh, nukadoko-fermented pickles, and koji-aged meats all share umami, yet this descriptor, which only recently entered Western vocabulary, can fall short of adequately capturing the boldness, intensity, and savory quality of these dishes. Rich Shih of Our Cook Quest, a culinary education website (see page 285), likes the term *umptious* to describe these flavors that leave a residual texture lingering on the tongue. Celebrity chef David Chang uses the word "Hozon," but at this writing it hasn't become the general term. In these pages, taking the lead from chef Jeremy Umansky (see page 184), we are going to refer to foods built on the backbone of miso as "amino pastes" or our own moniker, "tasty pastes."

It's About the Journey

If there were an elephant in the room, er . . . book . . . it is that the recipes in this book are almost plant-based, but not entirely. We aren't telling you not to eat meat, nor are we saying that you *should* eat meat; that is a personal decision. We think being plant-centered — that is, taking meat off the center and putting it on the side of the plate as a savory fringe — makes the most sense for us. A young man told us recently that he thought that opinions and judgments about people's food choices were becoming just as polarizing as politics and race. We do not wish to contribute to this polarization. This book is not about judgment.

This book is about flavor, and about ways to eat a little healthier for you and the planet. It is not about preaching; it is about feeling good in a busy and confusing world. It is about playing with your food.

Food is life. We are what we (with "we" being ourselves and our microbes) eat.

Food is incredibly personal, and so incredibly social. It is what binds us; we all need to eat. When we break bread with strangers, we leave the table friends.

In closing, foods — both ingredients and techniques — travel with people. Remember the telephone game, in which the story passes from one teller to the next, changing and evolving with each person's understanding until the story barely resembles the one that was first told? Food makes the same journey through kitchens. The form, presentation, and ingredients of our favorite dishes transform as they move down the road. Many of the ferments in this book are rooted in ancient China yet are solidly Korean, Japanese, or Indonesian now. These particular foods are more important than ever. We think that each and every one of us holds a piece of the puzzle to solve some of the challenges we all share as residents of one planet. This work seeks to honor ancient food traditions, while introducing and demystifying food that is beyond delicious, is lighter on the planet, and has the potential to alleviate some of the health crises many of us face. We believe that these foods deserve a wider audience. We want this to be the beginning of a conversation, a sharing, and a journey.

PART 1

Learning

This is a tale of two families: grasses and legumes. There are thousands of different species in each family, but we are going to zero in on the plants that humans have been consuming for thousands of years and fermenting for nearly as long.

The grass family (Poaceae) includes not only the grasses of your lawn but also those grasses grown for their edible grains — that is, cereal grains. The legume family (Fabaceae) includes beans, peas, and lentils.

While grain fermentation has been utilized worldwide for a long time, the same cannot be said for legume fermentation, which has been all but absent in much of the world until recent years. Why? It's hard to say. Modern Western cuisine has a long (melting pot) tradition of borrowing and modifying exotic foods and flavors — think bagels, pizza, or quesadillas. Could miso, tempeh, and natto be next?

CHAPTER 1

Fermentation Fundamentals

Whether you are new to fermentation or a veteran of sauerkrauts, pickles, and kombucha, legume and grain fermentation is in a league of its own. The cast of microbes is more diverse, their interplay is more complex, and the potential flavors they can create is richer than with other fermentation techniques. In this chapter, we will survey the state of food fermentation to put the techniques of this book in perspective.

Transformation through Collaboration

Microbes are with us all the time. They are with our food all the time. Okay, let's be honest, they are really running the show here on Earth, never resting, always eating, splitting, and transforming. We are only going to scratch the surface of the microbial world by discussing a very few of the players that affect our food: yeasts, bacteria, and molds. At the most basic level, fermentation happens when these yeasts, molds, and bacteria begin to break down food.

Let's back up a moment. What is fermentation? Fermentation is the chemical breakdown of a substance by bacteria, yeasts, or other microorganisms, often resulting in effervescence and the release of heat. In this book, we will expand the definition of fermentation to

include reactions induced by microorganisms or enzymes that split complex organic compounds into simpler substances. Put more plainly, these fermentation microorganisms — be they bacteria or yeasts or molds — digest these foods first and what is left after they are finished is more easily digested by the microbes in our bodies. The fermentation microbes do some of the heavy lifting so that the microbes in our gut can focus on taking up the vitamins and minerals for our bodies to use.

As you will learn, some foods are transformed by the cooperative effort of a complex community comprising all three types of microbes. The CliffsNotes (so you can sound smart in casual conversation) are as follows: Yeasts digest carbohydrates, including sugars, and make alcohol in the process. When we're using yeasts to ferment foods, sometimes we stop the yeasts before they make alcohol (as in the case of bread), sometimes that alcohol shows up in the end product (such as wine), and if allowed to continue the alcohol is just a player along the way. When you're making vinegar, for example, you use yeast to transform juice into alcohol and then let bacteria take over to turn the alcohol into vinegar.

Bacteria take center stage in many fermentations. To continue the example above, vinegar makers love acetic bacteria and vintners avoid it like the plague because they cannot allow the microbes to complete their natural progression. Lactic acid bacteria is used to make cheese, yogurt, charcuterie, sauerkraut, kimchi, and pickles. These bacteria acidify foods and lower the pH in the process of fermentation, usually through a series of lactic acid teams that take their respective turns in lowering the food's pH, creating an environment that is inhospitable to undesirable bacteria that we don't want in our foods. Some other kinds of fermentation bacteria work from the opposite end of the pH scale, driving up the pH and alkalizing food rather than acidifying it.

Molds, which are a type of fungi, are multicellular and characterized by growing in a cobweb of filaments (often called a filamentous fungi). They are also a tool in the culinary artist's toolbox. You're probably familiar with what mold looks like, but we're talking here about *edible* molds, many of which are highly prized in culinary traditions. We will introduce you to culinary molds to make tempeh and koji, not unlike the manner in which cheesemakers use mold (*Penicillium candidum*) to make Brie, in which the mold produces not only the tasty rind but also enzymes that soften the curd to the creamy texture. Tempeh is a one-step mold

fermentation. Once inoculated with mold spores (*Rhizopus*) and incubated, the resulting tempeh is the end product.

Koji is the Japanese name for the fungus *Aspergillus oryzae*. In contrast to rhizopus, koji spores are only the first step for creating many diverse ferments. Koji is grown on soy, rice, barley, and other legumes and cereals. Remember our yeasty friends that happily turn fruits into alcohol? Well, they don't do so well with grains and legumes because, unlike fruits, these foods don't have a ready amount of sugar for the yeasts. When koji grows, it releases enzymes that break down starches into sugars (a process called saccharification) like sucrose, which both yeasts and bacteria gladly consume. These more complex sugars are then further broken down into simpler glucose and fructose. Glucose is pure energy for bacteria and for humans. Koji sets the stage for ferments that are wonderfully complex with likely dramatic interactions between enzymes, yeasts, molds, and bacteria.

If the saccharification process sounds familiar to you, maybe it's because you know something about a more Western version of this process: making beer or whiskey from malt, or germinated cereal grains. Just like koji, sprouting the grain creates natural enzymes, and their role is to break down the starches in the grain into simpler sugar forms. In the final stages of malting, the grains are heated to a temperature high enough to stop the sprouting process while preserving the enzymes.

Many traditionally fermented foods have become victims of industrialization, subject to production shortcuts that reduce the time it might take microbes to transform our food from fresh perishable ingredients into delicious stable comestibles. Industrialized methods include

Top Four Roles of Fermentation

1. **Flavor.** As we will see later, fermentation changes the flavors, aromas, and textures of the food.

2. **Preservation.** Whether it be through lactic acid, acetic acid, or alkaline fermentation, in the end we ferment foods to prevent spoilage.

3. **Enrichment.** Fermentation improves the nutritional value of cereal grains, legumes, and vegetables by doing such things as increasing the protein content or increasing the availability of essential amino acids, essential fatty acids, and vitamins.

4. **Detoxification.** "Detoxing" our bodies has become a fashionable undertaking complete with celebrity-touted programs, but our bodies are actually designed to eliminate toxic wastes just fine on their own, thank you very much. Fermented foods repair and feed our gut biota, giving our natural detoxification abilities a needed boost.

adding enzymes, boosting temperature to speed up the process, or skipping fermentation altogether and achieving flavor through highly processed sugars or additives. However, we are now in a time of rediscovery and a food renaissance as we come full circle, embracing the microbes we so recently denigrated as "germs." Because although Louis Pasteur explained a lot about the microbes to avoid, we are learning now that "germ theory" isn't the full story.

Not All Fermented Foods Are Probiotic

In our fervor to embrace fermented foods, especially as probiotic, there is often some confusion about fermented foods that contain probiotics versus those that do not. Probiotics are *live* microorganisms, usually bacteria, that when eaten can help maintain or restore the beneficial bacteria we need in our digestive tract. Many fermented foods are rife with live probiotics. When we consume these foods, the microbes that can make it past our stomach acids (and we need to be honest here that the percentage is small) reach our gut — the center of our microbiome. Our microbiome is the entire community of microorganisms that live in and on our bodies. Understanding how these microbes (and there are many) that make up our microbiome affect our well-being is currently the most mind-boggling and exciting research on understanding our health. Even though relatively few microbes make it "still alive" into the gut, research keeps showing that eating foods rich in probiotics is good for our health, though we don't completely understand how it all plays out.

However, fermented foods are not beneficial to human health solely due to their probiotics. They are also important because fermentation transforms food in a way that makes it more digestible. In some cases, it even makes foods that were inedible now edible. Think about fermented foods that do not contain probiotics, like sourdough bread, coffee, and chocolate. These foods are produced by fermentation of their raw ingredients, but at some point in their production, they undergo processing — often the application of heat — that destroys any microbes in them. These foods are healthier, more digestible, and tastier because of the actions of the microbes, but they no longer contain live, active probiotics. (We feel you — oh for a world where a strong cup of coffee and a big chocolate bar provided us all the probiotics our bodies needed for the day. It just doesn't work that way.)

Tempeh, for example, does not have live probiotics yet it is touted as a food that helps strengthen digestion. It is not fully clear why this is so, but scientists agree that tempeh

Origins of Starch and Legume Fermentation

Shen Nung was a legendary mythical emperor who purportedly reigned from 2898 to 2698 BCE, which if you are counting is two hundred years. He is sometimes referred to as the god of agriculture. Age and celestial placement aside, he is credited with introducing the Chinese people to the five sacred grains. The term *wu ku,* "five grains," is believed to be the oldest recorded classification for grain. They are generally thought to be:

- ▸ Rice
- ▸ Wheat and barley
- ▸ Millet
- ▸ Glutinous millet
- ▸ Soybeans

Soybeans are a legume, of course, so why are they included with these grains? The reason probably lies in the translation; many scholars agree that the original intent behind wu ku was to list the five most important foods grown in the valleys.

Within these five grains, we might surmise the beginnings of culinary fermentations in the East.

As you can see, people have been eating rice for a very long time. Steaming became the preferred cooking method. As steamed rice ages, spores begin to thrive upon its surface, creating sugar, then yeast, then alcohol. In ancient times, with that happy discovery, this mold might suddenly have become a good thing, something to be kept and tended and improved over the years in the making of rice wine. From there, it's easy to imagine someone (now lost to history) making a leap in wondering whether this mold could be used on other grains.

Soybeans in ancient times would have been prepared by boiling. If you haven't had the pleasure of tasting freshly boiled soybeans, it is not a taste that makes you happy to be alive; rather, it is one that makes you think life is hard and gray and joyless. Soy then would have been a perfect candidate for innovative fermentations, leading to new flavors and foods that would come to define the cuisine of China for thousands of years.

positively benefits gut health. Also, it is not eaten raw. It is easy to digest and contains heat-stable antibiotic agents that act against some diseases. In areas in Indonesia where there is a high consumption rate of tempeh, there is a low incidence of dysentery, despite constant exposure.

In this book, we will discuss both types of fermented foods: those with and those without probiotics.

Spontaneous (Wild) vs. Cultured Ferments

Microbes are so ubiquitous on our foods that in many cases when we leave food alone under ideal conditions — like setting up shredded cabbage in a slightly salty anaerobic brine — we will have fermented food in a week or two. This is called spontaneous fermentation, or wild fermentation, because we didn't add anything; we just set up the right environment to make the existing microbes feel at home. Technically, however, "spontaneous" is a bit of a misnomer, since the microbes don't spontaneously appear but most often were already present on the ingredients or in the air surrounding the foods in question. For this reason, many folks prefer to describe this type of fermentation as wild fermentation.

For many ferments, however, the type of microbe that does the work of fermentation is important, and for this reason we add a microbial inoculant (commonly called a culture) to the food to jump-start fermentation. These cultured fermentations harness the various microbes to create the results and flavors we want in a pretty consistent manner. Of course, despite our intent to control the fermentation, these are live foods that respond to the conditions around the ferment, so there is often batch-to-batch variation.

In this book we will explore a few simple spontaneous fermentations and many cultured ones.

When horse gram beans are left to ferment without inoculation, diverse microbes move in and sour the paste for Pone Yay Gyi (page 95). This wild style fermentation has worked reliably for thousands of years.

Meet the Maker
Betty Stechmeyer
GEM Cultures

The year was 1979 and Jimmy Carter was president, a total solar eclipse had passed over the Lower 48, and Bill Shurtleff and Akiko Aoyagi's book *The Book of Tempeh* had just come out (for more on Bill Shurtleff, see page 41). They were on tour promoting tempeh as a fantastic source of protein to feed the ever-growing worldwide population. Their work furthered the idea, introduced 10 years earlier by Frances Moore Lappé in her classic work *Diet for a Small Planet*, that our only chance at sustainability on this planet was to fully use protein, not to run it through animals first. Betty likened Bill and Akiko's passion and mission to that of Sandor Katz, calling him the Pied Piper of "fermentation fervor." Their audiences (mostly university students) were equally inspired.

When Bill and Akiko came to the University of Michigan, Betty Stechmeyer and Gordon McBride were in the audience. Gordon's contract as a botanist at the university hadn't been renewed, and he and Betty were looking to move back to his family's farm in Fort Bragg, California. Call it synchronicity, call it chance, call it simply a series of events, but this was the evening they would begin their long-term relationship with microbes.

During the talk, Bill shared his frustration that there was no reliable source of tempeh starter in the United States. Betty and Gordon thought that producing tempeh culture could be a way they could make a living in Fort Bragg without debt. Since the startup costs would be minimal (they converted a closet in an old farmhouse to an incubation chamber by painting the walls with smooth white paint), they knew they could quit anytime. The running joke was that they'd moved to Fort Bragg to grow mold in a closet.

Soon, Betty and Gordon started GEM Cultures (GEM being Gordon's intials). The timing was good, as Americans were exploring

Eastern thoughts and healthful diets from Ayurveda to macrobiotics. They were also moving "back to the land" and baking their own bread, and making tempeh a good fit for this ethos. For the first five years, GEM Cultures' only product was its tempeh spores. Interest was growing, and after a conversation with Westbrae, a producer of vegetarian food products, Betty and Gordon began to import *koji-kin* (*Aspergillus oryzae* spores) to resell as well. Soon they added a culture for making tofu, and then they began to think that people might be interested in some of the cultures that their own family had tended to for years, including a family sourdough that had been kept alive since 1868 and a viili-soured milk starter from Finland that came over with Gordon's family more than 100 years ago.

Betty and Gordon hit on a trend. To this day, GEM Cultures, now run by their daughter Lisa, still produces tofu coagulants, sourdough starters, and dairy cultures, as well as soy cultures, water kefir, and kombucha starters (though no longer tempeh starter).

Fermentation and Safety

When basic production standards are followed, fermented foods in general, and the ferments in this book specifically, are quite safe. That's more impressive than it might sound because many of these foods have historically been produced by people with no formal training in microbiology or chemistry and often in unhygienic environments. Still, if you are like us, you want to go a little deeper than that simple reassurance. For that, we need to first understand what is happening in the different methods of fermentation. In this book, we focus on lactic acid, alkaline, and mold fermentations. Let's look at how each of these works so that you understand how, alone or together, they provide the protection and safety we want.

Lactic Acid Fermentation

Lactic acid fermentation is all about lactic acid bacteria (LAB) converting fermentable sugars into lactic acid, which lowers the pH of the food. This type of fermentation is responsible for yogurt and cheese, sauerkraut, kimchi, and so many other vegetable ferments, as well as salami and cold cured meats. In this book, a few of the spontaneous ferments that sour quickly, such as injera, rely on lactic acid bacteria to do their magic. Lactic acid fermentation is also a player in everything from soy sauce and kombucha to beer and other alcohols. For example, in its purest form, a lactic acid vegetable fermentation lowers the pH level to 4.0 or below and creates an anaerobic environment below the brine, both of which safely preserve the food.

Dosas are a good example of lactic acid fermentation working simultaneously on legumes and grains.

Alkaline Fermentation

Alkaline fermentation is all about the bacteria of the genus *Bacillus*, and specifically *Bacillus subtilis*, which is commonly found in the upper layers of soil and in our guts. This bacterium secretes enzymes that break down the protein in legumes into peptides and amino acids, while also producing compounds that inhibit the growth of pathogens. Ammonia is created and released in this process, which raises the pH very quickly to levels of 8.0 or higher. Most pathogens and spoilage microbes, like *salmonella* and *shigella*, cannot survive at these pH levels. *E. coli* is a little tougher but will die at pH levels of 9.0 or higher.

Mold Fermentation

In this book, mold fermentation undergoes a very similar process to alkaline fermentation. A substrate, be it beans or grains, is soaked in water and then boiled — again, the acidification and softening of the skins is important. This time, a species of edible mold is introduced — a rhizopus species for tempeh or aspergillus for koji. These molds take in oxygen and transpire carbon dioxide. Sounds like us, right? That CO_2 inhibits other microorganisms. To outcompete other mold species, the rhizopus or aspergillus species produces a lot of filaments, which quickly cover the beans to such a degree that the beans are protected from any other molds. Finally, as if just to make sure, the rhizopus or aspergillus expresses some antibacterial activity

Natto is an alkaline ferment that creates a biofilm that also helps keep other microbes away.

Another key to safety in alkaline fermentation in legumes is that the legumes often have a hard outer shell that must be soaked for many hours and then boiled for many minutes to break this armor. The soaking creates a lactic acid fermentation first, to combat spoilage, then the boiling kills those lactic acid bacteria and other bacteria and molds. At this point *Bacillus subtilis* can move in — either naturally from the air, as you'll see in cheonggukjang (page 123), or with a culture, as in natto (page 118). *B. subtilis* is competitive and tough — it can take temperatures over 100°F/38°C — so it begins its work when other bacteria cannot, effectively giving it the head start it needs to outcompete all others.

As the mold moves across the substrate, it knits it together and transforms it.

to inhibit other molds. If the temperatures are kept in the right range and the mold is strong, it will also inhibit other microbes, like bacteria and yeast, from growing. However, all of these microbes are opportunistic, and if the balance is thrown off — with the most common cause being overheating — the molds will die off and *Bacillus* will move in, in this case where it is not wanted.

Keeping House

When you are preparing any type of food, it's important to keep a certain level of cleanliness. With many of the ferments in this book, it's important that you take cleaning to the next level — sanitation — to avoid spoilage as well as cross contamination if you're making a number of different recipes. This isn't as difficult as it may sound. We have made all of the ferments in this book in our farmhouse kitchen, using the same fermentation chambers to make different recipes with no cross contamination.

As you will read, the most important control you have in the incubation phase is the temperature. If the ferments are warm enough and don't get too hot, they will thrive, outcompeting the competition. You'll want to keep your incubation chamber sanitary, as well as the vessels and utensils that you use. We often use the same casserole trays for koji, tempeh, and natto. The trick is to clean them well. Given that many folks have dishwashers, that first step is easy. Some ferments will require an extra step of sanitization, which can be as simple as using a spray of 190-proof alcohol (like Everclear) and allowing to air dry, a brewer's sanitizer (like Star San), or good old-fashioned hot water. In many cases, pouring boiling water from a kettle into your dish and keeping it there for 30 seconds is perfect.

Finally, it's important to monitor the temperature of your beans or grains during the period leading up to inoculation. The microbes need specific temperature ranges for inoculation, and your substrate will be the "cleanest" as it moves into that temperature range. If you allow your substrate to cool before you inoculate it, you will give other microbes a chance to move in. Don't fret, however. This is supposed to be fun, and there is some leeway. Remember that folks have been fermenting throughout the world for thousands of years.

Salt

Misos, tasty pastes (fermented foods built on the backbone of the miso technique), and amino sauces depend upon a relatively high salt content (3 to 20 percent), depending upon how long the ferment will be in the crock (or jar), to control the breakdown of protein and protect the food from putrefaction. The longer the ferment sits at room temperature in a crock, the more protection it will need. That's why white miso, which has a relatively short fermentation from a few weeks to 3 months, has much less salt than soybean hatcho miso, which has a long fermentation of 2 to 3 years or more.

As you explore modern misos, or tasty pastes, you will see that salt percentages may be much lower than in traditional recipes. This is because traditional fermentations must take into account the potential vagaries of the weather. Outside of temperature-controlled spaces, salt is key in controlling the fermentation because the temperature is out of the maker's control. The ferment is traditionally started in the fall, when temperatures are cooler, to give the yeasts and bacteria a slow thriving start. They then begin to become more active in the spring and get very busy in the summer, when it is warm. Modern, small-scale makers are making miso in the fairly steady ambient temperatures of a home or commercial kitchen, so the salt can be adjusted according to taste preference. (You will read more about this in chapter 9.)

We also want to note that the super-high salt levels of some of the ferments of the past can dull the nuanced and exciting flavors that are produced by the yeasts and bacteria, creating a flat saltiness. But ultimately the value of saltiness is a tool and personal decision. We heard one chef extol the benefits of a high-salt amino sauce as a much cleaner flavor — so it depends what you are after.

We also depend on salt to arrest the growth of the koji fungus to prevent it from reproducing through spores (which add dank, musty, or bitter flavors) while still preserving the active enzymes that produce amazing flavors by breaking down the proteins, fats, and carbohydrates of the legumes or grains being fermented. (Though as is the case with everything else, there are exceptions; Chinese *douchi*, or fermented black beans, for example, allow koji to sporulate to take advantage of those flavors.)

Not too long ago, salt choices were limited to mainly table salt, kosher pickling salt, and sea salt. Now we can taste our way around the planet with the salts of the earth and sea. Salts of all colors and flavors are available in even the most common grocery stores. We can choose white, red, pink, or black salts; finishing, milling, or cooking salts; and small crystals, flakes, or large blocks. Because not all salts are the same, their density (and therefore weight) will vary according to their mineral content and coarseness. The measurements in this book are based on regular finely ground salt, like you would find in most table salts, and are given in grams for consistency.

We use natural salts in the recipes in this book and that's our preference for all ferments, be they vegetable, legume, or grain. Natural salts come from the earth or sea, are relatively unprocessed, and have healthful trace minerals and lower levels of sodium chloride. If you have a favorite rock salt, pink salt, sea salt, gray Celtic salt, or something similar, go right ahead and use that. We tend to use Redmond Real Salt fine grain because of its high mineral content, discernable sweetness in taste, and wide availability in the U.S. Himalayan Pink salt is a good choice, but make sure it is from Pakistan and does not contain added fillers and ingredients.

We avoid most refined salts because the minerals are removed in the refinement process, which increases the percentage of sodium chloride and we think leads to a sharper salty taste. Avoid salts that have added ingredients such as anti-caking agents and iodine (the trace amounts found in natural salts are fine), which in higher levels can inhibit fermentation. Most inexpensive sea salts are refined (instead of evaporated), but they work well if there are no additives. We love evaporated sea salts, especially the grey sea salts which are harvested from the bottom of the evaporation pools, for flavor and mineral content, but they have become of victims of our times. These salts can also contain heavy metals and microplastics due to due to high levels of pollution in our oceans.

The ability to salt food properly is the single most important skill in cooking.

THOMAS KELLER

CHAPTER 2

Fermentation Equipment

WHAT YOU NEED

Unlike simple lactic acid fermentation (think sauerkraut, pickles, and kimchi), where the ambient temperature of your kitchen counter is usually within the wide range needed for successful fermenting, the types of fermentation in this book require a narrower range of conditions for success. Because of this, each type of ferment requires a controlled environment. In this chapter, we will help you navigate the options for keeping your microbes happy as they work for you. Remember that people have been fermenting with the simplest of tools and vessels, figuring out what works best for them, for thousands of years. We invite you to adopt a similar adventurous spirit as you explore the tools and vessels available to you.

Creating the Perfect Microbial Home

Every ferment needs an incubation space — a spot where conditions are right to foster the microbes that will do the fermenting. We like to think of it as a fermentation womb, but you can call it an incubation chamber if you like. Here we present an overview of many of the options you can use to create this space, so that you can start to think about what will work for your own situation. We will go into detail about each microbe's specific requirements in part II.

The Merriam-Webster definition of *incubate* (right after "to sit on eggs") is "to maintain something (such as an embryo or a chemically active system) under conditions favorable for

hatching, development, or reaction." For the incubation of bacteria, yeasts, and molds, success depends on getting these conditions right. Truly, your success and your fun depend on a reliable system that works for you and promotes the development of the microbes you desire. Since each microbe has preferred conditions, you can keep the right microbe growing — and the wrong microbes out — by controlling the temperature and humidity of the environment. For example, *Aspergillus oryzae*, the fungus we use to make koji, needs a comfortable 88°F/31°C to thrive. *Bacillus subtilis*, the bacteria we use to make natto, needs a much warmer climate. If the koji gets too hot, the aspergillus dies and the bacillus, now in its ideal environment, begins to thrive.

There are a lot of ways to control the environment of the ferment, on the cheap and not-so-cheap. We will share the insights we have gained from our experiences in order to help you find a system that works for you. The frustration of not being able to maintain a cozy environment can make what could be an easy process unsatisfying and difficult — and it all comes down to your incubation setup. Learning how to keep a consistent temperature-controlled, humidity-controlled home for these ferments is the difference between thinking fermenting is fun and thinking it isn't at all.

Some of these setups may require a modest investment. We get it; you may not know if you are going to become a regular natto or tempeh maker. You may just want to give it a try. Clearly it would be silly to invest in a special piece of equipment just for a curious flirt with these fermentations. Luckily, many incubators are multifunctional, like bread proofing boxes and dehydrators, so you can use them for more than one purpose. We've also included popular fermentation hacks for items most people already have, like a heating pad or oven, that we've discovered from others and used ourselves. All of these methods have their drawbacks, and we don't want you to give up before you even know whether you take pleasure in the process or the foods you might have created. So please, be patient and enjoy the journey. Trust us: once you have the right incubation setup, the rest is pretty simple.

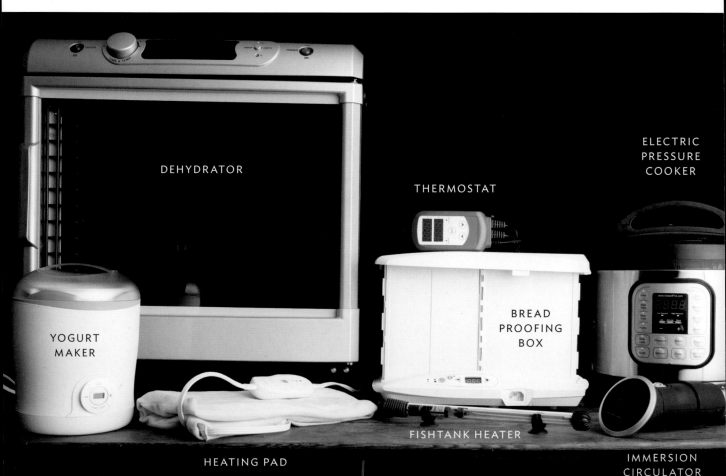

DEHYDRATOR

THERMOSTAT

ELECTRIC PRESSURE COOKER

YOGURT MAKER

BREAD PROOFING BOX

FISHTANK HEATER

HEATING PAD

IMMERSION CIRCULATOR

Your Oven

The good news is that you likely already have one and they are made to hold heat. An oven with good temperature control (or an oven operated by a patient person) can be made hospitable to all the ferments in this book. Most ovens can't be set as low as is needed for fermentation, but there are workarounds. For example, the culinary molds (tempeh and koji) will need to be held at a temperature between 85°F/30°C and 90°F/32°C. Ovens that are turned off but have an interior light left on and gas ovens with a pilot light often come really close to these temperatures. The downside is that your oven is off-limits to baking while you are fermenting. To save yourself the barrage of profanity that will surely issue when you preheat the oven to bake cookies and smell the melting ferment, put a piece of painter's tape over the knob after setting your ferment inside. You will be surprised by how many times you go to preheat your oven and look at the tape and wonder, "Now why did someone leave that there?" before you remember your ferment within.

Before you start fermenting in your oven, test it. Turn on the oven light (or, if you have a pilot light, do nothing) and place a thermometer where you would place a tray. After 15 to 20 minutes, check the temperature. Do this again a few times over the next hour. These readings will give you a sense of how warm the oven stays. If it gets too warm, you can turn off the light periodically or prop open the door with a wooden spoon. If the temperature is too low, you could try putting a heating pad or a pot of hot water in the oven (periodically replacing the hot water, as needed). Keep taking readings after making adjustments until you have established a good system for keeping the temperature where you need it.

Be aware that the oven light can cause one part of the oven to be hotter than another part. We used to ferment yogurt in our oven, for instance, and the jars nearest the bulb always got too hot. Another issue is that the heat from the bulb or the pilot light can be drying; it may dry out your substrate before the fermentation can happen.

Heating Pad

A heating pad is the most common setup suggested to beginners. We are going to share tips for making it work, but we must say right off the bat: it is cheap but frustrating.

The first source of potential frustration is that heating pads vary in quality, and newer pads have a safety feature that makes them shut off automatically after a certain period of time. If you are going to try using a heating pad, find the old-school variety that doesn't turn off if left on for hours.

Another option is to use a seed-starting mat. This type of mat doesn't get as warm as a heating pad, which in our experience meant we didn't overheat anything, but in some cases, it had trouble maintaining the temperature we needed.

Set up your heating pad or mat in a box or, for better insulation, a cooler. Set your fermentation inside, and wrap everything with thick layers of towels and blankets. If you can add a thermostat to that cooler, place it in the wrappings near the ferment, and you will have more control and get better results for the more particular mold ferments. (See photo on page 26.)

We have found that it is possible to make all the ferments using a heating pad, but we had the most success with natto and much more frustration with tempeh and koji. That said, many a first tempeh has been made inside a cooler with a heating pad.

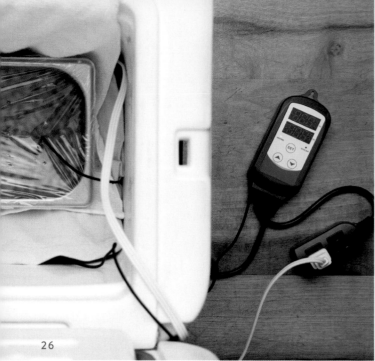

The heating pad is set in a cooler and controlled by a thermostat that is tucked into the ferment.

Yogurt Maker

A yogurt maker usually operates steadily between 110°F/43°C and 115°F/46°C, which is too hot for koji and tempeh. It can be a little high for natto, too, though we have gotten good results anyway. There are some yogurt makers that are designed to also make natto, and some natto makers that claim they also make yogurt. If you already own a yogurt maker, it's worth your time investigating if it will work for you.

Electric Pressure Cooker

An electric pressure cooker or Instant Pot is a wonderful invention when it comes to cooking beans quickly — especially soybeans. We initially bought our Instant Pot thinking we would use it as an incubation chamber, but we soon realized that while it has a number of great settings, it didn't quite have manual temperature controls for the specific temperature needs of many of these ferments. The Instant Pot

yogurt setting has three different fermentation temperatures:

▶ **The "Less" mode** runs right around 95°F/35°C and is for fermenting glutinous sweet rice (jiu niang, page 213). It can also be used to keep a dosa or idli batter warm. If doing this, use a lid from another pot or a cloth to cover — this will help keep the temperature in balance.

▶ **The "Normal" mode** for making yogurt runs right around 109°F–115°F/43°C–46°C. This is a little warm for natto, but it seems to work well nonetheless. It is not quite warm enough for amazake.

▶ **The "More" mode** is for pasteurization and is set to 180°F/82°C, which is too high for fermentation.

The cooker has proven invaluable in preparing beans and grains for fermentation.

Dehydrator

Dehydrators are a natural choice for a fermentation incubation chamber. The biggest challenge is that the fans are constantly blowing, which can be drying; you will have to make sure that your ferments are covered and protected yet still provided with some airflow. This is as simple as covering the ferment with aluminum foil or plastic wrap with perforations. Take care when you make the holes (see photo on page 149) — if your holes are too large or too numerous, your ferment will dry out. Sometimes adding a bowl of water to the dehydrator can help keep the humidity a little higher. We know folks who have found ways to cover the fan, which is usually encased in a metal mesh safety cage of some kind, in order to stop the airflow without impeding or damaging the fan.

Lab Incubator

A lab incubator is similar to a dehydrator, but it doesn't have a fan drying out your ferments. The downside is that it is a considerable investment; a small one is in the same price range as a high-end dehydrator, and it doesn't give you the option of using it for dehydrating. We have not used one, but we have heard good things about these and assume that the temperature control is much more advanced than in a consumer dehydrator.

We have found that the cabinet-type dehydrator with shelves, a door, and an electronic thermostat control has more consistent heat than the type that has trays that need to be shuffled. We have also found that they are well insulated and don't blow hot air throughout the room, which makes them more efficient. While dehydrators have temperature settings, many do not have a thermostat, but you can hook one up in order to control the temperature. Some dehydrators are preprogrammed to specific temperatures and do not allow you to choose other temperatures, which can be a problem for ferments with narrow temperature ranges. Like with the oven (page 25), you may want to use a thermometer to check your dehydrator's and ferment's actual temperature and consistency. It is disheartening to set the temperature for 85°F/30°C only to come back and find that your ferment is over 100°F/38°C. An external thermostat can be a workaround.

We have used our cabinet dehydrator with a built-in thermostat for all of the ferments in this book. However, it is a large and loud piece of equipment. Ours happily lives in the fermentation kitchen away from our home space, but if you have tight quarters and plan on fermenting in your living space, consider whether the ambient white noise of a dehydrator running through the night will work for everyone in the house.

Bread Proofing Box

A little folding bread proofer will work for many of the ferments in this book. The design is quite impressive, allowing you to fold up the box for flat storage. The chamber is not airtight, so we have not found a lack of airflow to be a problem. It is a bit of an initial investment, but it is considerably less expensive than a good dehydrator.

Along the bottom is an aluminum heating plate, which is the sole source of heat for the device. Note that many proofing boxes also function as slow cookers. Be sure to use the proofing setting for fermenting. This setting makes use of the wire rack that fits above the heating plate, which heats the surrounding air through convection. The proofer should come with a shallow aluminum water pan, which you can fill with a small amount of water and place in the center of the heating plate and below the wire rack to increase the humidty. Another reason to use the proofing mode is that you can adjust the temperature in 1-degree increments, so making natto at 106°F/41°C is no problem (the slow cooker mode has temperature increments of 5 degrees). Be sure to always place your ferments on the wire rack and not directly on the heating plate.

The downside is that some proofing boxes aren't big enough to hold large fermenting vessels (in our case, casserole dishes full of natto). In a pinch, we have stacked up to three smaller containers by placing some grating between each layer and offsetting them as much as possible to help with airflow.

Ferment Terrarium

For this setup, you are just using an aquarium heater and bubbler inside a cooler to create a nice humid incubation environment. You'll fill the cooler with water, and the aquarium heater will keep it warm, the bubbler will help circulate the water so that the temperature stays even, and the insulation of the cooler will make the whole thing very efficient.

A ferment terrarium is perhaps the easiest system for managing mold ferments because the water bath enveloping the floating fermentation tray not only keeps the substrate warm in the beginning but also works at cooling the growing mycelium as it begins to grow and create its own heat. It is inexpensive to set up and perfect for making small quantities. Because the humidity is perfect, you can avoid the single-use coverings of plastic wrap, or in the case of tempeh, plastic bags. The downside is that you are limited by the size of the stainless-steel pan that will fit in the cooler. The great thing about the aquarium heater is that it won't overheat the mold ferments (it was built to keep fish alive, after all); however, this also means it won't generate enough heat for natto or other warmth-loving ferments.

Gunter Pfaff, husband of Betsy Shipley of the company Betsy's Tempeh (see page 132), pioneered this method of fermentation for tempeh. Betsy spent a lot of time researching more information for us, for which we are extremely grateful, and we are honored to share Gunter's method. You will find Betsy's recipe for tempeh panini on page 355.

PRO TIP

If you have hard water, add ¼ to ½ cup white vinegar to the water bath of your ferment terrarium. This will keep your pump or circulator from calcifying and the sides of the tub or cooler from having a white bathtub ring of residue. We've found that the water can be used for two batches of tempeh in a row or one batch of koji before the water must be changed and the cooler or basin cleaned.

Build Gunter's Ferment Terrarium

Materials

- Cooler (see note below)
- Aquarium heater
- Aquarium bubbler (air pump and stone)
- Small tube of silicone kitchen and bath caulk

Tools

- Electric drill
- 1¼-inch drill bit
- Permanent marker
- Duct tape or electrical tape

Note: For the terrarium, you'll need a large insulated container big enough to contain your preferred tray and available to be a forever ferment terrarium. An 8-inch-deep Cambro insulated food carrier works well. Note that the more insulated the cooler, the less condensation on the inside of the lid. You can use an upcycled old picnic cooler.

Keep the stone that comes with your bubbler wet. If it dries out, minerals will deposit in the small holes and not let air through. If your bubbler doesn't come with a stone, see the hack on page 30.

For fermenting in the terrarium, you'll also need pans that can fit inside the cooler. Ideally you will find one with a high rim that can be floated on the water; this is your fermentation tray. You can use full- or half-size steam table food pans with 4-inch sides that float and not need to use supports. We also tested glass casserole dishes, which to our excitement not only floated but produced some excellent koji and tempeh. Alternatively, if you find you want or need support, your fermentation tray can be held up by four 6-inch water glasses.

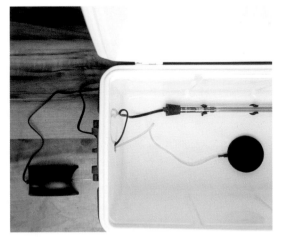

1. Drill a 1¼-inch hole in the side of your cooler, 2 to 3 inches from the top. The hole should be big enough to allow the cord for the aquarium heater plug as well as the air pump tubing to pass through.

2. Place the heater in the bottom of the cooler and pass the cord for its plug through the hole. Place the air pump on the outside of the cooler, pass the air tubing through the hole to the inside of the container, and connect the tubing to the stone. Secure both the heater and stone, if necessary.

Instructions continue on next page

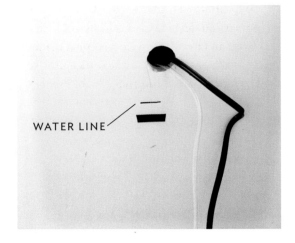

WATER LINE

3. Set a small piece of tape so that its top edge is 2 inches below the bottom of the hole. Fill your cooler with water to the top of the tape.

4. Place your empty fermentation tray (the pan with a high rim) in the incubator. Add more water to the cooler until the tray floats up slightly (about 1/10 inch). Slowly remove the tray. When the water is calm, set a new piece of tape so that its bottom edge marks the water line. Drain the cooler, then draw a line with a permanent marker where the final water line was. This will be the water level from now on. Remove both pieces of tape. Use the caulking to fill up the hole.

———— PRO TIP ————

If your air pump does not come with an air stone, you can crimp the end of the air tube and poke it with holes, about every inch or so, using a pin. Arrange the tube at the bottom of the cooler, weight the end with a rock, and place another rock about 12 inches down the line.

That's it — now you have a proven and reliable fermentation incubator. Fill the cooler with water to the line, put the lid on it, set the heater to your desired temperature, and let it run for 24 hours, taking regular readings, to make sure the water temperature stays constant and that the heater is accurate. (Our aquarium heater keeps the water at a consistent 89°F/32°C when it is set to 84°F/29°C.)

Immersion Circulator

An immersion circulator is basically a device that heats and circulates water, keeping the temperature at a precise number. It works just like the ferment terrarium. As the price of immersion circulators has dropped in recent years, more and more folks are discovering them for home cooking.

You can attach an immersion circulator to the side of a pot or pan filled with water to create a very hospitable fermentation environment. Then you put your ferment on a stainless-steel tray and float it in the warm water. Be sure to cover the entire tub with a lid. **Note:** You can get specific tubs and lids for this purpose, but these lids create a lot of condensation and can drip into your ferment. We found that placing a clean kitchen towel over the tub and pulling it tightly under the lid will fix this. This is less of a problem with the insulated lids of a upcycled cooler. This setup (like the terrarium) works well because the temperature and humidity are constant. The ferment is kept warm while it is nascent, and once it starts metabolizing, the steady temperature of the water will act to keep it from overheating.

Natto works well in an immersion circulator incubation setup (and it's the only setup that avoids the use of one-time plastic and aluminum foil). We have found that it is especially effective for making tempeh and koji in a tray. It is also amazing for any of the amazake or koji-type marinades when done under sous vide. In sous vide (French for "under vacuum"), food is dropped into airtight plastic bags and placed in this controlled chamber of circulating water for long and slow cooking. For fermentation, vegetables or meat are joined with a particular microbe, like amazake or shio koji, and the cozy "chamber" of warm water keeps the microbes

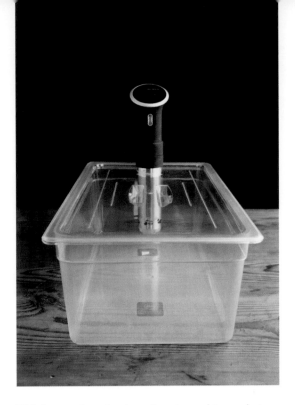

This immersion circulator is set on a bin ready to float a tray for fermentation.

happy and perfectly preserves the active enzymes. Some people have used this method to speed up a tasty paste (see nut tasty pastes, page 280) or modern miso.

The shortcoming of this system is that the batch size is limited — you can usually only make one tray at a time. But again, storage is easy, so its size can be an advantage in small living situations.

Retrofitted Refrigerator

This is for the do-it-yourselfer and those who want to make big batches of ferments for family and friends. Keep your eye out for a nonoperational refrigerator (there are a lot out there). The smaller dorm-style refrigerators as well as wine refrigerators are perfect for a small incubator. Using a full-sized fridge may be your best, most inexpensive option for large batches, as you can fit a lot of trays of koji, natto, or tempeh in this setup. You are basically recycling a very well-insulated box with shelves. You can also build something similar with a cooler.

The simplest way to convert an old fridge is to install an incandescent lightbulb and a thermostat to turn the lightbulb on and off to regulate the interior temperature. You will also want to drill a small hole in opposite corners to allow for natural passive circulating airflow. (For example, if you drill one hole in the front right bottom corner, you should put the other hole in the top left back corner of the fridge.)

This setup works well for most ferments. It may work for amazake if you can bring the internal temperature up to the necessary level.

	Natto	Tempeh	Koji	Amazake
Oven	x	x	x	x
Heating Pad	x	x		x
Yogurt Maker	x			
Electric Pressure Cooker	x			
Dehydrator	x	x	x	x
Bread Proofing Box	x	x	x	x
Immersion Circulator	x	x	x	x
Retrofitted Refrigerator	x	x	x	x
Ferment Terrarium		x	x	

Fermentation Vessels

The vessels that are used to make the recipes in this book range from flat open trays to wooden vats, ceramic crocks, and glass jars. You will see that once again this ancient art uses humble tools that work.

Some kinds of ferments require you to set a weight on top of the fermenting ingredients. For traditional brine-based ferments, the weight keeps the ingredients submerged in brine. For the misos and tasty pastes in this book, a weight is used to press out liquid and gases. For these ferments, an important question when selecting a vessel is: Will your vat, crock, or jar allow you to get a significant amount of weight on top of the paste? In an ideal world, your weight would be equal to, or even up to 25 percent more than, the weight of your paste. In other words, if you are making 8 pounds of miso, you need to weight it down with 8 pounds, or even better 10 pounds, of weight. This can be difficult to achieve with a small batch, but as you will read below, with some creativity you should be able to weight down your ferments.

Casserole dishes. Inexpensive glass casserole dishes work well for the incubation phase of making natto, koji, and tempeh. The best size is dependent on your batch size and incubation space.

Crocks and lids. Crocks are the traditional vessel of choice for fermentation because they are sturdy, they easily accommodate weights,

CROCK AND LID

GLASS JARS OF
VARIOUS SIZES

STEAMER PAN

CASSEROLE DISH

and the ceramic material has micropores that let the ferment breathe a tiny bit. They work well for ferments that need a longer fermentation and weights on top, like miso and tasty pastes. A drop-style lid that drops down into the crock allows the weight to push the lid down through the crock at any level. In the same vein, a crock with straight sides will ensure that the lid and the weights stay level as they drop in the crock.

Jars. Amazake and shio koji ferment well in glass canning jars. It is a little more challenging to ferment miso or tasty pastes in them because the glass does not have micropores that let the ferment breathe, and fitting heavy amounts of weights on top can be difficult. However, jars will work in a pinch, though they may take some stirring and a little more monitoring than a crock. Be sure to choose jars that are much larger than you need so that you have room to pile on weight. Glass canning jars are great for storing finished ferments like miso, stinky tofu, and natto in the refrigerator.

Pans. We're talking here about the stainless-steel steamer pans used to hold hot buffet items. You want the 4-inch-deep pans, which usually measure 12 inches by 20 inches (full size) or 12 inches by 10 inches (half size). The half size should fit in a bread proofer or immersion circulator tub, and both sizes should fit in a full-size cooler that has been retrofitted to be an incubator.

Plastic bags. Quart-size BPA-free ziplock bags are one option for fermenting tempeh. Vacuum-sealed bags can be used for koji, shio koji, and amazake marinades.

Weights. Weights go hand in hand with crocks and jars. As we mentioned above, weights are very important when you are making miso and tasty pastes. Weights will help push out the excess liquid and gases produced by the fermenting carbohydrates, preventing the ferment from becoming too sour or burdened with undesirable ethanol flavors. Weights can be just about anything. Traditional vats are often piled with stones. We have used marbles, ceramic or glass weights, and salt-filled plastic bags. Your goal is to pile on as much weight as you can.

Other Equipment

Aluminum foil and plastic wrap. To make some of the ferments in this book, and especially natto and koji, you will need to control the humidity in your fermentation vessel with some kind of covering. Perforated aluminum foil or BPA-free plastic wrap will do the job in most situations.

Meat grinder or other masher. This is not a necessary piece of equipment, but a grinder of some kind can be very useful for making miso, especially if you want a smoother, pastelike texture. Some misos are best ground before being aged, and others are blended when they are done fermenting.

Digital kitchen scale

Scale. For the types of ferments in this book, a gram scale is nice to have, especially if you start creating your own recipes. It will make calculating salt percentages and koji-to-bean ratios much simpler when you are making miso and tasty pastes. Many ingredients, most notably salt, will vary widely in weight depending on type (a tablespoon of one type of salt may be half the weight of a tablespoon of another salt, for instance). For this reason, we give you both volume and weight measurements for our ferment recipes. Digital scales are inexpensive and easy to find, and most can switch from measuring ounces to measuring grams quite readily. Be sure that your scale will allow you to calculate and deduct the weight of your measuring container (called the *tare*).

Thermometer. All of these ferments require a certain temperature range. Most of the incubation systems have some sort of temperature control, but in our experience, all of them are

Measuring pH

Acidity and alkalinity are measured by pH, which stands for potential of hydrogen. Simply put, pH compares the relationship of hydrogen ions with hydroxide ions. Acidic substances have a pH between 0 and 7, alkaline substances fall between 7 and 14, and distilled water is in balance with a pH of 7. It is easy to tell by tasting if a ferment is acidic or alkaline, so in general, we let taste, not measurements, be our guide, especially since we are after flavor anyway.

However, if you want more precision or are just curious, you can measure pH quite simply by using pH test strips. These reactive strips change color according to the pH of your solution. Or use a pH meter, which requires dipping into a calibration solution to work properly.

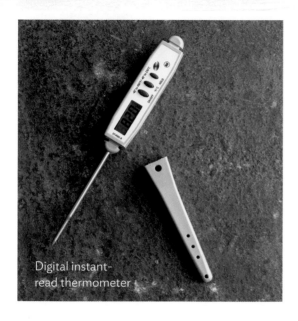

Digital instant-read thermometer

off in some way or another. It's for this reason that we recommend that you have a secondary thermometer. Remote-read probe-type thermometers work best because you can leave them in place and read them without disturbing your ferment. They have become quite common and inexpensive.

A basic instant-read thermometer will also work quite well and captures the temperature inside your ferment. The downside is that every time you check the temperature, you run the risk of disturbing the environment and the ferment as you poke into it.

Another popular choice is an infared thermomenter that works by aiming a "gun" at the outside of the ferment. It is quick and doesn't disturb the ferment in any way, but it only gives you the temperature of the surface of the vessel, not what is going on inside. And it doesn't work well on shiny metal surfaces.

Thermostat. A thermostat is a device that controls the temperature, turning on and off the electricity that powers the heater in order to maintain the temperature you set. Some incubation setups come with a built-in thermostat. If yours does not, you may want to plug your heat source, be it a lightbulb or heating pad, into an external thermostat to help you keep a stable temperature. Make sure you find a thermostat that fits your needs — some are more precise and can be set to temperatures to within one-tenth of a degree, while others are accurate only within a couple of degrees.

Pressure cooker. Both electric and stove-top pressure cookers will make cooking beans and grains prior to fermentation so much more energy efficient — not to mention easier and faster. The electric cookers allow for maximum control and are generally foolproof. If you do use a manual pressure pot, it is a good idea to have your cooker checked once a year. Most university extension offices will do this for free.

CHAPTER 3

A Guide to Legumes and Cereal Grains

FOR FERMENTATION

This chapter is just the beginning of what is possible with the fermentation of some legumes and cereal grains. There are so many more varieties out there, but we hope this overview not only will give you plenty to work with but will pique your curiosity for the vast world of bean and legume nutrition.

If you think that *legumes* is a fancy name for beans and that beans are boring, we hope this chapter will change your perspective. There are 13,000 species of legumes, but humans have chosen to eat only about two dozen of them. All beans are legumes, but not all legumes are beans. Legumes include the entire bean family plus other ferment-friendlies like chickpeas, lentils, mesquite, soybeans, and peanuts — all covered in this book. Soybeans are the main traditional legume in most of these fermentations, so we'll dig a little deeper into this bean.

If all you can think of when you hear "fermented grains" is beer, well, fair enough, that's a pretty popular one. We humans are pretty selective; there are more than 50,000 known edible plants on this planet, yet two-thirds of our plant-derived nutrition comes from only three plants, all of which are cereals: corn, wheat, and rice. Rice has traditionally been the dominant grain for many Asian fermentations. However, that hasn't stopped us and many others from exploring many more cereal grains worthy of adding to your fermentation crock.

Why Eat Legumes?

Legumes are good for the soil and good for our bodies. They are suited for growing in a wide range of climates. They give back to the soil by pulling nitrogen from the air and burying it into the soil through their roots. Legumes are an excellent source of protein, carbohydrates, fiber, vitamins, minerals, and phytochemicals. That's a lot to claim, so let's unpack each one of these benefits.

Protein. When they are eaten with grains, legumes provide us with high-quality complete protein, meaning that they provide us with all of the nine essential amino acids our bodies cannot create themselves.

Carbohydrates. Legumes are 25 percent or more energy-giving carbohydrates by weight. These are complex carbohydrates, meaning that they are made up of simple carbohydrates (sugars) that are woven together in chains. It takes our bodies a while to process them and release the sugar, so we don't get blood sugar spikes like we do with refined carbohydrates. Some of the bean starches are so resistant to digestion that

Resistant Starches

We were introduced to the concept of resistant starches, a type of prebiotic, by Dr. Art Ayers, a longtime friend who happened to also be a biomedical researcher. We wrote about them in our second book, *Fiery Ferments* on page 221:

You may have seen (and been confused by) the word prebiotic. Looks like probiotic, sounds like probiotic, but it's not. A prebiotic is a nondigestible (by us) food component that promotes the growth and health of beneficial organisms. An example is resistant starch, found in foods such as potatoes, rice, and plantains — but only when they are cold. What? We know you don't want your potatoes cold, but stick with us. When this coil-shaped starch is heated, it melts into a blob, but as it cools it slowly recoils, becoming once again resistant to digestion. And why, you ask, is this a good thing?

When resistant starch lands in the large intestine, it meets a group of bacteria waiting for dinner. As they break it down, they release small carbohydrate molecules (waste) that neighboring bacteria consume. They then secrete even smaller waste molecules called short-chain fatty acids. The most important of these fatty acids are butyrates, which are happily absorbed by the colon's lining, encouraging blood flow, keeping tissues healthy, and providing an energy source for your body.

Resistant starch has three main health benefits: First, it delivers fermentable carbs to bacteria in the colon to help prevent colon cancer. Second, it helps us better regulate the levels of fat in our blood, which aids in the prevention of heart disease. Third, it helps us absorb minerals, which otherwise would just pass through us and give us no benefits. Beans are an important source of resistant starch. Fermented beans are 8 to 15 percent resistant starch, making them an excellent source for us and a much more palatable and enjoyable option than cold potatoes!

they make it all the way to our colons, bringing plenty of carbohydrates for our resident micro-flora. In other words, these resistant starches (see box on page 38) are prebiotics, and they are just as important to us as probiotics.

Fiber. At some point in your life, someone has likely recommended that you eat more fiber — even if that someone was just an actor wearing a generic white lab coat and pitching some fiber mix in a pleasing fruit flavor. The advice is good, but there is a better way to get your fiber. When dietary fiber reaches our colon, our beneficial bacteria happily digest it into short-chain fatty acids, which are associated with a reduced risk of colon cancer. A diet rich in fiber also reduces the risk of diabetes, heart disease, irritable bowel syndrome, and obesity. Most Americans get less than half of their daily recommended dietary fiber, which is 25 grams for women and 35 to 40 grams for men. Legumes are one of the best sources of soluble dietary fiber, and many of the recipes in this book will easily give you 15 or more grams of fiber with a single serving.

Vitamins and minerals. Legumes have substantial amounts of vitamins A, B, and E. They also contain a number of essential minerals, including significant levels of potassium, which increases insulin secretion in the pancreas. All this and they have low sodium levels as well. Note that while legumes do contain iron, zinc, and calcium, these minerals are not fully available to us due to other legume constitutents, called phytates, which are especially prevalent in soybeans (we'll talk more about them on page 43). The good news is that fermentation makes these minerals more available.

Phytochemicals. Lastly, legumes are a good source of phytochemicals called phenols, which protect the plants from oxidation and do basically the same thing for us when we eat them. Studies have shown that consuming foods that contain these phenolic compounds is correlated with a reduced risk of cardiovascular disease, inflammation, and death by cancer.

In summary, beans are really good for us. There are just a few issues: They can be hard to digest, they are not always so tasty, and much of the nutrition is locked away by those pesky phytates. What is to be done, you ask? Fermentation, we answer. Fermentation turns these healthy foods into top-notch, stellar fare that helps prevent diseases while boosting our health.

Out of the thirty thousand types of edible plants thought to exist on Earth, just eleven — corn, rice, wheat, potatoes, cassava, sorghum, millet, beans, barley, rye, and oats — account for 93 percent of all that humans eat, and every one of them was first cultivated by our Neolithic ancestors.

BILL BRYSON, *AT HOME: A SHORT HISTORY OF PRIVATE LIFE*

Soybeans

This bean goes by several names, including soybean, soya bean, or just plain soy. Soybean (*Glycine max*) has the highest protein levels of any plant and grows in many soil conditions. It has been consumed by people in Asian countries for centuries, perhaps playing a key role in their lower incidence of chronic diseases. And it's controversial on many levels.

Through the three thousand years that humans have been consuming soy, we have both changed, though the soybean has arguably been altered more by us than we have been changed by it. Some argue that when you weigh the economic, medical, and political factors, no other food plant has coevolved to support mankind as much as the soybean, which is interesting for a bean that is not easily digested in its natural form.

Soybean is the legume most improved as a human food by fermentation.
SIDNEY W. MINTZ

There are many varieties of soybeans, but these three are the most available.

STANDARD SOYBEANS

BLACK SOYBEANS

NATTO SOYBEANS

Bill Shurtleff

It was the end of summer, yet the day was hotter than it should have been. Distant mountain fires filled the air of the San Francisco Bay Area with oppressive smoke as we drove up to the SoyInfo Center, which is also Bill Shurtleff and Akiko Aoyagi's residence. Bill met us outside and ushered us in, quickly closing the doors to keep the cooler, more comfortable air inside. We sat in the living room, surrounded by shelves heavy with books. More books and files crammed the upstairs library. While the SoyInfo Center supports a rich database of free downloadable information, it was awe inspiring to see so much knowledge represented in material form.

People write cookbooks for different reasons. Some are following a passion, some want to preserve and share their heritage through recipes, and others are mired in obligation, paychecks, or whatever. For Bill, who with Akiko wrote the cornerstone works that introduced the United States to powerful traditional soy foods — *The Book of Miso, The Book of Tempeh, The Book of Tofu,* and others — the reason was a steadfast commitment to solve world hunger. He was initially influenced by the 1971 work *Diet for a Small Planet,* in which Frances Moore Lappé argues that world hunger results not from a lack of food, but from poor food policy and an inefficient use of food. Rather than running about 85 percent of the world's total crops through animals that we will eventually eat, Bill argues, there is a more efficient way to get our protein: from soybeans and grains grown not for herbivores but for

humans. By bypassing the animal, we avoid the dramatic environmental toll that global livestock production makes upon our planet.

Decades before Internet resources, Bill and Akiko went to the countries where these fermented food traditions and techniques could only be learned by experiencing them in person. Rather than capturing techniques in photos, they chose, through Akiko's talent, to hand illustrate these processes in a simple yet eloquent way.

Bill and Akiko's books, which have been (and likely still will be) updated as new studies are published, present soy foods as a practical, low-cost source of protein for underdeveloped countries. "I am not a seller of soybeans — I am a seller of vegetarianism. I want soy food to be the most important protein in the world," Bill told us emphatically. "It's not about me. I am just a conduit."

Judgment of the Bean

We need to come clean about something. During our vegetarian/vegan years we were big consumers of soy, and especially tofu, but then something happened: We read about the dark side of unfermented soy. And for a while thereafter, soy left our diet completely.

You may have heard about some of the dark side: Rain forests in Brazil are being leveled in the country's race with the United States to become the world's largest grower of soybeans. Soybeans contain antinutrients, which from their name says it all, as well as phytoestrogens, which can impair the thyroid and cause fatigue, infertility, and even breast cancer. Finally, there is the fact that the majority of soybeans grown in the United States are genetically modified.

As we eventually learned, fermenting soybeans makes these legumes not only safe to eat but incredibly healthy. And organic regulations now prohibit genetic modification. So, eat your soybeans — fermented!

Trypsin Inhibitors

As we've established, the soybean is extremely nutritious. The problem for humans is that it also contains antinutritional factors that prevent us from properly digesting it and absorbing the nutrition it has locked within. Take, for example, the soybean's high levels of protein. In order for our bodies to use protein for tissue growth and body maintenance, we need to break it down into amino acids, which our bodies can absorb. We do that through enzymes like trypsin, which are excreted by our pancreas into our small intestine. Once there, trypsin activates other enzymes that digest protein into small amino acids.

Now imagine that you just ate a tofu salad or drank a glass of cold soy milk. Your body's digestive process kicks in and the trypsin is there and waiting, but unfermented soybeans contain trypsin inhibitors. The inhibitors glom on to the trypsin, preventing it from processing the proteins into simpler amino acids. The proteins pass through in a form our bodies cannot easily metabolize, and we lose the nutrition. The good news is that a thorough soak and boil or steam, followed by fermentation, deactivates most of these inhibitors.

GMO Soy

More than 99 percent of the soybeans grown in the United States are genetically modified organisms (GMOs), and they have one of the highest pesticide contamination rates of any crop. While we found research suggesting that GMO soybeans are safe for human consumption, we also found research that suggests quite the opposite. In the end, you need to decide for yourself what you believe to be true. In our case, we only purchase USDA-certified organic soybeans, which are pesticide- and GMO-free, because we believe there is enough evidence to warrant it.

Phytates

To understand what phytates are, we need to start with the mineral phosphorus, which is a vital nutrient not only for plants but also for animals like us. In fact, it comes in second behind calcium as the most abundant mineral in our bodies, making up 85 percent of our bones and teeth. Phosphorus keeps our kidneys functioning and our teeth and bones strong, and it helps bring energy to our body's cells. It is even in our DNA and our RNA. Plants store phosphorus as a compound called phytic acid. So far, so good. If phytic acid kept to itself the story would end here, but it doesn't do that. Phytic acid is a strong chelator, which means it binds with other minerals. Once those minerals are bound, it becomes a phytate and those minerals, including the original phosphorus, are locked away inside. To free them we need a key, and that key is a phytase. Phytases are enzymes that break apart the phytate, freeing not only the original phosphorus but also other minerals like calcium, magnesium, and iron. Go phytases!

Plants naturally have some phytases and we have phytases in our guts, though at pretty insignificant levels. To break down our food into its nutritional components, we need the help of microbes, which are nature's most important source of phytases. Processing our food becomes a relay race — first the phytases in the plants go to work, then the phytases in the microbes, and finally those in our guts as we digest the food. Along the way phytates are broken up in a process called phytate dephosphorylation. Let's look at each leg of the race and what is happening:

Soaking beans at room temperature for 18 to 24 hours reduces phytates in two ways. First, because phytates are water soluble, many will dissolve into the soaking water. This is why it is important to drain the water and replace it with fresh water a few times during the soaking period. The second thing soaking does is to provide a supportive environment for the plant's natural phytases to work on the phytates. Research has shown that the optimal environment is water that is at a pH of 5.0–6.0. This pH happens naturally over a long soaking period as lactic acid moves in. In a pinch, it can be reached by the addition of a tablespoon of vinegar to 4 to 5 cups of soaking water.

Germinating beans is simply following the plant's design for their seeds — to prepare them to be plants. Naturally occurring phytases break down the phytates to release phosphorus, which the beans use as energy for their germination process.

Cooking beans doesn't have as much effect on phytates as you might think, and that's due to the fact that most phytates are heat stable. Our friends the phytases, however, like all enzymes, are heat sensitive, so when you cook your beans, their lap is done, and you are taking those plant phytases out of the race.

Fermenting further degrades the remaining phytates and provides the microbes in our guts, which represent the last leg of our phytase relay team, a big lead to finish the race.

Allergens

Food allergies have likely been with us since the earliest days of humankind, but they have gotten worse in recent years. According to the World Allergy Organization's latest *White Book on Allergy*, it's estimated that the number of people globally with food allergies has risen from between 1 and 2 percent in the early 2000s to 3 to 7 percent in 2017, or between 240 million and 550 million people.

Of the 170 foods that have been documented to cause food allergies, soybeans are one of the eight most common, according to the Food and Agriculture Organization of the United Nations. The other seven are cow's milk, eggs, fish, crustaceans, tree nuts, peanuts, and wheat. Approximately 2 percent of all food allergies are soybean related. We will do the math for you. In the United States, approximately 196,000 to 457,000 Americans are allergic to soybeans. That's somewhere between the population of Grand Rapids, Michigan, and Colorado Springs, Colorado. Another way to visualize the statistic is to line up a thousand people, in ten rows of one hundred people. The last person in the last row will be truly allergic to soybeans. However, if you ask this group how many people feel that they are allergic to soybeans, one-quarter to one-half of the front row might raise their hands. What these people likely have is an intolerance or sensitivity, not a true allergy. Food allergies are an abnormal response by our immune system to components, typically naturally occurring proteins, in our food. People have one of two reactions: immediate (within minutes) or delayed (a day or more), and the severity of the reaction is typically much stronger for the immediate reaction.

With food intolerances and sensitivities, symptoms can be quite uncomfortable but are not as serious. Milk is a good example. Some people have lactose intolerance, which means they lack the necessary enzymes in their small intestines to adequately process lactose. While lactose intolerance can cause uncomfortable gastrointestinal symptoms, there is not an immune response from the body. By contrast, people with a milk allergy need to avoid it completely, because even small amounts will cause their bodies to take systemic actions that can be quite serious.

Some good news for allergy sufferers is that fermentation appears to make a considerable difference. In one study conducted in 1993, where soy-allergic adults were exposed to a number of soybean products and tested for the immunoglobulin E (IgE) antibodies in their blood, researchers found that "fermentation appears to be particularly effective in reducing allergens present in a soy product" and "fermentation may alter or destroy allergenic epitomes."[4] It is unclear to us why this is the case, but research in this area continues. While this is promising research, if you are an allergy sufferer, please consult your health care practitioner before exploring soy-based ferments. And don't worry — everything in this book can be made soy-free.

The food of one may be the poison to another.
LUCRETIUS

Growth Inhibitors?

You might have heard that soy can stunt your growth. Our third son was born during our heavy soy consumption phase, and his formative years were fueled by that diet. Today, as a young adult, he is 6 feet 8 inches and an NCAA athlete. The average height of our four adult children is 6 feet 4 inches, so in our admittedly limited trial we certainly didn't see this stunting effect, or if we did, it kept us from having a family of NBA centers and busted door frames in our farmhouse.

Soy has gotten this reputation due to the presence of a blood-clot-promoting substance called hemagglutinin. In combination with trypsin inhibitors, it has been shown to limit enough nutrition in certain populations of children that it has stunted their normal growth. However, both substances are deactivated by — you guessed it — fermentation. It's a theme that we hope you are picking up on and why we are so passionate about the power of this process to unlock the good and remove the bad from legumes and grains.

Phytoestrogens

Soybeans contain the isoflavones genistein and daidzein, which in the plant world only occur in legumes. Isoflavones are functionally and structurally similar to the estrogens in our bodies, so they are called phytoestrogens. When soy is fermented, a chemical process breaks down the isoflavones (which are glycosides) into a form (aglycones) that is more easily absorbed by our bodies.

Research has linked a diet rich in isoflavones with beneficial effects on heart disease and some types of cancer, improved bone strength, and obesity prevention.[5] There is a lot of research to wade through, and it's sometimes a bit contradictory, but certainly there is general consensus on the anti-inflammatory effects of isoflavones in the body. Several studies show that phytoestrogens relieve many menopausal symptoms as well. A diet rich in soybean isoflavones and protein has been shown to alleviate some symptoms associated with type 2 diabetes, including enhancing the body's insulin secretion and reducing insulin sensitivity. Finally, foods derived from fermented soybeans help decrease the risk of onset and the progression of insulin resistance and type 2 diabetes.

What final judgment do we make on the soybean? It has been consumed for over three thousand years in Asian countries, is central to traditional Asian diets, and is predominantly eaten fermented or with meat. Studies show that such a diet contributes to lower incidences of heart disease and type 2 diabetes, as well as stronger bones, relief from symptoms of menopause, and a decreased risk of breast and prostate cancers. Like most topics, the benefit of soybeans is not black and white, but the evidence seems strong for finding ways to include soybeans in our diets — if they are fermented.

Mountains of Beans

Soy takes center stage in many traditional versions of fermented foods, but it is by no means the only legume that can be successfully fermented. Following is a quick tour of some more legume varieties to put in your crock. We don't even begin to cover all the legumes that lend themselves to fermentation.

Adzuki

These shiny little dark red beans with a white split are often boiled, mashed, and combined with sugar to make a sweet paste that is prominent in many doughy treats in Japan and China.

Adzuki beans (*Vigna angularis*) lend themselves to other more exotic sweet treats. When Christopher worked in Singapore, he frequented the hawker stands for lunch and became addicted to the red bean ice cream. His favorite versions were those something like a snow cone with an adzuki-based sweet syrup, along with a sprinkling of nuts and other unidentified toppings. Adzuki beans also lend themselves to delicious misos and tempehs.

Chickpeas

Chickpeas (*Cicer arietinum*) are known by many names, including garbanzo beans and gram. Besan, a flour made from chickpeas, is popular in East Indian cooking and used in the recipe for Myanmar-Style Fried Tofu on page 93.

Chickpeas have a strong nutritional profile. They provide significant amounts of most of the essential amino acids, and the missing amino acids can be picked up by adding grains or another source of protein like eggs, meat, or dairy. They are rich in protein, fiber, vitamins, and minerals.

Chickpeas, like many beans and grains, are more easily digested when germinated before cooking.[6] The flour can be called chickpea/garbanzo/gram/besan — all are the same thing. Fermenting the flour makes the protein more digestible.[7]

Common Beans

There are at least 30 varieties of the common bean *Phaseolus vulgaris*, including some of our favorites like Anasazi, black turtle, pinto, cranberry, orca, Marcella, Rio Zape, Santa Maria pinquitos, Mayocoba, Good Mother Stallard, and Moro.

Common beans are a good source of protein, carbohydrates, fiber, minerals, and vitamins. When working with these beans, the most important thing to note is that each variety has a different starch-to-protein ratio, which can give you wildly different results with ferments.

Though all great agricultural societies have their own staple starch — wheat in the Middle East and Europe, rice in Asia, corn in the Americas — beans are perhaps the one food common and indispensable to us all. Because of their ubiquity, beans are one of the few foods that serve as a unit of analysis and comparison across space and time. They are also among the few foods so avidly traded and transplanted across the continents throughout human history that today few people apart from botanists can keep all the species straight.

KEN ALBALA, *BEANS: A HISTORY*

Cowpeas/Black-Eyed Peas

Cowpeas (*Vigna unguiculata*) go by many different names (and varieties), including honey beans, African sweet beans, Sea Island red peas (see recipe on page 276), black-eyed peas, pink-eyed peas, and Southern peas in the United States; *frijol caupi*, *caupi*, *frijol de vaca*, and *carilla* in Spanish-speaking countries; *lobia* in India; *niébé* in the French-speaking countries of Africa; and *feijão* frade in Portugal.

In Africa, this legume is the primary source of protein for millions of people. Traditionally eaten boiled, fried, or steamed, they provide high levels of protein and other nutrients. There is only one major drawback: flatulence. That is, unless you ferment them!

SCIENCE SAYS . . .

A Fighting Flatulence Competition

In 2012, Dutch scientists fed 18 healthy non-smoking volunteers (7 women and 11 men) traditionally cooked black-eyed peas fermented using various means (yeast, fungus, or bacterium) to see which fermentation process had the most powerful effect on flatulence.[8] Basically, a cowpea porridge was prepared and served every morning for 11 days to these folks, with each porridge being one of nine formulas. The porridge varied only in terms of how it was fermented — with bread yeast, rhizopus fungus of tempeh fame, or *Bacillus subtilis* of the natto world.

Cowpeas were used because they have some of the highest levels of carbohydrates that resist digestion through most of our gut. When they finally meet their microbial match in our colons, they are broken down, but that process produces several gasses (hydrogen, carbon dioxide, and methane) and that, in a nutshell, is flatulence. The key to reducing this process is to break down those carbohydrates before we eat them.

At this point you might be wondering: Just how did they measure the participants' gas? The worst graduate student project ever? No, thankfully, they took breath measurements from the study subjects, both with their prebreakfast empty stomachs and then again every hour for 12 hours.

The results? They varied based on the birth country of the participants. There was generally less flatulence produced from the lactic acid bacteria ferment, but the degree varied. Participants native to Europe produced less abdominal gas than those born in Asian countries, who produced less than those born in African countries. Although all participants reported eating mostly typical European diets, those from Africa and Asia also noted that their diets still consisted of meals from their home countries. This is likely evidence of the different gut microbiota resulting from our early childhood conditions and current diets.

After reading hundreds of scientific studies over the last few years, we have come to realize that most studies have no intention of improving culinary uses of legumes — everything from flavor to texture seems to be beside the point. We use black-eyed peas to make natto and serve it as a loosely defined Hoppin' John (see the recipe on page 120). These legumes also make a good substrate for a tasty tempeh.

Horse Gram

Horse gram (*Macrotyloma uniflorum*), also called Madras gram, kulthi bean, huruli, and even soybean (this is how it was introduced to us in Myanmar, though it is visibly different), has been in use as a food since at least 2000 BCE. Its common English name comes from its use as a staple food for livestock, specifically racehorses, due to its high protein content.

This dark brown, reddish, or even yellowish legume thrives in difficult growing conditions such as drought, low soil fertility, and salinity. In fact, horse gram thrives even in soils contaminated by heavy metals. So, we can add to horse gram's benefits the fairly unique ability to serve as a soil reclamation plant.

Horse gram is an excellent source of protein (18 to 25 percent by weight), carbohydrates (52 to 60 percent by weight), and dietary fiber. In Ayurvedic medicine, the seeds of horse gram have been used to treat urinary stones and diseases and, when cooked and spiced, to treat the common cold, throat infections, and fevers. Recent analysis shows that horse gram has high levels of antioxidants and free radical scavenging activities, providing a scientific basis for these traditional therapies.

Horse gram beans are fairly hard to find; however, you can order them online (see the source guide on page 388) if you want to try Pone Yay Gyi (page 95) — the fermented bean paste of Began, Myanmar.

SCIENCE SAYS...

Living on Fermented Black Beans

In 2006, two researchers at Simón Bolívar University in Venezuela examined what happens to black beans when they are fermented.[9] They bought black beans at their local grocer and fermented them with a laboratory microfermenter, which is capable of continuously fermenting multiple samples under controlled conditions. They cooked the fermented beans and then made a mash with cornstarch and oil, which became the sole diet of 6-week-old rats.

They discovered that dozens of different yeasts, molds, and bacteria were present on the beans during the fermentation, but none of the molds or yeasts survived the fermentation. Of the bacteria, *Lactobacillus* species won out, becoming the dominant species and lowering the pH. When the researchers inoculated a later round of dried black beans with liquid from a previous fermentation (a process known as back-slopping) and then fermented them, it supercharged the *Lactobacillus* team, further reducing the soluble dietary fiber and other flatulence-producing sugars compared to the samples that were fermented with only the microbes occurring naturally on the black beans.

After two weeks of weighing the rats and their feed every other day, as well as collecting their waste, the scientists discovered that the rats that were fed the fermented diet digested more protein than the rats that were fed unfermented black beans. The research made no mention of differences in flatulence between the two groups of rats. It's too bad; I think we all want to know that one.

Black beans have a shiny black skin and a white center, and they are smaller than a pinto bean and less curved than a kidney bean. We especially like them fermented in tempeh (see Choose Your Own Tempeh Adventure, page 142).

Meet the Maker

Steve Sando
Rancho Gordo

Steve Sando has more passion for, and knowledge of, heirloom beans than anyone else you are likely to meet. We discovered his heirloom bean business, Rancho Gordo, when fellow fermentista Karen Diggs told us that we needed to talk to "the bean guy" and introduced us through e-mail. Still unsure of who "the bean guy" was, we googled his website. The Rancho Gordo website is beautiful and spunky, and it makes you want to dive in and put on a pot of bubbling beans for dinner. Talking to Steve makes you want to do the same thing. Rancho Gordo is a producer and purveyor of unique heirloom beans — beans that have been lost to time and commercial production, or never lost but found growing tucked in mountain valleys of Mexico. Steve's business model is about fairness and compensating the farmers who grow these beans and educating consumers not only about the beans and where they came from, but also how to cook them.

It has been more than 17 years since Steve threw it all in to become a backyard farmer, spending long hours at local farmers' markets selling his heirloom tomatoes. He realized that first year that he would need something to sell earlier in the season, before his tomatoes came on, so he turned to beans. A man motivated by flavor (he become a tomato farmer out of frustration that his local high-end grocer only offered flavorless Dutch varieties), he was hooked on heirloom beans at the first taste.

"I wasn't a very good farmer," Steve confessed to us. "I was really good at the beginning of the season and by the end it's like, oh, will this thing ever end," he said. Instead of growing the heirloom beans himself, he pivoted to finding farmers who were already growing heirloom beans. He took many trips to Mexico, spending time on the ground to build relationships and trust with local farmers. He could have more easily told them what he wanted them to grow, but he did the more difficult thing of learning about what

they already knew how to grow well and then building a market around those beans.

Steve summed up his knowledge thusly: "The key is how you create opportunities, so people can do what they do best. That's really what I have tried to focus on."

Lentils

There are six species of lentils, but only one, *Lens culinaris* has been cultivated. Lentils vary in color from black to brown, gray, green, yellow, and red, and sometimes they are even speckled. If you are lucky enough to have access to Middle Eastern or Indian markets in your area, look for these different varieties.

Lentils take the prize as the oldest legume crop. Archaeologists at the Franchthi Cave in southern Greece found evidence that humans were consuming lentils in 11,000 BCE. It was probably one of the earliest of all crops domesticated in Africa, Europe, and Asia.

Lentils are about 25 percent protein, 56 percent carbohydrates, and 1 percent fat by weight, making them a rich source of cheap and easily available protein. That is, if you ferment them. Fermentation reduces the trypsin inhibitors, improves the digestibility, and reduces or eliminates some off flavors of lentils. We have made a number of nice lentil tasty pastes (see the section on making your own modern tasty pastes, page 268). They also work quite well in tempeh.

Runner Beans

Varieties of runner beans (*Phaseolus coccineus*) include scarlet runner beans, runner cannellini beans, the almost garish Christmas lima, and some of our favorites: thick-skinned ayocote negro and ayocote morado.

Runner beans have a starchy, potato-like quality. We weren't sure how they would do in miso, but we made a few excellent misos from this family. We also love the unique flavor and texture of runner bean tempeh.

Tepary Beans

Tepary beans (*Phaseolus acutifolius*) are the only legumes in this book that originated in North America. Where exactly in North Amerca is a bit of a dispute. There is clear evidence of tepary beans in the Tehuacán Valley in Mexico 2,300 years ago. The name *tepary* (*tepari* in Spanish) is thought to have originated from the Tohono O'odham Native Americans of southern Arizona, where the beans are believed to have been a staple food for over five thousand years.

Tepary beans were very popular in the early 1900s, but around the beginning of World War I, they lost ground to other common beans in the battle for Americans' palate. However, they are enjoying a comeback in recent times, and we think that this is because they compare very favorably to soybeans in terms of protein levels (with around 20 grams of protein per 100 grams of dry beans), yet they have lower levels of antinutrient factors like trypsin inhibitors, phytic acid, and oligosaccharides, which are known to produce flatulence. When compared to other common beans, unfermented tepary beans have a stronger odor and take longer to cook. However, when they are fermented, we found their flavor to be as good as or superior to that of other legumes, and they are the only legume that can match soybeans' fermentation characteristics when it comes to making natto. If you are looking for soybean alternatives in a ferment, consider tepary beans, a North American original.

A Guide to Cooking Beans

You can find a lot of information out there on cooking beans, and it often seems contradictory. To soak or not to soak? To add baking soda, salt, acid, or special cultures? What's a cook to do? The following instructions are the result of our research combined with a practical approach to cooking that doesn't get hung up on the details. We include step-by-step cooking instructions for soybeans because they cook up differently from all the other types of legumes.

There's Nothing Like a Good Soak

Let's talk about soaking and why it's a very good idea. The first reason to soak your beans is to reduce their cooking time, which also saves you fuel. Soaked beans take less time to cook because the cellulose of the cell walls has been broken down, which makes the hard pectins more soluble and susceptible to the heat of cooking. Remember that the humble bean is basically a baby plant protected by a husk. The trick to getting a tender, good-tasting legume is to break down this tough outer coat.

The second reason to soak your beans is to reduce the antinutrients inherent in them. The simple act of allowing the beans to bathe quietly on the counter in a pot of water will remove most of the main culprits — phytates and tannins. Remember that these antinutrients are water soluble and will be leached into the soaking water, so it is a good idea, if you can (no need to run home from work or wake in the middle of the night), to change out the water once or twice during the soaking time.

So now that you see why it's important to soak your beans before cooking, how do you go about doing it?

What to soak them in. You may have read about different water solutions. Some recipes say you should always soak your beans in purified water, while others say salted water, or water with baking soda, or water with an acid like whey or apple cider vinegar. We tried all of these methods and found the best results using fresh water and a long soak, which lowers the pH. We found that vinegar sometimes gave our tempeh off flavors, and we learned that it wasn't necessary to lower the pH — the pH came down naturally because lactic acid was already present. Sound like fermentation? It is. This is important for two reasons. First, as we discussed earlier in this chapter from the Dutch flatulence study, a lactic acid fermentation is one of the most effective ways to break down the carbohydrates our body has difficulty digesting. Second, remember from the soybean section earlier in this chapter, phytates are broken up by phytases, and phytases do best in an environment where the pH is between 5 and 6.

Soaking affects your beans in ways that you might not expect. When soaked right, your beans emerge from the cooking pot with their skins intact and just as tender as their creamy insides. When not done right, your beans emerge from the pot looking pretty beat up, skins torn and hanging off the exposed insides like they went a few rounds with a wild animal. When this happens, it's likely that the beans' skins weren't softened up enough before they were cooked. Instead of the water easily penetrating the skins and equalizing the pressure within and without, the skins maintained resistance, pressure built up inside the beans, and boom! — or rather lots of tiny booms! — the bean skins exploded.

How long to soak. While a surprising amount of research has been done on this topic over the past 30 years, the conclusions vary depending upon the legume. We recommend soaking beans for 18 to 24 hours, though beans can be ready in as little as 12 hours, which should be enough time for even the toughest beans. However, hardness varies quite a bit. This can be a result of age, as older beans are harder and can take longer to soak and cook. Also, the variation is seen between varieties — for example, the tiny tepary bean often takes a full day. Find what works best for your favorite beans and stick to that.

The Best Ways to Cook Beans (All but Soybeans)

There are several options for cooking beans. You can cook them in water or steam them in a steamer basket, either in a pot on the stove or in a pressure cooker. A pressure cooker is an efficient way to cook or steam the beans, but it is by no means necessary (in fact, we find it's best not to cook beans for tempeh in a pressure cooker, as you need them to be al dente and it's easier to monitor the cooking on a stovetop). We have found, however, that for many of the recipes in this book, steaming produces a better end product than cooking directly in water. Cooked beans should be tender — they should crush easily between your thumb and ring finger and have a good mouthfeel — yet they should not be so soft that they are falling apart and splitting.

Cooking beans on the stovetop. Put the beans in a pot and add enough water to cover them by 2 inches, then place the pot on the stove over medium-high heat, cover, and bring to a boil. The beans will usually generate a lot of foam and scum. Skim this off. Turn down the heat and keep the beans at a simmer until tender. You want to avoid boiling from here on out, as the beans tend to fall apart in the rough waters. The timing will be different depending on the bean. Dehulled dal beans or split peas for tempeh are usually done shortly after boiling; check after 5 minutes, then again at 10. For most beans, simmer for at least 45 minutes. Some varieties, like black-eyed peas and lentils, will be tender at this point. Other varieties, like tepary, will take longer, so check every 30 minutes. **Note:** Beans that have been stored way too long (before they ever got to you) can be challenging, and in the worst cases, never soften.

Cooking beans in an electric pressure cooker. After soaking your beans, pick off any split beans or loose hulls. These could clog the vent. Place water in the bottom of the cooker. For steaming, you'll need just 1½ to 2 inches. For boiling, you'll need to make sure the beans are covered by 2 inches of water. Then add the beans. For steaming, place the beans in a steamer basket. Seal the pot with the lid. Follow the manufacturer's instructions for cooking beans in your pressure cooker.

Soaking and Cooking Soybeans

Soybeans can be a challenge to cook, taking significantly longer than most beans. We have found steaming produces a better end product than cooking them directly in water.

Rinsing and soaking. Once you have measured the soybeans needed for the recipe, place them in a bowl and run water over them. You will notice bubbles rising when you add the water; these are the saponins, which are very different from the saponins in quinoa (see page 161 for more information). We generally run our hands through the beans (it's fun), pour off the water, then add water to cover the beans so that the water is significantly higher than the bean level. In general, you're looking for a 1:4 beans-to-water ratio. Soak them for at least 18 hours, then drain and rinse well. Change the water once during the soaking period if you can, as it will remove the first of the impurities that have been leached into the water. By all means, don't set your alarm to get up in the middle of the night to do this.

Cooking without pressure. Cooking soybeans without pressure can be time consuming. For split and hulled soybeans, and other beans, this stovetop method gives you the most control of the texture. If you are cooking your whole soybeans in a pot without pressure, make sure they have plenty of room in the pot. Add water to cover the beans by at least 2 inches, then cover with a lid and bring to a boil over high heat. Skim off any foam that may come to the surface of the water. Allow to boil for 15 minutes and reduce the heat to medium-low and simmer for 1½ to 4 hours, or until the soybeans are soft. You may need to crack the lid to keep the pot from overflowing. Check the water level every hour or two and add warm water as needed to keep the beans covered by a couple of inches. For cracked and hulled soybeans, see the instructions in the tempeh chapter on page 137.

Steaming without pressure. You can also steam soybeans without pressure. Pour a few inches of water into a large pot and set a steamer basket inside it. Add the beans, cover with the lid, and bring to a boil over high heat. Then reduce the heat to low and let the beans steam. Check the water level frequently, and add more water as needed to keep the pot from burning dry. This will take a number of hours.

Steaming with pressure. We find that cooking the soybeans under pressure works particularly well when making natto and koji ferments, as the beans have a superior texture. The texture of the beans is less of a concern for miso, and one must be careful to not overcook beans for tempeh.

When steaming beans in a pressure cooker, first pick out any split soybeans or loose hulls after soaking. These could clog the pressure cooker vent. Pour a couple of inches of water into the pot, set the steamer basket in place, add the beans, and close it up. Cook for 30 minutes at 15 pounds of pressure or for 45 minutes at 10 pounds of pressure. If you have an electric pressure cooker, use the steam or bean setting for 45 minutes on high pressure. Soybeans soaked and hulled for tempeh take 10 minutes when steamed under high pressure. We don't recommend boiling beans in a pressure cooker because the hulls and foam get all over the pot, and despite precautions we had trouble with some batches.

We're not going to lie, we love our electric pressure cooker. We cooked a lot of beans while writing this book, and buying a programmable electric pressure cooker was a game changer. We have a 6-quart pot, and we have the best results with 2 pounds or less of the normal-size soybeans.

Soaking and Steaming Beans
in an Electric Pressure Cooker

1. Rinse the beans thoroughly.

2. Let the beans soak at room temperature, replacing the soaking water with fresh water once.

3. The beans will be ready to cook after 12 to 24 hours of soaking, depending on the bean.

4. Drain and rinse the beans.

Instructions continue on next page

56

5. Steam under high pressure. For al dente, steam hulled soybeans for 10 minutes, hulled common beans for 8 minutes, and whole common beans for 18 minutes. For soft beans, steam common beans and natto soybeans for 35 minutes, and steam soybeans for 45 minutes.

6. The beans are ready! The cooked beans should maintain their shape and be soft enough to easily crush between your thumb and ring finger.

Zero-Waste Tip:
Don't Toss the Bean Water!

Aquafaba is the fancy name for bean water — the cooking liquid that comes from boiling beans. In this book, you will use some of it to get the consistency of your miso just so, but it also has many other uses. You can use this viscous water as an egg substitute in all kinds of creations, like meringues, for example. The thicker and goopier, the better. Soybean and chickpea aquafabas are particularly thick (think about the goo in canned chickpeas), but any bean water can be used. Your aquafaba should be the consistency of runny egg whites; if it's not, simply boil it to reduce it. The Internet will then give you more recipe ideas than you will know what to do with.

Grains: Another Nutritional Powerhouse

Why eat grains? In both developing and developed countries, cereal grains are often the main source of people's carbohydrates, protein, fiber, vitamins, and minerals. When consumed in their natural whole-grain form, many have even been found to help prevent chronic diseases. While these grains can be consumed raw, their digestibility and flavor are usually improved through cooking. Fermentation takes both to another level.

Fermentation drastically lowers the level of antinutritive compounds (phytates, tannins, and polyphenols) found in grains that would otherwise prevent our bodies from absorbing protein and minerals. Fermentation also makes the vitamins in grains more available to us.

In recent decades, the issue of celiac disease and nonceliac gluten sensitivity (NCGS) has led to a great deal of interest in gluten-free grains like rice, amaranth, quinoa, teff, and millet varieties. However, there may be something else at play for those in the nonceliac group besides gluten. Researchers at Monash University in Australia wondered whether FODMAPs (fermentable oligosaccharides, disaccharides, monosaccharides, and polyols) could be a factor. Basically, FODMAPs are carbohydrates that humans can't digest too well. When they pass through our small intestine undigested, they become food for the bacteria in our colon, producing a lot of gas and other symptoms, including irritable bowel syndrome (IBS). The question is: Could FODMAPs, which are present in high levels in many of the same grains as gluten, be the cause of NCGS symptoms?

In a 2013 study in Australia,[10] 37 participants with self-reported NCGS started a controlled diet that was gluten-free and low in FODMAPs. Later they were divided into three groups, and each received varying amounts of gluten but stayed at the low-FODMAP level. The researchers found that only 8 percent of the participants who believed they were gluten sensitive showed symptoms when gluten was introduced to their diets. This is a really big deal for the roughly 15 percent of the U.S. population who identify as gluten sensitive. It may be that gluten is not actually their problem.

For those for whom FODMAPs might turn out to be the trigger, fermentation is, again, the hero because it dramatically lowers FODMAP levels in beans and grains.

One final but culinarily important point: fermenting the following grains enhances their flavor by forming several volatile compounds that become the aromas you will enjoy while making our recipes. It's time to go beyond bread and cereal.

Rice

CHRISTOPHER WRITES: *In the fall of 2014, I made the first of a number of trips to Arkansas to work with four rural community colleges. Having grown up in the neighboring state to the north, Missouri, I expected to see fields of corn, soybeans, or sorghum. But as I made my way along winding rural roads, I kept seeing sunken fields and towering grain bins that reminded me of those I had driven through so many times before in Northern California, home of the organic rice juggernaut Lundberg Family Farms. I had always assumed California was the rice heartland of the United States, so after driving past so many rice fields, I asked my colleagues that first evening, "What's up with all the rice?" They gave me a quizzical* *look that said, "Where have you been for the past hundred years, our newfound unenlightened West Coast friend?" That year, Arkansas produced nearly 12 billion pounds of rice — more than the next five states combined. At least I wasn't in Arkansas to share my agricultural knowledge.*

Asian rice (*Oryza sativa*) is one of only 2 cultivated and 21 wild species of rice. While African rice (*Oryza glaberrima*) is grown only on a small scale in West Africa, Asian rice is grown around the world. It has two main subspecies, *japonica* and *indica*, whose names give you a hint about their countries of origin. The *indica* variety is thought to have been first domesticated in India, as well as other countries east of the Himalayan mountains, like Myanmar and Thailand. From there, *indica* dispersed throughout the tropics and subtropics. The *japonica* variety traces back to southern China, from where it eventually moved northward, becoming the species best adapted to temperate climates.

How important is rice to humans? Well, rice is the primary staple for about half of the world's population, so for a few billion people, it's pretty important. Rice is now grown within a band stretching from the 55th parallel north to the 36th parallel south, or if you like, think between Denmark and New Zealand. Humans have been selecting for improved varieties for a very long time, resulting in about 120,000 different varieties. More than 90 percent of the world's rice production goes to feed consumers in Asia. Rice is so important in Japan that it is sometimes called *shushoku*, which means "principal food." In fact, among all the grains, something unique about rice is that nearly all of it goes to feed people, not livestock.

The number of rice consumers is growing, and in many regions of Asia the urban sprawl to house those consumers is eating up the rice fields, causing the situation we are in now: increasing populations depending upon rice and decreasing capabilities to grow more. Some people are looking to discover new varieties that produce more, while others are trying to preserve and rediscover the unique traits, including lost flavors, of some of those 120,000 varieties in the world. We explored the latter track, looking for rice varieties that are traditionally used and then experimenting with others to see what is expressed in the fermentation process. In addition to flavor, all the thousands of varieties are distinguished by, among other things, length, grade, glutinous levels, and color.

Rice varieties range in color from brown to gold to red and even black. White, the color we most commonly associate with rice, is missing from this list because it comes not from nature but from processing. When we remove the bran layer, what is left is the white kernel. We process rice to increase its shelf life, make it more convenient to cook, change its cooked texture or flavor, or simply for appearances.

Without rice, even the cleverest housewife cannot cook.

CHINESE PROVERB

Carolina Gold Rice

One of the varieties making a successful comeback is Carolina Gold rice. This variety, thought to be the ancestor of long-grain rice in the Americas, went all but extinct after the Depression. Fortunately, the USDA seed bank had stores, and in the 1980s, Dr. Richard Schultz brought this beautiful golden rice back to the fields of the coastal wetlands around Charleston, South Carolina — and back to our plates.

Parboiling. For most rice, the first step in processing is milling, which is like mining away successive outer layers of the rice seed. The first layer to be removed is the rice husk, which is good because we can't eat that, even though the husk does contain nutrients. That's where parboiling comes in. Also known as hydrothermic rice treatment, parboiling helps protect the nutrients that would otherwise be stripped away by milling, and it improves our sensory experience of the final product. Half of all rice grown in India and a quarter of all the rice grown worldwide is parboiled.

Rice that is parboiled is soaked, steeped, and steamed in the husk. As this happens, the water-soluble vitamins, minerals, starches, and proteins are driven deeper into the grain and compacted, making it much harder for them to be removed by the milling process to follow. Enzymes (like lipases) and biological processes like germination are inactivated as well. When the parboiled rice is dried and then milled, it has higher yields, better quality, a much longer shelf life, and a desirable glassy or shiny appearance. When cooked, parboiled rice has higher nutrition. Finally, cooked rice that was parboiled seems to be easier for us to digest than rice that was not parboiled. That's all the positives of the process, and there are a lot of them.

On the negative side, parboiling rice that is destined to have its bran removed, turning it into white rice, actually decreases some of the vitamins and minerals as compared to regular brown rice that is not parboiled and does not have its bran removed. And then there's the possibility of odor, which Kirsten noticed the first time she parboiled rice to make idli and dosa. If the steeping step goes too long (more than 8 hours) or occurs at too high a temperature (above about 150°F/66°C), then amino acids that contain sulfur split, releasing hydrogen sulfide. Yep, you are remembering correctly from your high school chemistry days: that rotten egg smell. It isn't in all parboiled rice, though, and you might not come across it, as it can be completely avoided with modern equipment.

Milling. Milling does a number of things, in this order: First it removes the husk, which is typically used as fuel for boilers but can also be used in gardens or, as you will learn in this book, to make ferments! With the husks removed, we have brown rice (or red, purple, black, et cetera, depending on the variety). The next step is to remove the bran, at which point we have white rice. As you might have guessed, we can also make use of the rice bran in fermentation, in this case in a nuka pot (page 317).

Polishing. Polishing picks up where milling leaves off: with the white rice. Abrasive substances (including long strips of sandpaper) and/or water polish the rice, resulting in very smooth and white kernels that better expose the starch hidden within. Polishing also removes the essential proteins and amino acids, vitamins, minerals, and fiber — basically, everything that is good for us — from the rice.

Why would anyone do this? First, polishing removes fats, which are responsible for rice going rancid. Removing the fats extends the rice's shelf life from at best a few months to years. Second, polished rice requires half the time to cook. Third, tradition — many traditional Japanese fermentation recipes call for polished rice. White rice signified wealth and brown rice, which contains all of the before-mentioned parts, is not suitable for koji because the natural surface of the rice kernel is too hard for a strong mycelial growth, especially for sake. There are many traditional brown rice ferments also.

Barley

Barley (*Hordeum vulgare*) is the oldest known cereal grain. Only up to about one-quarter of the worldwide harvest goes to food production (and most of that goes to making malt), which is a shame because barley is an excellent source of vitamins, minerals, and complex carbohydrates. The problem with barley is that it requires a relatively long cooking time, and even then, its minerals are not that available to us. That changes dramatically if the barley is sprouted and then fermented. Then the carbohydrates become more digestible and the minerals more available, effectively improving everything barley has to give us.

Barley has been bred for malt production for beer for a very long time — varieties were optimized for better yields and disease resistance, not flavor. But that is changing. Breeders at Oregon State University are cross-breeding ancient varieties with current ones with a focus on the flavor rendered in beer, which is something innovative for beer makers who felt that barley was barley and malting was where the magic happened. Imagine a day when these same breeders, or the next generation, are breeding barley here in the United States for koji? By the same token, in Japan they have a *koji tane* (starter) that is specifically for barley. You will find recipes for barley koji on page 203 and we also use it in tempeh on page 160.

Corn

Corn (*Zea mays*), also known as maize, ranks with wheat and rice as one of the top three crops produced in the world. It was originally domesticated in Mexico about nine thousand years ago.

Varieties of corn are called races by the farmers who grow them and the botanists who study them, and there are well over a hundred of them worldwide. It's likely that they were all the result of farmers or breeders finding the right combination of kernel color, shape, size, starch content, and cob configuration to match their needs.

Nixtamalized corn kernels have been soaked and/or boiled in an alkaline solution (usually lime or lye) in order to improve their digestibility and alter flavor compounds. Nixtamalization transforms the corn in very much the same way that fermentation transforms its substrate. This process is important, as it makes available the nutrients in maize that our bodies typically cannot access. The ancient process is still used today in the southern United States, Mexico, and Central America to transform corn kernels into tortillas, tamales, hominy, grits, pazole, and pozol (a nonalcoholic beverage made and enjoyed in southeastern Mexico). We use nixtamalized corn in tempeh, where it provides wonderful flavor and texture. It is an ingredient in Ayocote Bean and Hominy Tempeh (page 174) and is the main player in Hominy–Pumpkin Seed Tempeh (page 176). We haven't yet had a chance to try growing koji on nixtamalized corn, but we intend to and encourage you to do the same.

Millet

Millet is a big family of small cereal grains, though the term *millet* is often knocked around to refer to small-seeded annual grasses. There are some great names to be found among the hundreds of species: bird's foot, finger grass, bulrush, foxtail . . . How about a Japanese barnyard millet to get your taste buds going? The variety of names gives us a hint as to their diversity and to their long relationship with humans. We go way back.

Millet is about 15 percent dietary fiber, and its outer layers have high levels of minerals like magnesium and calcium. As is the case for many cereal grains and beans, these nutrients are not very available to us when millet is raw or cooked. It also contains antinutrients like phytic acid, polyphenols, and tannins. This changes for the better with fermentation. In a study from the Sudan,[11] bulrush millet (also known as pearl millet) was naturally fermented at room temperature for 36 hours, with samples pulled every 4 hours and analyzed. The researchers discovered fermentation made the protein more digestibile and significantly reduced phytic acid and polyphenol levels. The effect on tannins was variable.

As recently as the Middle Ages, millet was far more prevalent than wheat in most cultures' diets. The supernutritious millet is waiting to be rediscovered. We hope to entice you with a few of the recipes in this book, like Ancient Grains Tempeh (page 169), Millet Koji (page 204), and Good Mother Stallard Millet Tasty Paste (page 273).

Oats

Oats (*Avena sativa*) are unique among the cereal grains for their healthful combination of function and nutrition. For at least 20 years, both the U.S. Food and Drug Administration (FDA) and the European Food Safety Authority (EFSA) have endorsed oats — specifically their soluble fiber in the form of beta-glucans — for their ability to lower blood cholesterol levels, which

A Culinary Fungus Strikes Again

Huitlacoche (*Ustilago maydis*), more commonly known as corn smut, is a fungus that grows on corn when rainwater gets into the husks and causes the kernels to grow grotesquely misshapen, many times their original size. It is a delicacy or a disease, depending on where you are standing. For a farmer hoping to grow corn for production, it's not so great. But for diners, it's a wonderful thing. The fungus significantly improves the health benefits of the corn. The transformed kernels have far more protein and much more lysine, an amino acid (and you know what we get with aminos . . . umami!), than the original kernels. Corn smut has a similar texture to mushrooms; it was eaten by the ancient Aztecs and is still a prized food in Mexico.

helps reduce the risk of heart disease. In fact, the first scientific papers reporting the significant effect of oats on blood cholesterol date back to the early 1960s. Oats are 10 to 20 percent protein, and compared to other cereal grains, they are rich in essential amino acids and have both dietary fiber and functional fiber.

Functional fiber is a term introduced in 2001 by the Institute of Medicine to help differentiate naturally occurring fiber from plants (dietary fiber) from fiber that has been extracted from the plant or has been synthesized (functional fiber). Because most of the U.S. population doesn't get enough fiber in their diets, food companies have taken to increasing the fiber content in food products with functional fiber. All this plus high levels of unsaturated fatty acids (lipids), vitamins, and antioxidants allow oats to have significant effects upon diabetes, hypertension, colon cancer, and inflammation.

Oats, like barley and rye but not wheat, have a coarse outer layer, which is actually part of their previous flower, that protects everything within. When you remove these hulls at the mill, you are left with oat groats, where the good nutritional stuff remains. You might have seen oat groats in the bulk bins of your grocery store, but it's more likely that you are familiar with rolled oats. Basically, rolled oats are oat groats that have been passed through large spinning drums with cutters and then steamed and squashed (to make rolled flakes) or pulverized (for quick-cooking oatmeal).

When oat groats are fermented with koji, some truly wonderful flavors present themselves. As they emerge from the fermenter, the smell is a mix of the distinct sweet smell of koji and a freshly baked oatmeal cookie. The first time we fermented oat groats, the smell was

so intoxicating that it set our minds exploring where koji oat grouts could be used, as you can see from the recipes in chapter 7.

You may be wondering: Why ferment oats at all? After all, oat bread and oatmeal raisin cookies seem pretty popular and contain much of the before-mentioned goodness. As is the case for other cereal grains and beans, the protein in the oats is easier for our bodies to absorb after fermentation. This is because fermentation degrades the proteins into peptides and free amino acids. Fermentation also enhances oats' antioxidant capacity. And so much simple sugar is released that you don't need any added sugar — how cool (sweet) is that?

Quinoa

You are most likely to see quinoa (*Chenopodium quinoa*) in breakfast cereals, as a grain bed for salads, or in gluten-free beers. It is not a true cereal grain but is technically classified as a pseudograin, alongside amaranth and buckwheat. (We don't use amaranth or buckwheat in this book, so quinoa will have to represent the pseudograins, which it does pretty well.) What's the distinction? Cereal grains are the seeds of grasses, whereas pseudograins are the seeds of broad-leaved plants. You would be excused if you couldn't tell the difference looking at the seeds — the difference is only obvious when you see the plants from which the seeds come.

Pseudograins are gluten-free. Quinoa is a bit unique in its protein structure because, like the soybean, it contains all the essential amino acids, making it a complete protein. This is a big deal because the other cereal grains are low to deficient in the amino acid lysine, so their protein is less available to us.

Quinoa has a nutty flavor. It can be used as a substitute for other cereal grains, especially rice. You prepare quinoa in the same way you would rice, with one important exception: you need to rinse quinoa well to remove saponins in its seed coat. The saponins are antinutrients to us and very different from the soybean saponins you'll read about on page 161. They make the seeds bitter and poisonous to animals looking to interrupt the plant's reproduction cycle. Rinsing the seeds removes the bitter taste and antinutrients. Simply place the quinoa in a mesh strainer and run water over it until the water runs clear. Some quinoa packages say that it's prerinsed, but we usually give it a rinse just to make sure.

The popularity of quinoa is undeniable. While its nutritional profile is strong when it is cooked, quinoa really shines when it meets *Rhizopus oligosporus*, one of the preferred fungi for tempeh. In a study conducted in Poland in 2016,[12] scientists made tempeh with quinoa and then compared it to traditionally cooked quinoa. They found that the quinoa tempeh had higher levels of soluble phenols and greater antioxidant capabilities. The fermented quinoa also had higher levels of protein, lower levels of carbohydrates, double the thiamine, and over five times the riboflavin compared to the cooked quinoa. You can enjoy these health benefits for yourself in the Choose Your Own Tempeh Adventure (page 142) or our Ancient Grains Tempeh (page 169). We have grown koji on quoina to a delightfully sweet result. We made a black bean quiona tasty paste as part of our research (see the guide to making your own tasty pastes on page 268).

Sorghum

Sorghum (*Sorghum bicolor*) comes in at number five on the list of the world's most important cereal grains. It's grown for food, forage, fuel, sugar, and brewing (especially gluten-free beer).

The trouble with sorghum is that it isn't easily digested, even when boiled (as for porridges) or baked (as for breads). Fortunately (you may at this point be getting tired of the refrain), fermentation improves sorghum's digestibility and nutrition. In the Sudan, for example, you'll find kisra, a fermented bread similar to dosa or injera. During this simple lactic acid fermentation, amylase enzymes reduce the starch in the sorghum to next to nothing, and the sugar content also decreases, especially during the first 3 hours of fermentation. This is interesting because the starches release sugars as they break down. As this happens, other microbes happily feed on those sugars. Also interestingly, and this may be counterintuitive, as the batter ferments, the fiber content doubles. Thiamin and niacin also increase, while some amino acids increase, and others decrease.

In an experiment conducted in 2011, Taiwanese researchers fermented sorghum with koji. They discovered that sorghum's digestibility goes from 66 percent when raw to 42 percent after steaming, but it is brought up to as much as 81 percent through fermentation with koji. What is even more interesting is that they could enhance the protein of the lees (the waste product) of the sorghum brewing process, making what once had no value into a nutritious source of protein.

Let Northern sorghum supersede New Orleans sugar, and Northern flax take the place of cotton, and Northern tobacco drive out Dixie's weed.

EDITOR, WISCONSIN'S *LA CROSSE DAILY REPUBLICAN* NEWSPAPER, 1861

Both koji and tempeh work well with sorghum. We especially like sorghum in our Ancient Grains Tempeh on page 169.

Teff

Teff (*Eragrostis tef*) might not be familiar to you unless you are Ethiopian, have eaten at an Ethiopian restaurant and inquired about the massive pancakelike bread, or have found it while looking for gluten-free grains. A reasonable estimate of the number of people who know and likely love teff is about a quarter of a million, but that's growing. Teff's rise to global awareness is clearly due to its protein structure — teff does not produce gluten when hydrated and therefore has become a popular ingredient in gluten-free recipes.

Teff varies in color based on variety, from white to creamy white, reddish, light brown, and dark brown. It averages about 11 percent protein and 73 percent carbohydrates by weight, of which nearly all of that is in the form of starch. These starch granules are very small (the word *teff* is thought to derive from the word *teffa*, which means "lost" in the Amharic language). Teff is consumed as a whole grain and is an excellent source of essential amino acids, including lysine, which is usually deficient in cereal grains. It's also higher in many vitamins and minerals than many of the other cereal grains.

The most widely known fermentation of teff is injera, which is a large, soft, spongy pancake with holes or "eyes" and a wonderful, slightly sour taste. As is the case for other grains, spontaneous fermentation of teff greatly reduces its phytic acid content. One study from Belgium in 2017[13] found a 66 percent drop in phytic acid levels within the first day of fermentation, and those levels kept dropping until the third day, which is good news, because this is just about as long as you would reliably want to ferment injera. The Belgian scientists also measured a substantial drop in pH, from 5.8 to 3.4, after that same three-day fermentation. This is evidence that our friends the lactic acid bacteria are hard at work creating a low-pH environment to outcompete other microorganisms. What's interesting is that it's not the lactic acid bacteria that is responsible for reducing the phytates. They are only creating ideal conditions for the phytase enzymes contained within the teff to do their thing, which is to degrade the phytates.

We didn't try to grow koji on teff, as the grains are tiny and cooked as a porridge instead of as individual grains. But you will find a recipe for injera bread (page 84), and we also sprinkled a bit into our recipe for Ancient Grains Tempeh (page 169).

Kernza

In the early 1980s, the Rodale Institute chose intermediate wheatgrass (*Thinopyrum intermedium*) as the most promising perennial plant to focus upon to ultimately create perennial wheat, comparable to annual common wheat but with a much better ecological footprint. The result was Kernza, which is so new you can barely get your hands on it. If you do, use it in tempeh or koji as you would wheat.

Wheat

CHRISTOPHER WRITES: *We were wheat farmers for a summer. It was on an extremely small scale — just a quarter of an acre fenced off from the barnyard pasture.*

We had no idea what we were doing when we brought home the only bag of seed wheat we could find at our local farm store. We all took turns walking barefoot over the freshly turned soil, casting out handfuls of golden-white seeds onto the poorly nourished ground. We watered, we waited, we wondered how we would store all of it.

Over the hot, dry summer our wheat field grew from patchy islands of sprouts to delicate little plants to long, strong stalks with seed heads beginning to form — at which point the goats, cows, and horses lost their minds. Layers of electric fence were laid in, which were breached regularly. Welded field panels were bought, hauled in, and wired together to form a new perimeter, which was then laid flat by the cows. Finally, we evicted all those on four legs to another pasture and cleared our collection of barricades to reveal a field that looked more like a maze than a swaying field of golden grain.

I think there is a reason you usually see a line of workers, not one lone person, scything grain fields in those big oil paintings of yore. It's hard, lonely work. The blade is long and sharp. You are repeatedly swinging it in great arcs that end uneasily close to your legs. Worse yet, it's just the beginning. Next there is gathering, stacking, threshing, cleaning, and winnowing the husks from the kernels. Did I mention we were wheat farmers for just one summer?

Humans have been consuming wheat since our hunter-gatherer days, when wheat was wild. In fact, wheat predates us, having started its own evolution before we walked the Earth. As humans began to develop an agricultural-based existence, evidence suggests that we began to select wheat plants to use as next year's seed crop based upon a couple of traits: the size of the grains and how well they hung onto the stem, or spikelet. During this time, whether through natural occurrence or human selection, the earliest species of wheat produced einkorn (*Triticum monococcum*) and emmer (*T. dicoccum*), both of which are sometimes categorized as ancient grains.

A couple of thousand years later, emmer met a wild grass (isn't this always the way these things work?), and from that union we got spelt (*T. spelta*), which would eventually lead to what we call common wheat or bread wheat (*T. aestivum*). Some early farmers also selected for certain traits, producing what we know today as durum wheat (*T. turgidum*) and giving all the pasta lovers something to sing about. So, we have cultivated four types of wheat — einkorn, emmer, spelt, and durum — for at least six thousand years.

Any of these varieties can be used as a substrate for koji. In this book, we call for the whole kernels of wheat, called wheat berries. The whole berries are cracked to give the koji something to sink its teeth into in the shoyu recipe (page 298). In the Choose Your Own Tempeh Adventure section (page 142) we use various whole cooked berries of varieties like farro, kamut, and emmer or bulgur, which are basically wheat berries that have been parboiled, just like parboiled rice (page 60). Finally, we use wheat flour in the Salt-Rising Bread recipe (page 77).

Preparing Grains for Koji and Tempeh

While we have included a few spontaneous grain ferments, most of the grain ferments you will encounter in this book call for preparing grains as a substrate for koji (*Aspergillus oryzae*) or tempeh (*Rhizopus oligosporus*). For this you'll need to cook the grains, and the most important thing to keep in mind is that the grains should be al dente — tender but still firm.

If you undercook the grains, they will be too hard and dry, and they won't contain enough moisture for the mold to form. If you overcook the grains, tempeh will become too wet while incubating, making a substandard product, and for koji, the spores will not be able encapsulate each grain, which makes a matted, wet koji. If the grain is quite overcooked, oxygen will not reach much of the grain, causing uneven coverage by the mold and decomposition.

The traditional method of preparing grains for fermentation is by steaming, as this keeps the final cooked grain from being overly wet. This is especially important with polished grains. We have found that a rice cooker works well for cooking single batches of many grains. Steaming in a stainless-steel basket over an inch of water in an electric pressure cooker also works well, and quickly, for grains. Barley, for example, comes out perfectly when steamed under pressure for 30 minutes for making tempeh or koji. Rice is done after 18 to 20 minutes.

The traditional practice is to rinse rice before cooking it to remove starch, which we talk more about on page 191.

If you don't have a rice cooker or a pressure cooker, you can easily steam rice on your stovetop (see instructions on the following pages). You can use a bamboo steamer over a pot or a dedicated steamer, or make your own by filling a large stockpot with a few inches of water, placing a small stainless-steel bowl upside down in the pot, then placing a metal steamer basket (without feet) on top of the metal bowl.

SCIENCE SAYS . . .

Gluten-Free Wheat Thanks to Microbes

In 2006 the University of Bari in Bari, Italy, reported a breakthrough in reducing gluten in wheat. By utilizing 10 lactobacilli species typical in sourdough fermentation and fermenting wheat for 24 hours in a very wet mixture, they were able to nearly totally hydrolyse the gluten — so much so that when they created baked products with the sourdough and fed them to confirmed celiac disease patients, those patients reported no adverse effects.[14] While their results were encouraging, they hinted in the report that they were working on something even better: combining the lactobacillus with fungus. In a study in 2010,[15] researchers refined the lactobacilli mix and added *Aspergillus oryzae* and *A. niger*; the result was a reduction in the gluten in the wheat flour from 103,127 parts per million (ppm) to 10 ppm. That is half the typical lowest detectable concentration level of 20 ppm and was shown to be digestible by participants with celiac disease with no adverse reactions. Researchers continue to refine the process and the team of microbes being used, but it gives those who suffer from celiac disease and gluten intolerance hope that their gluten-free baked goods will no longer necessarily equate to wheat-free.

Steaming Rice and Other Grains

A GUIDE TO LEGUMES AND CEREAL GRAINS FOR FERMENTATION

1. Soak the grain for at least 6 to 8 hours or overnight; soak whole grains or brown rice up to 24 hours, changing the water once about halfway through.

2. Pour the grain into a colander. Let it drain for at least 30 minutes, shaking intermittently to help remove any excess water, as water will cause the grain to cook instead of steam, resulting in lumpy grains.

3. Once the grain is "dry" — that is, you're no longer able to shake out any drips of water — line your steamer basket with butter muslin or fine cheesecloth. (Be sure to wash and boil the cloth between uses to santitize it.) Place the grain in the cloth and fold up the edges over the rice to form a bundle.

4. Start boiling the water over high heat. When the steam rises through the grain (not just up the sides), it is time to start the timer. White rice takes about 45 minutes and brown rice takes about 50 minutes. (**Note:** Rice can take up to 1½ hours if a lot of the steam escapes.)

While the grain is steaming, watch to make sure the pot does not boil dry. Midway through the cooking time, check a few grains to make sure they are not cooking more quickly at the bottom than in the rest of the bundle. If they are, pick up and rotate the bundle so that the grain will cook evenly.

5. The grain is ready when you can push your fingernail through it. The core should not feel any firmer than the rest of the grain, and the whole kernel should have an even texture that we think of as a bit rubbery. This is al dente.

Zero-Waste Tip:
Don't Throw Out the Rice Water

It turns out that the water that you use for soaking rice is seriously good stuff, full of vitamins B_1 and B_2 as well as lipids and starches. This nutrient-enhanced solution can be useful in a number of ways:

▸ **Use it as a broth for soup.** The colloidal property will make your broth smoother and the vitamins will make it richer.

▸ **Water your plants.** Give those nutrients to the potted plants, not the drain.

▸ **Water yourself.** Rice water is a good facial wash and hair tonic. It can balance and clear skin if used once or twice a week, and it strengthens hair when used as a rinse.

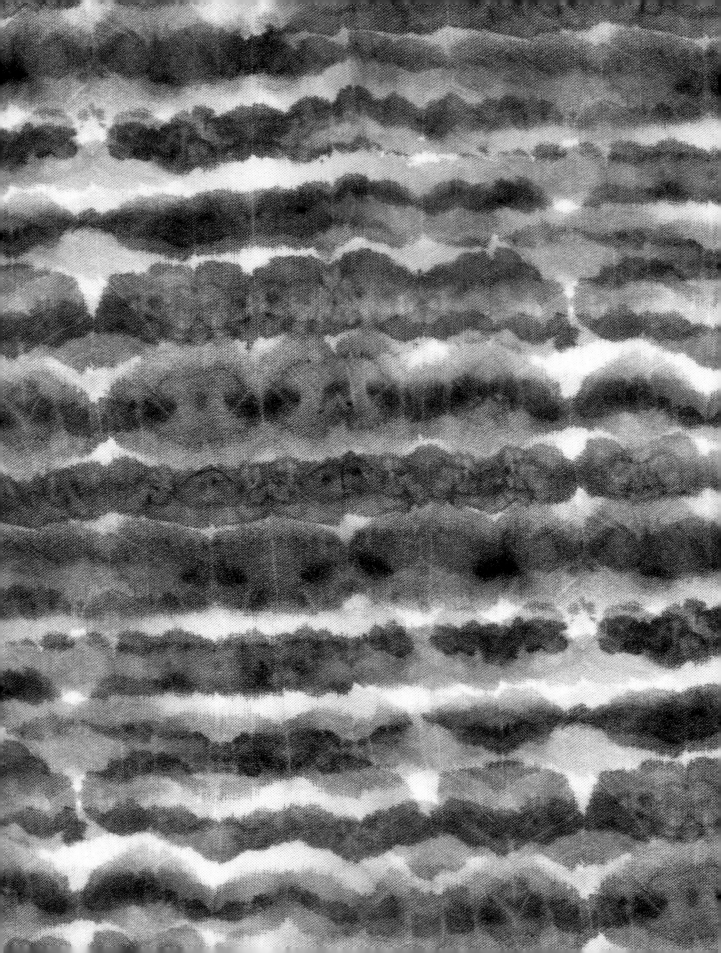

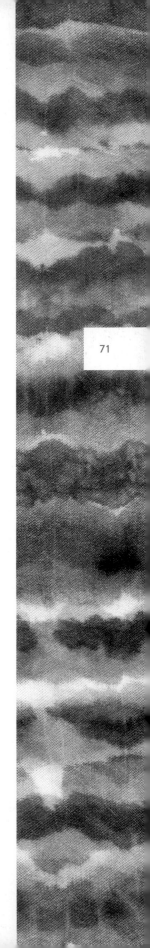

PART II

Making

There may come a time when you find yourself buying a lot of miso, tempeh, and amazake from your supermarket and making regular trips to your closest Asian market for natto and stinky tofu and to your local Indian market for dosas and idlis. It could be time to start making your own. Or maybe you live in a place where none of these foods are available, so to enjoy them on your plate, you have to make them yourself. This part of the book is about learning to do just that.

We will start with spontaneous ferments — things you can make in partnership with the wild yeasts and bacteria already on the legumes or grains. Then we move on to ferments that require you to introduce certain microbes, and finally, you will learn how to grow your own microbes to do your bidding. Get ready for an adventure.

CHAPTER 4

Getting Started: Spontaneous Ferments

A.K.A. WILD FERMENTATION

Most of the ferments in this book are made by adding a culture (specific microbes) to a substrate (a base like rice or soybeans) in order to produce a fermented product. If you have never worked with a culture before, it can feel intimidating. This chapter will let you test the fermentation waters, so to speak, with quick wild ferments that do not require cultures. If you want to start fermenting right now (no need to wait on cultures to arrive in the mail), this chapter is for you.

When you hear the term *spontaneous fermentation*, you might be thinking of something like spontaneous combustion. One moment there is nothing, then the next — boom! — fermentation. That's not quite what happens. The microorganisms responsible for the fermentation are already on the legumes or grains well before the fermentation starts. In fact, there are always a lot of microorganisms (desirable and undesirable) on our foods. The trick is to set up the proper conditions for the ones that produce the fermentation we want. To do this, we want to control the air, temperature, water, pH, sugar, and salt. Done right, the microorganisms we want to encourage gain the upper hand in this crowd.

Ingredient Amounts in Our Recipes

Because volume measurements are not as accurate as weight measurements and some people prefer the precision of weighing their ingredients, we've included both gram weights and volume measurements in part II. This gives you the option of making these ferments in the way that is most comfortable for you.

SALT-RISING BREAD

This bread is an American original born some two hundred years ago in Appalachia. Do you recognize the name? Either you are smiling at the thought of the last time you had this unique cheesy-tasting bread, or you are wondering how bread can rise using salt. Spoiler alert: it doesn't. The name likely refers to the traditional practice of keeping a bowl or bag of salt warm by a cabin stove through the night and nestling the bread's starter down into it for safekeeping. The very warm salt maintained a pretty consistent temperature, which is important because this bread is leavened by naturally occurring bacteria, rather than yeast. If we were naming this bread today, we would give it a more provocative and accurate name: bad bacteria bread anyone? But before you skip the recipe, please note that it is unique, delicious, and well worth knowing about, if not making.

The classic characteristic of a good salt-rising bread is its cheesy smell. It smells as if it were baked with equal portions Emmental Swiss cheese and Kraft Parmesan from the can. This likely comes from butyric acid, which is formed by the bacteria mix.

First, just a quick primer on how bread is usually made. When you mix flour, water, yeast, and some salt, multiple fermentation processes start. Enzymes wake up in this newly hydrated environment and begin breaking down the starches in the flour into simple sugars, which the yeasts hungrily gobble up, producing CO_2 and alcohol. From the yeasts forward it's the same basic process as making hard cider from apple juice or wine from grapes — only in this case, those millions upon millions of tiny CO_2 bubbles are trapped within a growing mesh of gluten proteins. It's the same for sourdough breads, except instead of adding a package of yeast, you add some starter that is already doing this starch-to-sugar-to-bubbles dance underneath the gluten.

Salt-rising bread switches teams, ignoring the services of yeasts for those of bacteria. The special instructions for this bread are geared to attract and capture warmth-loving bacteria, and to kill and then discourage the development of yeasts. Among the mix of bacteria, three players are key; two you will recognize, and one will surprise you.

Remember the different types of fermentation from chapter 1? Lactobacillus from the lactic acid fermentation camp and *Bacillus subtilis* from the alkaline fermentation camp are in this mix and likely have a symbiotic relationship. *B. subtilis* moves in early at the higher temperatures, and lactobacillus comes in a later, after things have cooled off a bit. The surprise third bacteria here is *Clostridium perfringens*, which often make the news when big outbreaks of foodborne illness occur. Usually *C. perfringens* outbreaks are caused by cooked food that is kept in a warm place for hours. While cooking kills the bacteria, it doesn't necessarily kill the spores of the bacteria, which wake up and multiply rapidly in high temperatures. An analysis[16] of the microbiology of salt-rising bread reveals that *C. perfringens* type A was present but did not produce any toxins, due to the act of baking. The recipe we share with you repeatedly captured the right bacteria, creating active starters and sponges and producing bread loaves with that lovely cheesy smell. Enjoy.

Where's the Sourdough?

Sourdough is a wild ferment, consisting of yeasts and lactic acid bacteria. There are as many as 50 species of lactic acid bacteria and 20 species of yeasts that can play a role in sourdough's microflora. Exactly which type is determined by what has caught a ride in on the flour, what is in the environment in your kitchen, and what is in the starter culture. Once the bacteria and yeast work it out — usually 100 bacteria for every yeast fungus — they become stable, and this relationship can last for decades or longer.

Our sourdough recipes have changed at least as often as our starters over the years, which is why you won't find our favorite sourdough bread recipe in this book — we simply don't have anything to offer beyond the plethora of recipes already out there. Unlike the majority of ferments in this book, sourdough is anything but obscure or poorly understood. We encourage you to explore sourdough if you haven't already.

Salt-Rising Bread

YIELD: 1 (9-INCH) BREAD LOAF

This bread is surrounded by plenty of lore about how unpredictable it is. James Beard even says that it is worth making but that "you may try the same recipe without success three or four times and find that it works the fifth time." When a leading culinary figure in America basically says, "I did my best, and good luck to you," you know there is a fair bit of magic involved. Of all the recipes we tested and modified, we got the best results from three-stage recipes that consisted of a starter used to create a sponge that became the dough. Begin the starter in the evening and you should have a warm loaf cooling on your counter the following afternoon.

Incubation tip: A bread proofing box works great for all three stages and lets you set the right temperature for each stage. Note that, like natto, salt-rising bread has a special smell, and through no fault of its own, the proofing box will not contain this smell.

FERMENTATION **18 hours for the first ferment, plus 2–4 hours for two risings**

---

1 tablespoon (10 g) cornmeal

1 teaspoon (3 g) whole-wheat flour

⅛ teaspoon (0.76 g) baking soda

½ cup (118 mL) milk

2½ cups (590 mL) warm water

3 cups (375 g) all-purpose flour

1 teaspoon (6 g) salt

Neutral cooking oil, for greasing the pan

1. Combine the cornmeal, whole-wheat flour, and baking soda, and sift into a small bowl.

2. Pour the milk into a heavy-bottomed saucepan and place over medium heat. Heat, stirring constantly, until the milk reaches 180°F/82°C, then keep it at this temperature for 15 seconds to scald it (see box on page 78 for why). Pour the scalded milk over the cornmeal mixture and mix well.

3. Cover the bowl with a cloth. It is important to cover with cloth, and not plastic wrap or a lid, because you want bacteria to find this starter and set up shop. Place the bowl in a warm spot (105°F/41°C to 115°F/46°C is ideal) and let it sit overnight or until it develops a healthy layer of foamy bubbles on top. The foamy bubbles are a sign that a healthy colony of bacteria is processing the carbohydrates in the starter and creating gas. The amount of time this takes will vary, but if no foam has developed after 18 hours, discard the starter and try again.

Recipe continues on next page

Salt-Rising Bread, *continued*

4. Warm a medium bowl in the oven at 200°F/95°C until it is comfortably warm but not too hot to grasp with your bare hands. (The bowl needs to be warm to maintain your new bacteria culture.) Add your starter and 2 cups of the very warm water, which should be at about 110°F/43°C or uncomfortable to the touch. Add 2 cups of the all-purpose flour and mix well to make a sponge. The sponge should have the consistency of cake batter. Cover the bowl with a cloth and leave in a warm place (at least 100°F/38°C) until it doubles in size (see above photo), 1 to 3 hours.

5. Warm a large bowl in the oven at 200°F/95°C until it is comfortably warm but not too hot to grasp with your bare hands. Transfer your warm sponge into this bowl. Add the remaining ½ cup

warm water (body temperature is fine here), the remaining 1 cup all-purpose flour, and the salt. Knead by machine or hand until the mixture begins to hold together as a dough, adding more flour if necessary.

6. Grease a 9- by 5-inch loaf pan. Place the dough in the pan, cover with a cloth, and set it in a warm spot (at least 100°F/38°C) until the dough has risen to the top of the pan (see above photo), 2 to 3 hours.

7. Preheat the oven to 375°F/190°C.

8. Bake the loaf for 30 to 35 minutes, or until golden brown. Then remove from the pan and let cool on a wire rack. Enjoy fresh; after a day or two, it's best when toasted.

Why Scald?

Search the Internet and you will find people who dismiss scalding as an antiquated practice. Before pasteurization, scalding milk was necessary to kill off bacteria, but we don't think that's the main reason why people scalded milk for bread recipes. More likely, it was to warm up the other ingredients to improve the rising process, to integrate other aromatic ingredients (especially for ice cream and custard recipes), or to deactivate or denature a whey protein called glutathione. Glutathione can interfere with rising, although in our tests we still got a rise without scalding. Still, by investing a little time in this process, you assure your bread the best chance of rising.

DOSAS AND IDLIS

Dosas and idlis are a wonderful combination of cereals (rice) and legumes (black gram/lentils) that ferment with just the microflora present on them, producing breadlike foods that are both more digestible and higher in nutritional value than the raw grains. They have similar ingredients but differ in how they are cooked. Dosas are cooked like pancakes or crêpes and lend themselves to being stuffed with savory fillings. Idlis are steamed, producing a light, spongy cake perfect for sopping up rich sauces for breakfast.

CHRISTOPHER WRITES: *I began working in southern India in 2001 as part of a team from Hewlett-Packard codeveloping solutions to problems faced by the rural communities in the states of Andhra Pradesh and Tamil Nadu. My responsibility was agriculture, so I worked with government officials, NGOs, and community organizations to improve rural people's food sovereignty. We spent a lot of time in sun-baked fields looking at withered crops and hearing about the dropping water table that the farmers were chasing with increasingly deep and expensive bore-hole wells — all problems that I suspected HP's printers and laptops didn't have a hope of addressing.*

Dusty, soaked with sweat, and perplexed by the complexities of the problems, we would always be ushered into a government building at dusk for a meal. At least a dozen of us would wait in white plastic chairs under glaring, bare fluorescent lights for a single slender man, who would enter carrying two swinging columns of stacked stainless-steel containers. The first time I saw him, I wondered if we would make it back to our hotel in time to hit the street vendors for a second dinner. I should not have worried. The copious dishes that emerged from those containers were mesmerizing and delicious. These evening meals always contained a couple varieties of dosas.

When we spent nights out in the field, we would awake to many roosters vying to be first to announce the day and the smell of fresh, piping hot tea with milk, alongside steaming idlis and several sides of colorful sambars and chutneys.

Both dosas and idlis have been made in these regions since at least the eighth century, according to Sujan Mukherjee, the executive chef at Chennai's Taj Coromandel Hotel, which was our home base in Chennai. That hotel is high-end, but I think the dosas and idlis were better from that little man in the village.

TUA NAO,
page 97

DOSAS,
opposite

MYANMAR-STYLE TOFU
(unfried), page 93

POHA IDLIS,
page 82

Dosa

Dosas are versatile for scooping up a variety of sauces and curries. Or you can dollop a bit of your favorite potato or coconut curry on top of the dosa, then roll it up and eat it like a wrap. The hardest part of this recipe is planning ahead so that you remember to soak the grains overnight.

FERMENTATION **12 hours**

1½ cups (270 g) uncooked white rice

¾ cup (150 g) split black gram (urad dal)

2 tablespoons (25 g) split chickpeas (chana dal)

1 teaspoon (4 g) fenugreek seeds

1½ teaspoons (9 g) salt

Neutral cooking oil or ghee, for greasing the pan

---

1. Place the rice in a bowl. Place the dals and fenugreek seeds in another bowl. Fill each with water, swirl with your fingers to gently wash, then carefully pour out the water. Repeat until the water is clear as you pour it out. Refill each bowl with fresh water to cover the rice and dal mixtures with 2 inches of water, cover the bowls with clean kitchen towels, then set aside to soak for 8 hours.

2. In the morning, drain the rice, reserving the soaking water for later use. Add the drained rice to the bowl of a food processor. Add ¼ cup of the reserved soaking water. Process the rice, adding a tablespoon of the reserved rice soaking water at a time as needed, until it reaches the consistency of smooth paste. It's good when it still feels a little gritty when you rub it between your fingers. Scrape into a large bowl.

3. Drain the soaked dals and fenugreek seeds, discarding the soaking water, and add them to the food processor bowl, along with ½ cup of the reserved rice soaking water. Process, adding

a tablespoon of the rice soaking water at a time as needed, until the mixture reaches the consistency of smooth paste. Scrape out and add to the rice paste.

4. Cover the bowl with a clean kitchen cloth and set aside to ferment for 8 to 12 hours. If the ambient temperature is below 80°F/27°C, consider placing the bowl in the oven with the light left on. It has properly fermented when it has doubled in size and is a little foamy.

5. Add the salt to ¾ cup of water and stir to dissolve. Pour this over the fermented batter and stir to incorporate together.

6. Oil an 11-inch cast-iron skillet or a nonstick griddle and set it over medium-high heat. Test the heat by placing a drop or two of water on the surface. It should immediately dance a bit before evaporating. If the surface is smoking or the water immediately evaporates, the surface is too hot.

7. Pour ⅓ cup of fermented batter into the skillet and spread with a ladle, using a circular motion to make a thin dosa. When bubbles start to appear, which takes about 2 minutes, sprinkle ½ teaspoon of oil over the top of the dosa. Loosen the edges with a spatula and then flip the dosa over. Cook for 20 seconds or so, then remove to a warm plate and cover with a clean kitchen towel.

8. Cook the remaining batter in the same way, lightly greasing the surface between dosas. Serve hot.

Poha Idli

There are many variations on idlis, as well there should be for a dish that has been in existence for at least 1,200 years. This recipe makes a soft and spongy idli that is perfect for absorbing rich, fermented sauces, giving you a double ferment treat for breakfast or snacks.

This version uses flattened rice, also referred to as beaten rice, chura, or jada poha. It is rice that has been flattened so that it is more like rice flakes. It also uses rice semolina (idli rava), which is rice that has been boiled, then dried and ground into a fine powder that resembles semolina flour. If you can't find both of these ingredients, a good plan B is to substitute the dosa batter from the previous recipe.

Finally, though we experimented with other containers to steam the idli, nothing performed as well as the tool made for them — the idli stand (pictured opposite, bottom of page). Idli stands are inexpensive and can be found in Indian grocery stores or online.

FERMENTATION **3 hours for the first ferment plus 8–12 hours for second ferment**

---

1 cup (59 g) rice semolina (idli rava)

¼ cup (46 g) beaten rice (chura or jada poha)

¼ cup (150 g) split black gram (urad dal)

1 teaspoon (6 g) salt

Neutral cooking oil, for greasing the idli molds

---

1. Combine the rice semolina and beaten rice in a medium bowl. Run cool water into the bowl while you gently swirl the rice with your hand. Let the water pour out of the bowl until the water run clears. Cover the rice with at least 1 inch of warm water.

2. Wash the black gram using the same process. Let the rice and the gram soak for 3 hours.

3. Pour off the water from the soaking rice into a small bowl. Add the rice to a blender. Process the rice, adding a tablespoon of the soaking water at a time as needed, until it forms a smooth batter. Pour the batter into a large bowl.

4. Pour off the water from the soaking gram into a small bowl. Add the gram to the blender.

Process, adding a tablespoon of the soaking water at a time as needed, until it forms a smooth batter. This black gram batter will be a little thicker than the rice batter; try to use as little water as possible. (Resist the urge to process the rice and black gram together — you won't get a good consistency.) Pour the batter into the bowl with the rice batter.

5. Add the salt to the batters and mix well. Cover the bowl with a clean cloth or lid and set aside to ferment for at least 8 hours and up to 12 hours, or until the batter is bubbly. If the ambient temperature is below 60°F/16°C, you might need to ferment longer. If the fermentation goes too long, however, it may become too sour for your taste.

6. Quickly beat the fermented idli batter by hand with a whisk for 10 to 15 seconds.

7. Disassemble the idli stand and oil each mold where the batter will go. Spoon the batter into the oiled molds, starting with the bottom one. As you fill the molds, add them to the stand until you have filled them all. Place the assembled idli stand in a steamer and steam for 8 to 12 minutes.

8. Pop the hot idli out of the molds and serve immediately with your favorite sauces.

INJERA — EAST AFRICAN BREAD

Teff is enjoying its day in the sun thanks to its gluten-free status, but we love it for its nutty taste. Injera is a large, soft, spongy, pancakelike bread from eastern Africa that is made with teff flour. Traditionally, it is made using a wild ferment that takes anywhere from 1 to 3 days to develop. In our tests, these ferments were not always as bubbly as they needed to be to get the characteristic holes and nice sponginess. We also wanted to be able to whip up some injera in just a day. We tested recipes that used baking powder or baking soda, but it felt like cheating and we could taste the leavening, which shouldn't be competing with teff's flavor. So we took inspiration from our sourdough starter and created a simple process for making a natural injera starter.

Injera

YIELD: ABOUT 5 INJERA, DEPENDING UPON SIZE

The traditional way to serve injera is to dollop on rich pastes and sauces and then eat the whole messy goodness with your fingers. If you are looking for a less messy appetizer, spread your favorite pastes or sauces on the injera, then roll up and cut them into segments, jelly-roll style.

Once you have gotten the starter going, you can just pull it out of the fridge anytime you want and make injera that same day. The starter recipe follows this recipe.

FERMENTATION **4 hours**

1½ cups (218 g) teff flour

1½ cups (355 mL) warm water (at body temperature)

¾ cup (177 mL) injera starter (page 85), preferably at room temperature

¼ teaspoon (1.5 g) salt

Neutral cooking oil, for greasing the pan

1. Combine the teff flour, water, and starter in a large bowl and mix well. Cover with a clean cloth and let sit until it becomes puffy and bubbly, about 4 hours. Give it a good sniff and you should smell delightful sour smells. If you took your starter straight out of the refrigerator, let the mixture sit for an extra hour to allow the microbes to warm up before they do their thing.

2. Oil a cast-iron skillet and place over medium heat. Pour the batter into the warm skillet, making a thin pancake approximately 8 inches in diameter. Cover with a lid, let cook for 3 minutes, then take a peek. The injera is ready when its top is full of popped bubbles and no longer looks wet. The edges may appear very dry and even a little curled up from the pan. Remove the injera from the pan to a warm plate.

3. Continue pouring and cooking injera until you have used up all the batter. After the first one, the total cooking time is usually just 1 to 1½ minutes.

Injera Starter

YIELD: ABOUT 1 PINT OF STARTER

If you have ever made your own sourdough or been tasked with keeping a friend's starter alive while they were away, this is going to feel very familiar. Basically, you make a slurry with teff flour and wait for wild yeasts and bacteria to move in — or at least until you accidentally neglect them to death, and then you have to start over.

FERMENTATION **3 days**

1 cup (145 g) teff flour

1 cup (237 mL) water

1. Mix the teff flour and water in a bowl or jar. Cover with a clean cloth and let sit at room temperature for 3 days.

2. After 3 days, check your starter. The surface should be puffed up and a bit dry looking, with holes here and there where gases are escaping, and if things are really rocking you can hear the gas bubbles popping down below the surface. If you don't see or hear any of this, start a new batch. You can keep the current one going for another couple of days, but mentally prepare yourself to let it go and use the new one.

3. Once you have an active starter, which is basically a colony of bacteria and yeasts, you will need to keep feeding them so they don't die. Every time you remove starter to make bread, replenish the main starter by adding an equal amount of a teff slurry (itself made with equal amounts of teff flour and water). Even if you are not making injera regularly, you still need to feed the starter regularly. The frequency with which you feed it depends upon the microbes' activity, which can be controlled by temperature. If you keep your starter refrigerated, activity slows way down, and you will probably need to feed it only every 2 weeks or so. If you are keeping it on your counter, you will need to replenish it every couple of days. Remove at least some of the starter every time you replenish it; if you are not going to use it to make bread, you can add it to your compost.

FERMENTED TOFU

Tofu is made from the curd of coagulated soy milk, like cheese is made from the curd of coagulated cow's milk. It is not traditionally fermented. If the soybeans used to make the tofu were properly soaked, then some of their antinutrients have been reduced, but their true nutritional potential is fully realized only through the transformation of fermentation.

We have included two fermented tofu recipes in this book: Chinese "stinky" fermented tofu (page 89) and Japanese tofu misozuke (page 319). Our method calls for a wild fermentation, which is the simplest way to make it; read on about the more complex methods using specific microbial inoculants.

Chinese Fermented "Stinky" Tofu

Setting out to discover the virtues of stinky tofu was not as straightforward as we had imagined. It is made in many different ways, varying by place and microbe. This fermented curd has a soft, runny, creamy center not unlike a bloomy rind Brie and a strong odor like aged cheese, is tasty like a sharp blue cheese, and is often pickled in brine like feta.

Stinky tofu is a catchall term for a wide range of types of fermented tofu. Some traditional names include *sufu, doufu-ru, chou dou fu, and mei dou fu* (the Chinese *chou dou fu* translates as "stinking tofu"). Let's face it, calling it like it is — smelly — is more fun. And it gives food writers a chance to compare nourishment to dirty socks, which share a sharp to downright acrid odor that can be off-putting (though enthusiasts proclaim that the stinkier, the better). Just between you and us, we don't find the smell offensive . . . but we like stinky cheese, so we may not be the best judges.

Written records show that the Chinese have been eating fermented tofu since the Wei dynasty, over 1,500 years ago. We came across many stories about the origins of this smelly treat, but perhaps the most colorful involves a failed student named Wang Zhihe of the Qing dynasty (later than the Wei dynasty, but who are we to let facts get in the way of a good story?). After failing the imperial exam, Wang Zhihe opened a tofu shop in Beijing. He needed to preserve some of his tofu. Some accounts say he made too much, others say the weather was too hot to keep it, one version says the entire fermentation was an accident. We are not convinced that he wasn't just curious — an early fermentation lab? Whatever the cause, tofu somehow ended up cubed in an earthen jar for some time. When Wang Zhihe retrieved it, he found it stinky and greenish, at which point, bravely, he put it into his mouth. Is it just us, or does that not point to experimentation?

Makers in different regions developed many methods of fermenting tofu. There are three core forms based on the microorganisms used: mold fermented, bacteria fermented, and lactobacteria brine fermented.

Mold-fermented tofu is a two-part process. First, the tofu is cubed, inoculated with a mycelial fungus (*Actinomucor, Rhizopus,* and *Mucor* are common), and left to sit for 3 to 7 days. At this point, the tofu, now called *pehtze,* is wonderfully fuzzy. In the second stage, the pehtze cubes are salted and dropped into a dressing or brine (such as rice wine, red koji rice, aromatic fermented vegetables, miso, soy sauce, milk, or stinky fish brine) and allowed to marinate and continue to ripen. This ripening period can vary from just a few days to weeks or months.

The most famous example of mold-fermented tofu is *chou dou fu.* Nighttime street vendors in Taiwan sell it, and they are likely to inoculate the tofu with *Actinomucor taiwanensis* and then age it in a brine made of fermenting vegetation (which most accounts describe as rotting). The brine ingredients vary by vendor or region but often include wild amaranth stalks, mushrooms, or even fermented black soy. The vendors make soup with it or serve the ferment as a deep-fried snack on a stick that one dips into chile paste or fermented sweet sauce.

We wanted to make chou dou fu, but we couldn't source *Actinomucor* spores. Knowing that this is a two-step fermentation where the first step is to use a microbe to break down the tofu, we deduced that perhaps we could hack the ferment and grow *Aspergillus* on the tofu and then brine it. Alas, the results were poor. So, the takeaway is that you need *Actinomucor* spores.

Mei dou fu comes from the Shaoxing region. The name translates as "fermented or moldy bean curd."[17] Like many traditional ferments, it gets its microbes from specific plants, in this case, pumpkin leaves or rice straw. Tofu cubes are placed on the leaves or straw to pick up *Mucor* and *Rhizopus* molds. Once furry, the cubes are salted and left to ferment for many months in clay pots with rice wine and other spices. During fermentation, it is the same old story: Enzymes generated by the molds and their microbial friends free the proteins and starches in the tofu to create amino acids and sugar. The yeasts that came in with the wine suddenly have a new source of sugar and — you know where this is going — you end up with drunken tofu. Traditionally, mei dou fu is served as an appetizer, a digestive aid (those enzymes), a side dish, or used as a flavor enhancer in dressings and sauces — which happens to be our favorite use (see the recipe for [Not] Blue Cheese Salad Dressing on page 310).

Bacteria-fermented tofu is also a two-stage ferment. The fresh curd is cubed and inoculated with *sufu* (*Bacillus* or *Micrococcus* bacteria) and left to ripen for about a week. At that point, you have *pehtze,* but this time it is secreting a yellow liquid. The pehtze cubes are salted and dropped into a dressing or brine to ripen.

Brine-fermented tofu is the most alluring to home cooks. Perhaps it is because they do not want their domiciles smelling like the *sufu*-producing districts of Taiwan. However, when we tried simply submerging unfermented tofu cubes in fermented brines, the end results weren't satisfying. The tofu soaked up some flavor of the brine, but it was clear that the proteins in the tofu hadn't been broken down — umami was missing.

One thing all fermented tofus share is that they employ microorganisms to process the curd to make it more digestible and layer on flavor. One common ingredient is well-fermented amaranth stalk, which is said to have some pretty disturbing odors. The odor may be what makes it so useful, though. Some fermented tofu makers take great pride in their well-guarded recipes for the stinkiest brine.

Stinky-Style Fermented Tofu

YIELD: 1 POUND

This fermented tofu has some funk and even some stink, but it is mild by stinky tofu standards. There is nothing rotten or putrid about the smell. The enzymes break down the curd, creating a creamy, cheesy flavor. The recipe is adapted from *Asian Tofu* by Andrea Nguyen. It is wonderfully simple and leaves plenty of room to play with flavors in the brine for the second stage of the fermentation. We suggest that you make the brine as the recipe suggests the first time. However, after you understand the base flavor, you can use different herbs in the brine to bring about unique variations.

We had the best results with a firm tofu. Extrafirm (very dense) tofu consistently tended to grow gray mold before it produced the desired yellow or orange spots during the first curing, and it didn't soften as nicely in the brine, which meant it wasn't spreadable. It is important to note that the softer the tofu, the quicker it will ferment, but also will be more delicate to pull out of the brine.

We truly enjoyed discovering this food; we hope you do, too.

FERMENTATION **2–5 days for the first ferment, plus at least 4 weeks for aging**

---

1 (16- or 14-ounce) block firm tofu

1–2 tablespoons (18–36 g) salt

½ teaspoon (0.80 g) chile flakes (optional)

½ cup (118 mL) water

¼ cup (59 mL) Shaoxing rice wine or mirin

¼–½ cup (59–118 mL) sesame oil, enough to create a seal on top of the tofu and brine

---

1. Cut the tofu into 1-inch cubes.

2. Press the tofu cubes (cutting the tofu into cubes before pressing allows you to press out more moisture). Place a cutting board on top of a dish towel to catch any extra liquid. Place a couple of paper towels on the cutting board and arrange the cubes in a single layer on the paper towels, leaving a bit of space between them. Place more paper towels on top of the tofu, then place a second cutting board or casserole dish on top. Set a sack of rice or some other heavy item on top to weight it down. Allow the tofu to drain, at room temperature, until it is just moist and feels quite firm, about 2 hours.

3. Transfer the tofu to a rectangular casserole dish large enough to hold the cubes in a single layer with at least ½ inch between all the pieces. A glass dish is ideal, allowing you to see what is going on. As Nguyen so aptly states in her recipe, you want to "create a Petri dish–like environment."

4. Cover the dish securely with taut plastic wrap. Poke six or so holes across the top with the end of a chopstick or skewer, so the tofu can breathe.

Recipe continues on next page

Stinky-Style Fermented Tofu, *continued*

5. Let the dish sit at room temperature for 2 to 4 days, depending on the temperature and humidity. (We found that at 58°F/14°C it takes 5 days and at 69°F/21°C a quick 48 hours.) The tofu is ready when it starts to look wet (you may even see a small milky puddle pooling below the cubes) and there are light yellowish orange mold spots on the tops and sides. It will have a strong smell. You may also get a bit of gray filamentous fungal growth. This is okay; just scrape it off. (In fact, we found that the tofu we made that was just starting to get to this gray stage was better than the tofu that we didn't allow to get this far.)

6. Combine the salt and the chile flakes, if using, on a plate. Coat each cube with this mixture, then gently stack the coated cubes in a glass jar. (A 1½-pint jar will work, or divide the cubes between two 1-pint jars.) Handle the cubes gently so they do not break apart.

7. Combine the water and wine in a small bowl, then pour the mixture over the tofu. The cubes should be submerged. (If you're experimenting with adding flavorings, now is the time.) Pour the oil over the brine so that you have a roughly ¼-inch layer of oil sealing the top.

8. Place the tofu in the refrigerator to age for at least 4 weeks. It will be ready to eat when it has a creamy texture and a cheesy sharpness. Store the tofu in the refrigerator, where it will keep indefinitely as long as you always use clean utensils when scooping some out. The tofu will mellow over time, becoming creamier and more remarkable.

TOFU MISOZUKE, page 319, wrapped with white miso and infused with shiitake powder

STINKY-STYLE FERMENTED TOFU, page 89

MYANMAR-STYLE SHAN SOUP, page 94

MYANMAR-STYLE FRIED TOFU, page 93

THE FERMENTED BEANS OF MYANMAR

A few years ago we were looking for a unique place to visit with our two youngest children before our nest emptied completely, and we chose Myanmar. We were enchanted not only by the land but also by the generosity and spirit of its people. From the first taxi ride in the middle of the night through the streets of Yangon, we were greeted with warmth and an openness none of us had experienced elsewhere. On our last night, our taxi driver wanted to take us by his home to meet his parents and see "how the people of Myanmar live." We had time, so we humbly accepted his invitation.

The busy streets of Yangon gave way to a maze of clustered homes divided by a road hardly wider than the compact car we were squeezed into. He brought us into a small, immaculately swept yard and gave us plastic chairs to sit on. He wanted us to taste how his mother had prepared the shrimp sour (a ferment he'd shown us the day before). It was delicious, and it was less salty and much milder than any of us expected, with none of the strong flavors we associate with other fermented seafood products, like fish sauce. For lack of a better comparison, its sourness was very similar to that of yogurt.

The four of us, all giants, sat awkwardly, but honored. The yard was lined with 20 or more small trees grown in plastic bags filled with dirt. He explained that they were seedlings from the star fruit tree we were under. He wanted to plant more because the weather just kept getting hotter; he said he didn't understand why, but he knew he needed to plant trees for more shade. It was all he could think to do. That was our last glimpse of Myanmar and its beautiful, kind people.

Burmese cuisine, referred to as "Myanmar food" by the local people, is rich with ferments, many of which are quite regional and little known to the rest of the world. Once we decided we were going to Myanmar, we researched its ferments. We went with the intention of learning everything about the national dish *laphet thoke*, or fermented tea leaf salad (sadly, not included in this book — tasty as it is, there are no fermented beans or grains in this one), and we came home having experienced many more.

Myanmar-Style Fried Tofu

YIELD: 2 POUNDS

We made a point of eating street food for nearly every meal while we were in Myanmar, and one of Christopher's favorites was crispy fried chickpea tofu, which had a crispy outside and a soft, creamy, almost doughy inside and was amazingly tasty when dipped in mysterious concoctions of savory and chile-fueled heat. We have seen many recipes online for this wonderful chickpea tofu, but most of them skip the fermentation step to save time. These recipes work, but you miss out on the fermentation benefits.

The thinner you slice the tofu, the crispier it will be. If you cut thicker slices, the middle will remain a bit creamy, which is also wonderful.

FERMENTATION **15 hours**

- 6 cups (1.4 L) water
- 2 cups (240 g) packed chickpea/garbanzo flour
- ½ teaspoon (3 g) salt (optional)
- ½ teaspoon (1.6 g) powdered turmeric (optional)

  Peanut oil, for coating the dish and for frying

1. Pour 2 cups of the water into a medium bowl. Add the chickpea flour and whisk until well combined. Cover the bowl with a plate or lid. Let the mixture ferment at room temperature for 15 hours.

2. After fermentation, add the salt and turmeric, if using. Stir well.

3. Bring the remaining 4 cups water to a boil over high heat. Whisk in the fermented chickpea mixture. Reduce the heat to medium-low and cook at a slow boil until the mixture is thickened and slightly reduced, 5 to 10 minutes. Stir continuously with a spatula to keep the mixture from sticking to the bottom of the pan.

4. Oil a small casserole dish, glass pie pan, or bread loaf pan. Pour the hot chickpea mixture into the pan. Drape a clean cloth over the pan and let it sit until it cools. Then place it in the refrigerator to set, about 6 hours.

5. When you are ready to fry the chickpea mixture, turn it out of its dish onto a cutting board. It should have the consistency of firm tofu. Carefully slice thin (about ¼-inch-thick) slices with a sharp knife.

6. Line a plate with paper towels. Pour about 2 inches of peanut oil into a wok (preferred) or heavy pan over high heat and bring it to 375°F/190°C.

7. Carefully slide tofu slices into the hot oil, leaving space between them. You will need to work in batches. Fry until lightly browned on the bottom, about 2 minutes, then flip the slices and continue frying until lightly browned on the other side, 1 to 2 minutes. These slices won't brown like fried potatoes, so don't try or they will fall apart. Just a hint of brown is perfect.

8. Using a slotted spoon, remove the slices and set on the towel-lined plate to drain. Fry the remaining tofu slices in the same way. If the oil begins to smoke or the tofu slices are cooking too quickly, reduce the heat to medium.

Myanmar-Style Shan Soup

YIELD: 4 GOOD-SIZED BOWLS

During our trip to Myanmar, we had planned to visit the region in the northern part of the country where tea leaves are fermented, which is home to many different ethnic groups, several with their own standing armies. We had to change our plans at the last minute because fighting broke out between Myanmar's government army and one of those regional armies, cutting off our access to the tea villages. We found tea plantations and the fermentation we were seeking, including this tofu, in other parts of the country.

In Burmese, this soup is called *hto-hpu nwe*, which either means warm tofu or hot tofu. We got various translations and, depending upon where we were eating it and which hot chile had been added, it did range from warm to very hot. This is a tasty base for some fun soup bowls.

FERMENTATION **12 hours for the first ferment and 3 hours for the second**

---

8 cups (1.9 L) water

2 cups (184 g) chickpea/garbanzo flour

1 tablespoon (15 mL) peanut oil

1 teaspoon (6 g) salt

½ teaspoon (1 g) powdered turmeric

Hot sauce (optional), for topping

Chopped cilantro, cooked rice noodles, chopped roasted peanuts, blanched greens, or finely sliced shallots (optional), for topping

---

1. Pour the water into a large bowl. Add the chickpea flour and whisk until well combined. Cover the bowl with a plate or lid and let ferment at room temperature for 12 hours.

2. Stretch a piece of cheesecloth across another large bowl and secure with a rubber band. Pour the chickpea batter through the cheesecloth into the bowl. This may take a little time and patience.

It helps to have a rubber spatula handy to periodically scrape the the cheesecloth to remove the chickpea sediment. Compost the chickpea sediment. Cover the bowl of broth and let ferment for 3 hours at room temperature.

3. Pour the oil into a heavy pot and rub it around to coat the bottom and sides. Stir the chickpea broth and pour all of it into the pot. Stir in the salt and turmeric.

4. Bring the mixture to a boil over high heat, then reduce the heat to medium-low and cook at a slow boil until thickened and slightly reduced, 20 to 25 minutes. Stir continuously with a spatula to keep the mixture from sticking to the bottom of the pot.

5. Remove the pot from the heat and serve the soup immediately, topped with a drizzle of hot sauce, if you like, and your favorite fresh toppings.

Pone Yay Gyi

MAKES ABOUT 1 CUP

Pone yay gyi ("bean sour") is a fermented horse gram (see page 49) paste that is found only in the Bagan region of Myanmar, where the climate is perfectly suited to growing the high-protein pulse. Travelers to Burma often bring this paste home to their loved ones, sort of like mouse ears from Disneyland. At the shop we visited, which was also a processing facility, you could buy small, brightly colored, multipack gifts expressly for this purpose.

The most interesting thing about this fermented paste is that it is made by reducing the bean residue in the cooking liquid (often called "pot liquor"), not the beans themselves. We suspect that this type of ferment could be made with the thick pot liquor of many beans, but we haven't tried it yet. Try making Horse Gram Tempeh (page 161) with the leftover beans.

To make a flavoring sauce for rice or noodles, mix 2 tablespoons pone yay gyi with some diced fresh chiles or red pepper flakes to taste and 1 tablespoon of peanut oil. Or make Pone Yay Gyi Shallot Salad (page 351).

FERMENTATION **24 hours**

---

1 pound (450 g) horse gram

¾ teaspoon (5 g) salt

---

1. Rinse the beans, then soak them overnight in a gallon of water.

2. Rinse the beans again, then place them in a pot with another gallon of fresh water. Bring the beans to a boil over high heat, then reduce the heat to low and simmer in a covered pot until soft, about 4 hours. Add more water as needed while cooking to keep the beans fully submerged. Remove the beans from the heat and let cool.

3. Working in batches, pour the bean liquid through a sieve into another pot. Use the back of a wooden spoon to press the beans against the sieve to remove all the liquid. Alternatively, gather the beans in butter muslin and squeeze out the liquid. The trick is to capture as much of the thick bean liquid as possible; some of the beans will become mashed, but there is no need to mash them on purpose. You can feed the leftover beans to livestock, put them in compost, or, as mentioned above, use them to make Horse Gram Tempeh.

4. Pour the bean liquid into a jar; you will likely have 1 to 2 quarts. Add the salt and stir to dissolve, then cover the jar loosely with a clean cloth. Place the jar in a warm place (80°F/ 27°C to 90°F/ 32°C is perfect) and let ferment for 24 hours, or until there are frothy bubbles on top of the liquid.

5. When the ferment is ready, skim off any bubbles. Pour the liquid into a pot. Bring to a boil over high heat, then reduce the heat to a low simmer and cook until the liquid reduces to a paste that has the consistency of overly dry refried beans, which should take an hour or more, depending on the size of your pot. As it cooks, stir the liquid with a rubber spatula, occasionally at first, and then constantly as it thickens and forms a skin on top.

6. Spread the paste on a plate in an even layer about 1½ inches thick. Let it sit in an out-of-the-way place to dry for 24 hours.

7. Transfer the dried paste to a jar, pressing out the air pockets, and seal. Store in the refrigerator, where it will keep for about 3 months.

Discovering Pone Yay Gyi in Myanmar

We landed in Bagan, Myanmar, early in the morning. We rented electric bicycles and suddenly we felt as though we were driving right through the pages of *National Geographic.* Bagan is that kind of place.

A few days into our stay, we hired a guide. He first took us through an indoor market in Nyuang U, the neighboring town to Old Bagan. There, we found all types of ferments — a type of preserved citrus, fermented mustard greens, fermented green mango, shrimp sour, tua nao, and many others. Despite our guide's insistence that they would make our Western stomachs sick, we went back later and bought some of the ferments anyway. While we understood his concern, we also knew that our bodies were used to ferments and that fermentation made food safer. We took the risk and did not get sick.

Next we drove into a neighborhood that we hadn't yet seen, where English colonial homes, cracked with faded grandeur, were surrounded by dust, tamarind trees, stray dogs, and curious children. These buildings were so out of place and so much a part of the place at the same

time. There, we found a small open-air facility that made and sold pone yay gyi.

Stepping into the facility was like stepping through a veil of smoky, sultry air, fire, and steam into another time. At the far wall, a stack 4 feet high and 20 feet long of gnarled gray pieces of wood awaited their fate in the two long and low brick ovens, which dominated the middle of the space. Inset along the tops of these long ovens were five wide metal bowls, like 3-foot woks, with beans bubbling away. Beans and bean paste were everywhere: in sacks, in pots, tumbling out of pots. The walls were bean-colored and matched the dirt and concrete floors. Electrical wires as thick as your thumb crawled like snakes along the rafters. Every now and then they dipped down to bite an electrical box or to just dangle in midair, waiting.

From this rustic, steamy, bean- and earth-toned enclave came a delicious bean paste that was packaged in garish red, shiny, single-serving packs and sold in the corner that was the shop. We bought several packs to take home.

Tua Nao

YIELD: 10–12 DISKS

Tua nao literally translates as "soybean spoiled." This fermented soybean paste, sometimes called *pè bohk*, is dried into thin, round disks that are used in much the same way as miso to flavor soups and sauces. Tua nao is found in many dishes in the lush, green, mountainous regions of Myanmar, as well as in northern Thailand.

The main microbe for this ferment is a strain of *Bacillus subtilis*, the bacteria that is also responsible for natto. As the soybeans break down during fermentation, they release a lot of ammonia. The upside is that this tips the pH scale high into a safe zone; the downside is that this ferment is known to be a bit inconsistent and can have a strong ammonia odor.

Once fermented, the beans are ground into a paste, rolled into the disks, and sun-dried for a few days. They are then fried in oil to make a base for sauces, eaten like crackers with the meal, or wrapped in a banana leaf and steamed to be eaten before a meal. This recipe is adapted from what we saw and learned in Myanmar, and the work of Naomi Duguid in her cookbook *Burma: Rivers of Flavor*. To ferment, you'll need an incubation chamber that is relatively humid and warm (around 106°F/41°C); see some of your options beginning on page 24.

FERMENTATION **2–3 days**

---

1 cup (175 g) soybeans

½ cup rice or rice hulls

Salt

Sesame seeds, red pepper flakes, pieces of Kaffir lime leaves, or minced lemongrass, ginger, or galangal, for flavoring (optional)

---

1. Rinse the soybeans and then transfer them to a large pot. Add enough water to cover them by at least a couple of inches. Let them soak for at least 6 hours and up to 24 hours, and then drain them.

2. Now it's time to cook the beans. You can boil them or steam them, following the instructions on page 54. When they are done, they should be soft and easy to squeeze between your thumb and ring finger.

3. Drain the cooked beans. Transfer them to a shallow glass or stainless-steel casserole dish and spread them in a layer about 1½ inches deep. Fill a small bowl with the rice and place it in the center of the dish, moving some of the beans aside to make room.

4. Cover the dish with aluminum foil, crimping the edges to seal tightly. Then, holding the foil taut, use a skewer or pointed chopstick to poke a series of air holes across the top about 1½ inches apart in a grid pattern.

5. To maintain a humid environment, place a small jar of water in the incubation chamber. Incubate the beans for 2 to 3 days. Check the ferment daily. In the first 24 hours, a white film will start to develop and you will smell a strong nutty aroma. You can check for biofilm by running a spoon across the top of the beans. The spoon should stick slightly and threads should form behind it. On days 2 and 3, check to make sure the beans aren't drying out.

Recipe continues on next page

Tua Nao, *continued*

6. After the fermenting period, the beans will have an ammonia odor and will likely be sticky. Transfer the beans to a food processor. Pulse to mash them into a smooth paste, adding water as needed. You want a thick paste that isn't too wet. You may need to process the beans in batches.

7. Add ½ teaspoon of salt for every 1 cup of mashed beans and mix well. Then mix in the flavorings, if using. You can now use the paste as is. Refrigerated in a sealed container, it will keep for 1 to 2 weeks. For a longer shelf life, continue on to make dried disks.

8. Form the paste into golfball-size balls (about 2 tablespoons of paste). Place each ball between two sheets of plastic wrap, parchment paper, or damp, clean cloth and roll into small, flat, thin disks. For an authentic feel, set the disks out in the sun to dry. Alternatively, dehydrate them in a dehydrator for 12 to 18 hours. Store the disks in a cool, dry place, like you would any dehydrated food. They will keep indefinitely.

CHAPTER 5

NATTO

AND ITS ALKALINE COUSINS

This chapter focuses on one of the most unique ferments and nutrient-dense superfoods out there: natto. Natto is a Japanese fermented soybean condiment that has traditionally been a breakfast staple. "Eat your natto" tumbles out of Japanese parents' mouths with the same it's-good-for-you tone as American parents might say, "Drink your milk" or "Eat your broccoli." Most Americans have never heard of natto, despite the love affair in this country with sushi and Japanese food. But we don't see that as a deterrent, as at one time most Americans didn't know what tofu was, and before that we didn't know about yogurt.

As is the case for all ferments, natto was discovered when food was left alone for a period of time. In this case, *Bacillus subtilis* went to work on soybeans. And honestly, some folks still think natto is more of an accident, or a contaminated ferment, than a proper intentional ferment. This is because, as with all microbes, when *B. subtilis* is in the wrong place, the results are ruinous. Acetic acid bacteria are amazing in vinegar, for instance, but they are a winemaker's nemesis; cheese and sauerkraut wouldn't work without lactic acid bacteria, but you don't want them in your cider. Similarly, *B. subtilis* is not welcome in miso houses or sake breweries, to the point where brewers are forbidden to eat natto during the brewing season, lest an errant spore finds its way into the sake.

While some of the ferments you will read about in this chapter are still made in traditional wild-style ways, in general, natto is cultured with specific natto spores. And if you would like to make natto *and* miso *and* tempeh at home, not to worry: we have been making all of them with some precautions in the very same spaces and have had no problems.

KIRSTEN WRITES: *When I was a cheesemaker I was warned against ever making blue cheese because, I was told, there would be no turning back. My cheddars, my fetas, my Goudas — all would become blue cheese. In fact, like the sake brewers who aren't allowed to eat natto, workers at our local creamery (that is world famous for its blue cheese) who have been in the blue cheese house during a workday cannot enter the cheddar facilities because they could be carrying spores from the blue cheese mold.*

Because of its spore-forming ability, B. subtilis, *the bacterium responsible for natto, is also tenacious and pretty indestructible. (Just wait, you'll see how tough it is on page 104.) I admit I was a little nervous about working with koji, miso, tempeh, and natto at the same time. I wondered: Should I even bring natto into my house? What if* B. subtilis *takes over all my fermentations?*

I did some research and talked to Ann Yonetani, microbiologist and natto maker at NYrture Food (you'll meet her on page 110). I learned that B. subtilis *is everywhere already — specifically in the air and the soil — so it's in the vegetables we bring in from the garden, the flowers on the table, our kitchen compost bucket . . . I should have learned this when I was able to successfully make cheonggukjang (page 123) and tua nao (page 97) — both of which require* B. subtilis *— without a culture before I ever ate or made natto in our home. The "extreme fear that never the two shall meet" may be more lore than necessity, Ann says. If you take enough care to make sure cultures are not carelessly thrown about, and you thoroughly clean up, I have found that cross contamination can be avoided.*

Ann told me, "In my own experience, for the first year of producing natto commercially, I worked in a communal incubator kitchen space and shared a fermentation room with other fermenters making huge vats of all sorts of lactoferments (kimchi, kombucha, kraut), and none of us ever had a problem. Of course, these are B. subtilis *competing with (primarily)* Lactobacillus. *That was totally fine. I have not personally tried doing fungal ferments (koji-based ferments and tempeh) in close quarters with my natto. However, I know at least two very well-established producers who make both natto and koji ferments in the same (small) facilities without a problem, and both certainly consume all types of ferments without restriction."*

I also learned that B. subtilis *can spend indefinite amounts of time in spore form, completely inactive. It takes the right conditions (in the case of natto, a heat shock and a food source) for them to wake up, produce enzymes, and contaminate foods. So yes,* B. subtilis *can be a contaminant for sure, but as with all fermentations, using* B. subtilis *is about our choosing and cultivating the microbes to "control rot" to our benefit.*

On Base: What Is an Alkaline Ferment?

Most of the fermentations people think of are alcoholic or acidic, in which the resulting alcohol and/or acid content (think pH here), along with the presence or absence of oxygen, controls which microbes survive, multiply, and carry out the fermentation. Alcohol ferments, like beer and wine, rely on yeasts as the dominant force (unless acetic acid bacteria move in and make vinegar, always a sad day for the cellarmeister). Acidic fermentations, like pickles and cheese, rely on lactic acid bacteria to lower the pH and make the food safe and delicious.

Alkaline fermentations are much less common in Western traditions. In this book we only cover some legume-based alkaline ferments, but there are many worldwide that are animal-based — for example the Mediterranean fish sauce garum. These ferments make use of the base side of the pH scale to control fermentation (and preservation) by increasing the pH to an alkalinity that is higher that 7, which makes the food inhospitable to unwelcome microorganisms. That said, alkaline ferments aren't as bulletproof as acidic ferments (more on that in a moment). Most legume-based alkaline ferments rely on the *Bacillus subtilis* bacteria for fermentation.

While this chapter focuses on natto, which comes from Japan, very similar versions of soy fermented with *B. subtilis* are found all over eastern Asia, southern Asia, and up into the Himalayas, as well as in central and west Africa. In truth, other than the fact that these other ferments rely on "wild caught" local *B. subtilis* instead of inoculation, there is very little difference between them and natto — with warmth and time, beans (mostly soy) become sticky, somewhat stinky, and infinitely more digestible.

In most areas where *B. subtilis* ferments are made, production is still done at the home or village level, like they have been for centuries. Since these regional ferments are wild, they also contain various other organisms alongside the dominant *B. subtilis*. These ferments are not stable over the long-term without some help. In villages that do not have refrigeration, they are cooked into soups or curries or preserved in some other fashion — like the sun-dried tua nao flavor disks made in Myanmar (page 97) and the traditionally buried ceramic pots of cheonggukjang in Korea (page 121).

In Japan, natto has evolved a little differently. It is inoculated with *B. subtilis* var. *natto*, a variety harnessed for consistently sublime flavor and success. Most of the production has become industrialized, whether by small

Kinema

Kinema is an alkaline soy ferment from Nepal and surrounding regions in the Himalayas. Traditionally this ferment is made only by women, and in these rural households, producing kinema provides the family with a small but crucial income.

To make kinema, the women soak and steam the small yellow or dark brown local soybeans, then crush them slightly, put them in a basket lined with fern leaves, and sprinkle them with wood ash. The airy fern leaves contribute to an aerobic condition, add microbial diversity, and serve as packaging, while the wood ash helps ensure the right alkaline conditions. Pretty cool, right? When fresh, this ferment stays viable for only a few days, so it is dried for a longer shelf life. The women sell it wrapped in fern, ficus, or banana leaves tied with straw. Kinema is usually fried in oil and added to onions, tomatoes, garlic, chiles, and other vegetables to make a thick curry that is served over rice.

artisans or large-scale producers, and is now refrigerated or frozen for long-term stability.

With lactic acid fermentations, acidic flavor or a sour pickly smell are both signs of a solid ferment. With these alkaline ferments, the beans need to be heavily populated (read: gooey, sticky, stringy) with the *B. subtilis* bacteria in order to ward off undesirable bacteria.

A weak alkaline ferment (underdeveloped threads and biofilm) is more susceptible than its acidic counterpoint to other microbes looking for food. Don't feel discouraged or wary, however — we will show you how to ensure that you have a successful ferment. Don't forget: people have been making alkaline ferments in their homes for hundreds of years.

Badass *Bacillus*

If we could vote for the most badass microbe, our choice, hands-down, would be *B. subtilis*. Here are our top five reasons.

5. They were on Earth before us, even before our humanlike ancestors from the Paleolithic era. Take note, people on a Paleo diet: *B. subtilis* species clearly count as Paleo because they undoubtedly were happily nestled in the guts of the early Paleolithic versions of ourselves.

4. Evidence of *Bacillus* species spans our globe; they are on every continent, and the oldest were found in salt crystals that are 250 million years old.

3. Many bacteria live in the soil, and *Bacillus* species were originally classified as soil organisms, but now scientists think of them also as digestive tract organisms. This means that their home is all types of creatures who dwell on and eat from the soil, from dinosaurs to insects to us, and they use the soil as a bus stop while they wait for the next host to happen by.

2. Speaking of hosts, *B. subtilis* cells have been found in the wee tummies of 40-million-year-old bees. *B. subtilis* is one of the most dominant bacteria species in honeybee stomachs and is believed to be key in the bees' ability to process nectar from flowers into honey.

1. Forget Earth for a moment. Scientists in the mid-1990s proved that the *B. subtilis* species could survive in space for years — dark space, with its killer radiation, ultra cold, and complete vacuum environment. These same scientists believe that the endospores of *B. subtilis* bacteria likely exist throughout the universe!

Why Natto?

Soybeans are the richest source of legume protein. When we add the bacteria *B. subtilis* var. *natto,* it produces enzymes that digest these proteins into simpler forms of peptides or even simpler amino acids, which is easier for us to digest. As the bacteria digest the proteins, they release ammonia, making the natto kind of stinky. The bacteria also digest the carbohydrates, to a point where the sugars are at near zero after 18 hours of fermentation.

The health benefits of eating natto regularly are clear and impressive, and the depth of the research is strong and convincing (we will get deeper into this shortly). If there is a true superfood, it might just be natto. But let's be honest, the health benefits need to be this impressive because, while natto devotees crave the earthy umami, slightly alkaline bite, and toasted nuttiness of these fermented beans, at first bite you may not be craving natto for its flavor — but we predict that will change.

Health Benefits

One of natto's most unique health benefits comes from the enzyme nattokinase. This enzyme appears to act as a natural blood thinner, helping to dissolve fibrin, an enzyme in the blood that can cause abnormal thickening of blood and blood vessel constriction. In other words, it works to prevent and dissolve clots and lower blood pressure. What's funny is that natto itself is sticky and goopy, and yet it works to release blocked or sluggish (sticky) blood. In fact, natto is all about healthy blood flow. Nattokinase is an enzyme that is secreted by *B. subtilis* into a tangle of natto strings. In 1987, Dr. Hiroyuki Sumi analyzed natto's goo and found it had highly active levels of this blood clot–dissolving enzyme.[18] Nattokinase is frequently used therapeutically to help prevent stroke and the advancement of cardiovascular disease. It has even been suggested that nattokinase can help with the prevention of diabetes and Alzheimer's disease by dissolving protein aggregates in the body, which are central to both disorders.

That stickiness is made primarily of polyglutamic acid and mucin, which sounds like *mucus* because it kind of is, but that is a good thing. Let us explain. Polyglutamic acid is a biodegradable and edible biopolymer that is produced by the bacteria *Bacillus subtilis*. Like other polymers (think plastics or our own DNA), it is made up of many smaller compounds that together give it unique physical characteristics. In this case, it's viscous and somewhat elastic. Mucin is a mucoprotein that coats the bacteria, protecting them from your stomach acid and allowing them to make it down to your digestive tract. Besides helping to deliver the probiotics safely through your digestive system, it may also lubricate as it goes, helping your immune system to keep strong and fight off viruses.

Natto is a complete protein that has all of the essential amino acids that our bodies need but cannot create on their own. It's a rich source of omega-3 fatty acids, but it really shines in the area of vitamins and minerals.

The vitamin and mineral content of soybeans varies widely depending on whether they are boiled or fermented. Let's start with minerals. While sodium, phosphorus, and zinc levels remain the same after fermentation, potassium, calcium, iron, and copper levels increase. Magnesium is the only mineral that is lowered by fermentation, though that decrease is by only about 10 percent.

Topping the list of soybeans' vitamins is K, a fat-soluable vitamin that comes in two forms. You might remember something about leafy greens and vitamin K — that's K_1. It's found in plant foods because plants require it for photosynthesis. So, eat your veggies and you get your K_1, which helps blood to clot. Vitamin K_2, on the other hand, isn't produced by plants but rather by bacteria, and it can be found in cheeses, sauerkraut, and fermented soybeans like miso and natto. Natto has by far the highest concentration — 15 times more than hard cheese, and over 200 times more than sauerkraut — which is good news for those of us looking to increase our bone density. K_2 is thought to build bone health as well as heart health, among other things. Studies indicate that K_2 decreases the incidence of bone fractures in women with post-menopausal osteoporosis.[19] Vitamin K_2 comes in many forms; menaquinone-7 (MK-7) is the form most easily used by the human body, and natto's vitamin K_2 is nearly all in this form.[20]

The *B. subtilis* in natto appears to play a unique role as a probiotic. Although it rarely shows up as a major player in analyses of microflora in the lower gut (the part of our digestive system that is easiest to sample), there is a wealth of evidence showing that people who eat natto or supplement with *B. subtilis* have a significantly different balance of gut microbes than

History of Natto:
War Beans or Monk's Delight?

Natto originated in ancient Japan, likely about a thousand years ago. Exactly when and how depends upon which legend or scholarly theory you want to believe.

According to one story, in the eleventh century, the northeast provinces of Japan were cold and crops often failed, though the government never failed to collect taxes, or at least try to. The people of the province of Oshu had had enough and were thinking of revolt. In 1051, a contingent of soldiers, led by samurai warrior Minamoto no Yoshiie, was sent northward to collect. One night, Yoshiie's camp was attacked just as he was finishing up boiling soybeans for the horses' dinner. Not one to waste perfectly good beans, Yoshiie tossed them into a straw bag and strapped it to one of the horses as he and his soldiers made their getaway. Hours later, Yoshiie opened the straw bag only to discover that a transformation had taken place in its warm, humid confines. Those soybeans were now covered in the threaded web of natto. He must have been hungry, because as the legend goes, he dug in and found it good. Thirty years later, during yet another military campaign, Yoshiie would bring the natto process with him, believing it made his soldiers strong in combat. Thus, natto became an important part of war rations.

A less violent origin story holds that the character pronounced "na" comes from the word *nasso*, which is a word for a temple's kitchen. The character pronounced "to" means "beans," so putting them together, we have "beans from the temple kitchen." And indeed, natto is a staple in the vegan Zen monastic diet. The image of monks preparing natto in their quiet temple kitchen is decidedly more peaceful than natto riding alongside retreating samurai in the night. You decide.

those who do not. *B. subtilis* seems to modulate our total gut ecosystem, helping our "good" bacteria species (*Bifidobacterium* and *Lactobacillus*) to flourish while helping to defend against "bad" bacteria (*Clostridium* and *Streptococcus*) by suppressing their growth. *Bacillus subtilis* isn't especially immune to the acidic bile ocean of our stomach; however, its spores do quite well.

Last but not least, the nutritional benefit of natto derives largely from the fact that fermentation counteracts the antinutrient properties of the raw soybeans. As with all the soybean ferments in this book, natto is a much more nutritious food than its nonfermented counterpart.

Natt(so) Much: Building a Relationship with Natto

We have traveled the country for a number of years sharing "sauered" vegetables and the gospel of fermentation. When we're on tour, every time we stand in front of a new group of folks we begin by asking, "Who here has fermented before?" The banter goes on as we figure out whether we are preaching to the choir or convincing the skeptical. When we imagine asking "Who here has tried natto before?," we picture ourselves in the fourth grade standing before the classroom as it erupts in "eeeew," "gro-oss," and general potty-type humor, with the class clown crying out, "Who cut the cheese?"

Have you tried natto? If you haven't, we don't want to dissuade you with this frank discussion of filaments. Yet discussing natto without mentioning the sticky, stringy, whispy film that coats the beans is like talking about a pickle and not mentioning sour. How do you feel about okra? In Japan, connoisseurs prize

foods that have the slippery texture called *neba neba* — foods like okra, sea urchin, raw egg, and slippery grated mountain yam. Natto is the ultimate neba neba food. The goo that coats natto is called a biofilm in the lab, or as Heidi Nestler of Wanpaku Foods (see page 125) describes it, "gossamer threads of goodness." No matter what you call it, this consistency is where the magic lies. It is natto's superpower, its "special sauce." There is a direct correlation between sauce goo and strong, happy probiotic bacteria and all the other health benefits that are products thereof — the stickier, the better. The caveat is that the goo factor can bring even the most adventurous eaters to their knees.

Okay, now that we have talked texture, let's talk flavor. What does natto taste like? This is a difficult question because we all taste food differently, and once you love it (we are thinking positively here; you *will* love it), you will likely pick up on flavors you didn't notice at first. The scent is part of the experience, and like all things fermented, it has funk and aroma. It has a toasted nuttiness to it. The flavor can be mild, with only a slight earthy ammonia (a young natto), to strong (well-aged natto), but honestly, for most people the flavor is not as disconcerting as the texture.

We like to remind people who are struggling with the stringiness that for some people, mozzarella is just as odd-looking as natto (cheese is, after all, milk "soured" by bacteria). In other words, it is just a matter of what you are used to. Non-dairy-eating peoples of the world are just as put off by the look of the strings of cheese that melt off the side of a hot piece of pizza as you may be by the sticky threads of natto. And while we are talking about cheese, we'll say that natto is kind of like a funky cheese and can be used as such. Its aroma is similar to that of washed-rind cheeses like Limburger, Chimay, Taleggio, and others, due to the role of fungal and bacterial aging on the surface.

Many fermented foods are acquired tastes. Coffee, beer, and wine, for example, all result from fermentation, and most of us don't necessarily like them at first, and then we love them. It has been our experience that most people who stick with natto end up loving and/or craving it (that is likely the microbiome talking); like many ferments, it finds a way into your heart (and gut).

Like many people, we didn't love natto the first time we tried it. We now realize that we just didn't know how to eat it. The hurdle to natto love can simply be its presentation — in other words, what it is served with, over, or in. We hope that some of our recipes entice you to try natto, or if you have already tried it, to see it in a new light.

And, of course, you may just love natto at first bite. Our 3-year-old granddaughter, of her own volition, requested "sticky beans" for a snack daily. After teaching you how to make natto, we share the traditional ways to eat it as well as a few nontraditional ways that may be more familiar to people in the West.

Why Make Your Own?

Like most foods, the industrial version of natto cannot compare to the small-batch handmade counterpart. The homemade version has more flavor and higher quality, from bean to bacteria. In the industrial version, the inoculated hot beans are plopped right into a little styrofoam package, a perforated bit of plastic is put on top, the package is sealed, the beans are allowed to ferment, and then the natto is frozen before

it heads out into the world. The difference between it and your homemade jar of natto is not unlike the difference between a hard, bland supermarket tomato and a sun-ripened juicy tomato picked right off the vine.

Fresh homemade natto has a richer, nuttier flavor than you will find in the packaged natto. Yet at the same time, it also has a gentle freshness. (If you want stronger flavor, just let it age longer in the refrigerator.) That said, there are a few artisan natto makers in this country whose product is delicious and high quality. If you are curious about natto but not ready to make your own, head to the source guide (page 388) and order some fresh natto from them.

Making Natto: An Overview

Unlike vegetable ferments, where good hygiene and clean tools are enough to ensure proper microbial colonization, natto requires some specific sanitation practices. This is because, as we mentioned, *B. subtilis* fermentation is alkaline, not acidic. And while alkalinity discourages other microbes, it is not nearly as strong a barrier to pathogenic microbial growth as a typical acidic ferment. Again, we refer to Ann Yonetani: "That said, if the fermentation goes well and the product is really super-fuzzy and gummy, you can rest pretty assured that you have a successful ferment, and the sheer overwhelming numbers of *B. subtilis* cells (and their biofilm) will effectively defend themselves from contamination."

We also take great care in washing our equipment afterward, not because we are worried about pathogens, but because we also use them to make koji and tempeh.

Meet the Maker
Ann Yonetani
NYrture Food

Meet Ann Yonetani, Ph.D., owner of NYrture Food, and one of less than a handful of folks* to make and sell natto in the United States commercially. NYrture New York Natto comes in four varieties: original, turmeric, black, and organic. This makes her a pioneer in the production of an ancient food, but her background is entirely contemporary — she holds a doctorate in microbiology and has spent years as a food science educator and biomedical research scientist.

Living in New York City, Ann had the same bucolic cheesemaker dream that many of us have had at some point. However, she wasn't looking to leave the city, and cheese making never seemed realistic. Somewhere along her journey, she realized that natto, a food she remembered from her childhood visits to Japan, had a lot in common with the chemistry of the cheeses she loved. And as a food scientist, she was spending a lot of time thinking and talking about how humans can move toward a plant-based diet, which is more important than ever for the planet. "Natto is a close as you can get to a vegan washed-rind cheese," she says, "plus it has that deep, funky, umami flavor that is missing in many vegan diets."

Like a fine cheese, natto is in the hands of microbes — humans aren't totally in control, and it can differ from batch to batch. Ann feels that wildness is part of the magic. "Food used to be fresh and dirty. We lived surrounded by nature. Nowadays, we arc exposed to too little microbial diversity," she says. This may be one reason why our collective microbiome seems to be deteriorating. Fermented foods may be a way to combat this loss.

Driven by a curiosity to learn more about this ancient food, she spent the summer of 2014 in Japan with her children exploring natto in its native habitat, eating all the varieties she could find and seeking out natto knowledge. She found a fifth-generation natto maker in Tokyo and soaked up everything she could about natto history and the art of making it. As soon as she returned to New York, it was game on: the experiments began (which she shared with her food science students). Ann knew that good natto starts with good soybeans that have been cultivated specifically for natto. She tried different organic soybeans — big and small, black and yellow — played with the cooking method, and worked on consistency, nurturing the *Bacillus subtilis* for the best flavor and the most superlative sauce. Soon NYrture New York Natto was born.

In the fall of 2015, Ann wanted to know: How probiotic was her natto? She let a batch age for a month in the refrigerator and then took it to labs at Harvard Medical School, where the microscopes are awesome. The first thing she learned is that the natto was still alive and well, doing what *B. subtilis* does — living, dividing, and sporulating (like making babies, but different).

The fact that the bacteria were actively growing and undergoing sporulation within the food was a big deal. Here's why: Most bacteria in food do not make it all the way to our gut microbiome. It is a numbers game, really; most of the probiotic bacteria in fermented foods don't do well under long storage conditions, and the tremendously acidic bath that is our stomach's digestive juices wreaks further havoc on the sensitive population. (This, by the way, is a good thing, because our acidic digestive bath is our first line of defense against any not-so-beneficial microbes in our food.) Here's where the spores come in. Natto's spores may be fairly impervious to the assault waged by the stomach's environment. If *B. subtilis* spores can reach the gut, and they may be more likely than most probiotic species to do so, they may move in and colonize, becoming the positive player in a healthy gut, as scientists believe them to be. For Ann, the news was great, because she found that even after several weeks of refrigerated storage in glass jars, her company's natto contained billions of *B. subtilis* cell-forming units alive and well!

*Other makers that we are aware of include Aloha Tofu in Honolulu, Hawaii; Megumi Natto in Sonoma County, California; Rhapsody Natural Foods in Cabot, Vermont; and Wanpaku Foods in Portland, Oregon.

Choosing the Right Bean

The best beans for natto are the small yellow soybeans that were specifically developed over centuries for natto. The normal (large and yellow) soybeans that you will find in most stores will, of course, ferment, but the flavor and texture will not be as good as if you use the smaller variety. We have found that the larger beans produce a more sharply flavored natto; however, we do use them to make cheonggukjang (see page 123). Edamame also will not work, as it is basically an immature soybean; it would be like making refried beans with the seeds in green beans.

If you have never tasted natto and are not sure if you will like it, we suggest ordering the smaller beans. The flavor is significantly better. Most of the beans that go to Japan for natto are actually grown in the United States. Two companies that sell GMO-free natto beans from family farms are Signature Soy and Laura Soybeans (see the source guide on page 388).

Can you make natto using other types of beans? Science hasn't rigorously tested the

REGULAR SOYBEANS

NATTO SOYBEANS

B. subtilis ferments on beans other than soy, so we don't know if the lauded health benefits are the same with, say, a pinto bean as they are with a soybean; we suspect not. While you can successfully substitute other beans for miso and tempeh, we and other natto makers have found that there isn't a perfect substitute for soy in natto. In most cases, the bacteria just do not build a good structure. We have found that the tepary bean makes a good natto (see page 119) and we had fun with our black-eyed pea natto (see page 120). Feel free to experiment, keeping in mind that the gooier, the better.

Natto Spores

Traditionally, natto was made by wrapping the hot soybeans in rice straw bundles, which provide a natural source of *B. subtilis*. These days, many strains are carefully cultivated for consistent flavors and aromas.

The most common strain you'll find for use as a starter culure is *B. subtilis* var. *natto*, which was first isolated from a batch of natto in 1906. The classification is important, because this strain does produce natto as we know it. Some manufacturers have gone on to create their own starters, which allows them to highlight certain traits, like resistance to low temperatures and thus low productivity of ammonia. We can all get behind such innovations.

We have had the most consistent natto using powdered culture. Spore cultures are widely available online (see the source guide on page 388); they may be called natto-kin spores or natto-moto. Different starter cultures are designed to be used in different quantities, so be sure to check how many batches you can make per package. (Though if you are making natto regularly, you may save some money by culturing with a previous batch; see page 113.) In general, natto starters will keep for 6 months in the refrigerator or freezer and for 3 to 4 weeks at room temperature.

Natto Fermentation Particulars

Natto is a quick and fairly easy ferment. In fact, it is a great introduction to bean ferments. It has a short turnaround time. Like many ferments, if things go awry, you'll know — a bad batch of natto has no goo or is quite unpleasant. (You can read all about natto mishaps in the troubleshooting section beginning on page 372.) Given that many people are unfamiliar with natto, we will try to walk you through the look, feel, flavors, and smells of this ferment, so that even if you have never tasted it before, you will feel confident that you have done it right.

Sanitizing. You will need to sanitize your fermentation vessel and utensils by boiling them for 5 minutes, using the sanitize setting on your dishwasher, or spritzing them with 190-proof Everclear alcohol or a no-rinse brewing sanitizer.

Soaking and cooking. The next step is soaking the beans, which is important for all ferments (for more on why, see page 52). Next comes cooking them. For best results, the soybeans should be neither overcooked nor undercooked. Overcooked soybeans can lead to a bland, almost flavorless natto. The simplest, shortest route to properly cooked soybeans is to steam them in a pressure cooker. The smaller natto beans take 35 minutes in a pressure cooker, instead of 45 for the common soybean. See page 54 for instructions.

Inoculating. You can use either powdered natto spores (see box on page 112) or a small amount of beans from a previous batch of homemade or commercial natto. We use about a tablespoon of homemade natto, or one-quarter of a 50-gram package of frozen natto. If using natto spores, follow the manufacturer's directions.

Combine a bit of just-boiled water and the natto starter in a sanitized bowl. This water will help disperse your natto culture and will be part of your final gooeyness. A small amount of water (less than ¼ cup) will produce strings that are the thickness of spiderweb filaments. If you use ½ cup or so, you will have a much thicker, viscous texture. Pour the warm steamed soybeans into the bowl. The beans are inoculated at a high temperature of around 175°F/79°C. The "heat shock" boosts the bacteria into action, not unlike some pine species that need a forest fire to trigger them to open their cones and release their seeds. The high temperature also reduces the likelihood that other microbes will have a chance to take up residence.

Since the steamed beans start drying as soon as you pour them into the bowl, you will want to work quickly to keep everything moist for even coverage by the culture. Use a wooden spatula or rice paddle to mix the beans and starter, taking care not to mash the beans. Then transfer the beans to your fermenting vessel, like a sanitized casserole dish. Spread the beans evenly and thinly — the thinner the better, as they benefit from plenty of oxygen. The beans should not be more than 2 inches deep.

To keep the environment nice and humid but still allow the natto some airflow, cover the dish with a piece of plastic wrap or aluminum foil through which you've poked a matrix of holes with a skewer or chopstick. If you're using plastic wrap, lay it right on top of the beans. When the spores start "waking up," they sense the environment around them, and the plastic wrap will give them a sense of a nice cozy home. It will also give the bacteria another surface to adhere to as they dig into their substrate. In our experience, natto made this way has a much more developed biofilm. (Fun fact: the bacteria also sense one another's presence and are able to judge the population in relation to their food source, slowing down reproduction when food is limited.)

The fermentation setup. The trickiest part in making natto is finding a place to incubate your ferment (see incubation setups in chapter 2). *B. subtilis* likes it tropical; a balmy 99°F/37°C to 113°F/45°C is perfect. When we make natto, we set the temperature right in the middle of that range: 106°F/41°C (note that some packages suggest a lower temperature, but this is the sweet spot we have found). Be aware that the life force of the microbes will create heat as the natto ferments. If you are using an incubation setup where you are monitoring the temperature manually, like a heat mat or yogurt maker, you may need to reduce the temperature of your device as the natto heats up. Heidi Nestler (see page 125) has seen the temperature of her natto spike to 127°F/53°C in a yogurt maker.

A bread proofer, yogurt maker, or water bath setup can be helpful because the environment stays more humid than in a dehydrator. You can use a dehydrator if you place a cup of water in with the natto (also make sure the holes in the cover are small). (Crunchy, chewy, dried-out "raisin" natto is not tasty.)

Incubate the soybeans for about 20 hours. When they're done fermenting, you'll see a white chalky film on top, or if you covered them with plastic wrap, they will look gooey on the surface. Run a spoon across the natto and see if you have plenty of sticky strings. If you do, success!

The smell. The first time Kirsten smelled natto fermenting, it reminded her of when she was a kid and her mom would come home from the salon with a new Afro perm (it was the 1970s). Natto is an alkaline ferment, so instead of a funky acidic or sulfur aroma, it produces an ammonia scent. All this is to say that if you or others in your family are sensitive to lingering odors, consider where you set up your fermentation space. A small batch doesn't produce a lot of odor and it may not be noticeable to everyone, but it is something to be aware of.

The taste. When your natto is finished, it shouldn't taste bad at all. Fresh natto is actually quite mild, with a toasted or nutty, earthy flavor. The longer the ferment, the stronger the flavor and smell.

As is the case for many ferments, bad is just that bad — poor viscosity, a dramatic display of off-colors, and composty smells will indicate that the batch is destined not for consumption but for the compost. Trust your senses; they will let you know. Good is funky and nutty. Bad is full-on rotten.

Aging and Storing Natto

As you are learning to make natto, you will discover how you prefer it to taste. Natto develops character and flavor as it ages. Taste your natto immediately after you make it. It will be just as sticky as ever, but the flavor will be quite mild. Then let it age in the fridge for 24 hours and taste again to see how the flavor has changed. You can continue to age it and test it at intervals to see if you like a mild young natto or a strong older natto. We like it aged for about 2 weeks.

There are varying opinions on how long natto keeps in the refrigerator. About a month is a good rule of thumb. If it is kept for several months (aging all the while), the main risks are that it will dry out or become contaminated with other microbes. We divide our finished natto into smaller airtight containers and place a piece of parchment paper on top to keep the natto from drying out. The smaller containers also reduce the risk of contamination that may come from dipping in and out of a larger container with utensils that may not be clean.

All imported commercial natto is frozen to slow its aging, although this may come at the expense of flavor because the act of thawing can trigger a secondary fermentation that can create off flavors. Importantly, freezing and thawing will also compromise many of the inherent health benefits of natto as well. However, if freezing works best for you, no worries. Freeze your natto after you have aged it in the refrigerator to your liking. When you're ready to eat it, thaw the natto in the refrigerator and consume it immediately.

STEP-BY-STEP

Let's Make Natto!

What You Need

- ▸ Natto soybeans
- ▸ Large pot or electric pressure cooker for steaming the beans
- ▸ Kettle for boiling water
- ▸ Bowl for inoculating the beans
- ▸ Glass or stainless-steel casserole dish for fermenting the beans

- ▸ Natto culture
- ▸ Aluminum foil or BPA-free plastic wrap
- ▸ Chopstick (if using aluminum foil) or large needle (if using plastic)
- ▸ Incubation chamber (see chapter 2)
- ▸ Spoon

1. Steam the soybeans, following the instructions on page 54. When the beans are almost ready, bring a kettle of water to a boil. Once the beans are ready, quick-release the pressure, if you used an electric pressure cooker. If you steamed them in a pot, turn off the heat, but keep the lid on the beans until you are ready to go. You will want to work quickly since the natto spores are activated by a heat shock.

Instructions continue on next page

2. When the beans are ready, pour boiling water from the kettle into a bowl and a casserole dish to santitize and warm them. Swoosh the water around, let it sit for 30 seconds, then pour into the sink.

3. Add ¼ to ½ cup of just-boiled water to the bowl, then add the natto culture and mix thoroughly.

4. Add the hot steamed beans to the inoculated water and stir gently, taking care to avoid crushing or breaking the beans, until well mixed. You want the beans to be evenly coated.

5. Pour the inoculated beans into the warm casserole dish.

6. Spread the beans in a thin, even layer, preferably from ½ to 1 inch deep, and no more than 2 inches deep.

7. Stretch plastic wrap or aluminum foil across the top of the casserole dish, crimping the edges to seal tightly. While holding the plastic wrap or aluminum foil taut so that it doesn't crush the beans, poke a series of air holes. If you are using plastic wrap, lay it directly on the beans after you poke the holes.

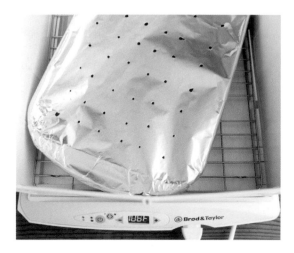

8. Place the natto in a relatively humid incubation chamber at 99°F/37°C to 113°F/45°C for about 20 hours.

9. The natto is ready when you can see a well-developed web of *B. subtilis* on top. Check the growth of the stringy biofilm. Properly fermented natto features a strong network of sticky threads. It should be very gooey, like taffy, and not thin, like a few spiderweb filaments.

Natto

YIELD: 4–5 CUPS

This is a basic recipe for classic natto. There isn't a whole lot of space for creativity. As we developed a taste for the biofilm, we discovered that if you use a little more water when inoculating the beans, you will end up with a little more biofilm in your natto.

PROCESS **natto (page 115)**	FERMENTATION **20 hours (30 hours for a more pungent natto)**

2½ cups (450 g) soybeans, preferably the small natto-style beans

Natto starter (*B. subtilis* var. *natto*; see note below)

Boiling water, for warming the containers and inoculation

Note: Use the quantity of natto starter specified by the manufacturer of your starter. It's usually in the range of ⅛ to ½ teaspoon (the variation stems from whether the starter is pure spores or spores dispersed in rice flour).

1. Soak the soybeans for 18 to 24 hours. Rinse thoroughly.

2. Steam the soybeans according to the instructions on page 54.

3. Pour a bit of boiling water into a bowl and a shallow glass or stainless-steel casserole dish to sanitize and warm them; let sit for 30 seconds and then pour the water out into your sink. Pour ¼ to ½ cup of boiling water into the bowl, then add the starter culture and mix thoroughly. Add the hot soybeans to the starter culture and mix carefully.

4. Transfer the beans to the casserole dish and spread in a thin layer, preferably about ½ to 1 inch deep, and no thicker than 2 inches.

5. Place a sheet of aluminum foil or plastic wrap across the top of the casserole dish, crimping the edges to seal tightly. While holding the aluminum foil or plastic wrap taut so that it doesn't crush the beans, poke a series of air holes across the top, about 1½ inches apart in a grid pattern. If you're using plastic wrap, then lay it directly on the beans.

6. Place the natto in a relatively humid incubation chamber (see page 114) at 99°F/37°C to 113°F/45°C; we have had the best results at 106°F/41°C. Incubate for 20 hours, or until you smell a nutty, alkaline aroma and you see a white film across the top of the beans. Run a spoon across the top of the beans. The spoon should stick slightly, and threads should form behind it.

7. Place the natto in a sealed container and refrigerate. Ideally, let it age for 1 week for deeper flavor. It will keep in the refrigerator for 4 to 5 weeks.

Variation: Black Soybean Natto

We love black soybean natto. The shiny black beans look like jewels, and we find the flavor of the natto made from them milder and sweeter. While we couldn't procure the small black soybeans that are specific to making natto (without going to Japan), the ones we did find online were quite tasty (see the source guide, page 388). They have a surprisingly beautiful green interior. Soak for only 6 hours, as they can be prone to sprouting. Make it exactly as you would regular natto.

Variation: Tepary Bean Natto

Any kind of tepary beans work quite well for natto. They develop a strong biofilm, yet the ammonia quality isn't as pronounced as with regular soybean natto. These native American beans come in white, black, and brown. Each variety has a slightly different flavor. Natto made from the white beans looks most like traditional natto. The brown tepary natto has a bit of an earthy, almost smoky flavor. The black smell quite beany but have a pleasant, mellow, nutty flavor. We feel the black tepary beans are a perfect substitute for black soy natto, given that we cannot source the smaller black soy natto. Tepary beans require a long soaking time (20 to 24 hours) and take much longer to cook (upwards of 2 to 3 hours on the stovetop), which makes us wonder if this because they are drought-tolerant desert beans. We've found that the texture is the best when the beans are pressure-cooked for 40 to 45 minutes. The beans should be soft enough to yield when squeezed between your thumb and ring fingers, but not mushy. Follow the instructions for regular natto.

Variation: All Other Beans

In our experience, black-eyed peas formed a decent biofilm, though not as strong as the soybean or tepary natto. Many other beans do not form a strong enough biofilm to truly be a natto, or they just don't taste good. As you experiment with bean varieties, remember that successful natto is sticky, gooey, and stringy. Follow the instructions for making regular natto.

Hikiwari Natto

This traditional variation is made from beans that have been roughly chopped before being cooked. We tried a few ways to get a similar ferment but found that trying to split up dry soybeans with average kitchen equipment is not realistic. Cooking the beans whole and and then roughly chopping them before fermentation was doable with our equipment but made a very strong-smelling natto. The best way to get that texture is to ferment them whole and then chop the finished natto, as you will see in our recipe for Tapenatto (page 346).

Bacillus-Fermented Hoppin' John

YIELD: ABOUT 1½ CUPS

This recipe is a fun twist on the simple classic Southern dish Hoppin' John. The peas are fermented and then seasoned similarly to the traditional dish, but instead of producing stewed beans, the result is a mellow side salad that we like to serve with, well, collard greens, of course.

This recipe came about when Christopher was researching the relationship between beans and flatulence (see box on page 47). We knew that bacillus ferments are very common in Africa, and that these peas stem from that part of the world.

We ferment the black-eyed peas in the same way we make natto, but the results are different. The peas are starchier, and the texture and flavor are different. The biofilm is respectable, but not nearly as abundant. This recipe yields enough to serve as a fun side to a Southern-inspired main dish with plenty of collards, a little mac and cheese or grits, and perhaps fried green tomatoes.

PROCESS **natto (page 115)**	FERMENTATION **20 hours**

1 cup (175 g) dry black-eyed peas

Natto starter (*B. subtilis* var. *natto*; see note)

Boiling water

AFTER FERMENTATION

3 tablespoons (18 g) thinly sliced scallions

1–2 teaspoons (5–10 mL) apple cider vinegar

Salt

Sour cream or Koji Cultured Cream (page 311), for a creamy version (optional)

Note: Use the quantity of natto starter specified by the manufacturer of your starter.

1. Soak the beans for 18 hours. Rinse thoroughly.

2. Boil or steam the beans, following the instructions on page 52, until al dente, 20 to 30 minutes (you don't want them mushy and overcooked).

3. Pour a bit of boiling water into a bowl and a shallow glass or stainless-steel casserole dish to sanitize and warm them; let sit for 30 seconds and then pour the water out into your sink. Pour ¼ cup of boiling water into the bowl, then add the starter culture and mix thoroughly. Add the hot beans to the starter culture and mix carefully.

4. Transfer the beans to the casserole dish and spread in a thin layer, preferably about ½ to 1 inch deep, and no thicker than 2 inches.

5. Place a sheet of aluminum foil or plastic wrap across the top of the casserole dish, crimping the edges to seal tightly. While holding the aluminum foil or plastic wrap taut so that it doesn't crush the beans, poke a series of air holes across the top, about 1½ inches apart in a grid pattern. If you're using plastic wrap, then lay it directly on the beans.

6. Place the beans in a relatively humid incubation chamber (see page 114) at 99°F/37°C to 113°F/45°C; we have had the best results at 106°F/41°C. The fermentation time will be right around 20 hours. You will know when the ferment is done by checking for good gooey string development.

7. Serve the dish at room temperature or cool, adding the scallions, vinegar, salt to taste, and sour cream, if using, immediately before serving. Leftovers will keep for a day in the fridge.

CHEONGGUKJANG
(Chongkukjang)

This Korean comfort food is similar to natto and undisputedly a cousin. We've often read that the biggest difference between the two is usage, but we have found that isn't the full story.

Commercial producers of cheonggukjang use a starter, but many traditional makers still make it by catching wild *B. subtilis*. To make it in the traditional method, cooked soybeans are placed in wicker trays lined with sheets of cotton muslin and then placed on stone floors warmed from underneath by fire. Small bundles of rice straw (for inoculation) are placed on top of the beans and then covered with another cotton sheet. A thick blanket is placed on top to tuck in the ferment and keep it around 106°F/41°C. *B. subtilis* moves in, and the biofilm forms along with the strong ammonia smell. Sounding familiar? From here, this condiment takes a different path from natto. Natto ferments for 20 hours, but cheonggukjang ferments for at least 2 days, and more often 3 days. As you can imagine, this ferment also develops a stronger aroma than natto and the texture at this point is no longer as stringy.

After fermentation, the fermented beans are pounded into a chunky paste using a deep mortar. Salt and hot pepper powder (*gochugaru*) are mixed in. In times past, this paste was made in the fall after the soybean harvest, pressed in a clay onggi pot, and buried in the ground for winter storage. Now it is made year-round and refrigerated.

Cheonggukjang is not traditionally eaten raw but is instead made into a simple thick stew (*cheonggukjang jjigae*).

This food is as soulful in Korea as chicken soup is for many in the West. And for good reason. As an affordable protein, it has kept the poorest populations fed for generations. In a very short fermentation period, you can take cheap soybeans and convert them to a high-quality functional food. Studies have shown that the amount of retinol, vitamins B_1 and B_2, and niacin increases with fermentation, and the digestibility of the soy improves by 30 percent. And as is the case for natto, the consumption of cheonggukjang can improve blood pressure and keep blood clots from forming. It has also been shown to suppress cancer cells. In a study conducted in 2013,[21] three anticancer factors were explored in three traditional soybean ferments: cheonggukjang, shuidouchi (from China), and natto. The researchers found that while all had remarkable effects, cheonggukjang was most effective in decreasing the growth of respiratory and digestive tract cancer cells and increasing cancer cell death rates.

Cheonggukjang

YIELD: ABOUT 6 CUPS

In this recipe, the *B. subtilis* is coming from an inoculant of organic rice or rice hulls, which are placed in a bowl among the soybeans. We also tried making cheonggukjang using a wild fermentation because we wanted to make it as traditionally as possible. Both methods worked (remember that *B. subtilis* is an easy-to-catch, ubiquitous microbe), but we did get a slightly better ferment with the addition of the rice hulls. In the name of tradition, we also tried setting up the ferment with a cotton cloth on a warm surface. We draped half of a sanitized damp towel over a sanitized glass casserole dish and positioned a small bowl of organic rice hulls in the middle of freshly steamed soybeans. We then covered the beans with the other half of the towel. We placed this on a heating pad insulated with thick towels. After 2 days, the beans were solidly sticky and stinky, just as they should be. We were pleased with the results — until we tried to get the beans off the towel. We now use the bread proofer.

PROCESS **natto (page 115)**	FERMENTATION **2–3 days**

2½ cups (450 g) soybeans

¼ cup organic brown rice or rice hulls

AFTER FERMENTATION

Salt

Gochugaru or another chile powder

1. Soak the soybeans for 18 to 24 hours. Rinse thoroughly.

2. Steam the soybeans according to the instructions on page 54.

3. Place the beans in a shallow glass or stainless-steel casserole dish and spread in a thin layer, preferably about 1½ inches deep and no thicker than 2 inches. Fill a small dish with the brown rice and place it in the center of the casserole dish, moving some of the soybeans aside to make room.

4. Place a sheet of aluminum foil or plastic wrap across the top of the casserole dish, crimping the edges to seal tightly. While holding the aluminum foil or plastic wrap taut so that it doesn't crush the beans, poke a series of air holes across the top, about 1½ inches apart in a grid pattern. If you're using plastic wrap, then lay it directly on the beans.

5. Place the beans in a relatively humid incubation chamber (see page 114) at 99°F/37°C to 113°F/45°C; we have had the best results at 106°F/41°C. Incubate for 2 to 3 days. Check the beans on the first and second day to make sure the environment stays humid. If you need to, you can place a small jar of water inside the chamber. A white film will develop in the first 24 hours and you will smell a strong, nutty, ammonia-like aroma. Run a spoon across the beans. The spoon should stick slightly, and threads should form behind it. On day 2, check that the temperature is constant and that the beans are not drying out. Ideally this ferment will go for 3 days, but if the smell is too strong for you, you can pull it.

6. When the beans are finished fermenting, combine with salt and chile powder to taste in the bowl of a food processor. Process in batches until mashed into a chunky paste.

7. Divide the paste into 1-cup portions. Press each portion together into a ball and place in an airtight container. Store in the refrigerator, where it will keep for 1 week. Tightly wrap any portions you don't use in plastic wrap and freeze them. Each portion is a perfect base for bowls of soup later.

Eating Sticky Beans

When we first started working with natto, we were determined to find recipes to help folks begin an enjoyable relationship with this ferment. After spending time with Heidi Nestler (see page 125), we wondered whether putting natto in more familiar Western food could be a gateway to enjoying natto in other more traditional dishes. We had some successes, like our natto energy bars (page 371); they are amazing, and you would never know natto is there. Our best efforts can be found in part III of this book. But we also tried natto in recipes where it did not belong (and you won't see those recipes in part III). Hiding natto in foods like lasagna, for instance, would make the whole dish taste like natto funk. Not that *we* would have tried that . . .

The truth is that whipped natto on toast, with a few of your favorite toppings, is pretty awesome (see page 345). It's nutritious, it's inexpensive, and it fills you up. Just wait . . . natto toast will soon be the new avocado toast.

Another great simple way to enjoy natto is in a grain and protein bowl to add nuttiness and umami. Ann Yonetani has come up with other novel ways of pairing natto with Western dishes, which she displays beautifully on her Instagram feed @nyrture (check out her natto ice cream!).

The live bacteria in natto will begin to die off at the same temperatures as lactic acid bacteria, at around 105°F/41°C, and those same high temperatures can deactivate the enzyme nattokinase. However, *B. subtilis* is a little different. When it sporulates (make more spores), the spores are quite heat resistant and will remain viable regardless of the temperature, making it to your gut. So you can still reap the gut health benefits of this probiotic bacteria when it is cooked, because once they reach your gut, those spores will germinate, form new cells, and form new biofilm.

That said, we ourselves are all about flavor, which means that we think sometimes it is okay to sacrifice live cells for an amazing dish. We believe in balance and sanity and not getting hung up on all the small "rules." We believe that food should be prepared and eaten with joy and love, and that this is ultimately the most nourishing.

Bon appétit.

Meet the Maker
Heidi Nestler
Wanpaku Foods

If natto has a Pacific Northwest publicity agent, it's Heidi Nestler. Not just an aficionado, she has spent many years teaching, coaching, and cajoling others to love it, even before she started making it commercially. She has taken natto to many festivals and events where she lovingly serves up samples of nori-wrapped natto and rice garnished with scallions. (She shares her recipe for these hand rolls on page 344.) At one point, she told us, "I'm out to save the soybean and I am not sure why."

While we'd each tasted natto before, the 2 days we spent with Heidi learning to make and appreciate natto were our true introduction to this food. It was natto boot camp. Heidi and her son, Ranmu, served us natto in so many traditional ways, from the simplest — seasoned natto on rice — to natto served with okra (that's right: sticky biofilm squared).

In addition to whipping the natto with chopsticks, Heidi also taught us another less common way to activate its goodness: she placed it on a cutting board and used a chef's knife to chop the beans until they were the consistency of a roughly chopped olive tapenade. This created a very different texture from traditional natto, and it is perhaps more appealing to the uninitiated Western palate. In fact, it was this method that inspired us to make Tapenatto (page 346) for this book.

Heidi also organized a natto tasting party for us, where she invited friends to taste nattos made with a variety of different beans. Along with the soy natto, there were nattos made from black turtle, kidney, cranberry, chickpea, adzuki, black soy, red lentil, black-eyed pea, mung, and orca beans. We all agreed that some of the flavors were not pleasing, others were okay, and some were just plain interesting, like the red lentils, which almost turned into the consistency of hummus. As we mentioned on page 119, different beans can be fun to try!

CHAPTER 6

TEMPEH

AND OTHER INDONESIAN FERMENTS

Tempeh is different from the rest of the fermentations in this book for a couple of reasons. It originates from Indonesia (more specifically Java), making it stand out from the other soybean ferments traditional to China, Japan, and Korea. And it can easily swing between being either a condiment (fringe) or a main dish (core), or both.

Tempeh is a white mold–covered cake made from the interaction of a fungus upon soybeans (or other legumes, grains, or seeds) that have been hulled, soaked, and then partially cooked. These al dente legumes and grains are then acidified during the soaking period (which acts as a short fermentation) and finally inoculated with rhizopus spores. The resulting fungus is not just on the surface but is woven throughout the cake. When you slice tempeh, you will see this clearly: the legumes are knit together in a dense mat of white mycelium.

Lightly boil or steam tempeh in a mixture of water and soy sauce (or other aminos), then slice and drop into hot oil for a few minutes and you will experience a savory flavor and satisfying texture that can easily double for meat in many dishes. It is a bit of a flavor chameleon, taking on the flavor of the marinade or flavored oil. Because of the enzymatic breakdown of its proteins, tempeh adds umami to dishes. For vegetarians, vegans, those on a budget, or anyone looking to eat less meat, it can easily become a protein-rich go-to. But it is also more than that. Tempeh is quick to ferment, quick to prepare, and highly versatile.

A Brief History of Tempeh

All the experts seem to agree that tempeh originated on the island of Java, one of over a thousand islands in the Indonesian archipelago. How it began is less clear, but we can assume that it likely grew out of either trade or war with a ferment-loving neighbor like China or Japan.

We know that the Chinese mastered the use of koji to produce soy sauce over a thousand years ago and would have brought it as a staple on long trading voyages to Indonesia. We also know that one of the names for soybeans in West Java is *kachang jepun*, which translates to "Japanese bean." The Javanese word for soy, *kedele*, shows up for the first time in the folklore tale of Banyuwangi, thought to have been written in the twelfth or thirteenth century. The story is one of lust, misunderstanding, betrayal, and, of course, a cast of gods and goddesses. While very good, the tale isn't really about tempeh, but it does establish it as a food many hundreds of years ago. It is very likely that the Indonesians were introduced to soybeans, koji, and the process of fermenting soybeans with koji by the Chinese or Japanese, and that the Indonesians adapted the process to produce the tempeh we enjoy today. From Java, tempeh spread to the rest of Indonesia and Malaysia, and the name became more generalized to refer to any fermented legume or cereal that mycelium had penetrated and bound together.

Fast-forward to the United States in the 1970s, when the awareness of the ecological implications of our carbon-based lifestyle began the environmental movement. Back-to-the-landers and food crusaders had hope that plant-based diets and a better utilization of Earth's resources would feed burgeoning populations. There was a lot of momentum after the 1973 oil crisis, and in 1977, Robert Rodale wrote, "Before long [tempeh] will be eaten widely and lovingly across this land of ours." William Shurtleff (see page 41) and Akiko Aoyagi, authors of *The Book of Tempeh*, crisscrossed the nation giving hundreds of presentations and workshops about the wonders of tempeh. Participants inspired by their teachings formed dozens of companies to make tempeh, cultures, or both, furthering the movement.

More than 40 years after Rodale's prediction, we can say that tempeh is lovingly eaten by consumers who are mostly in California, the Pacific Northwest, or the East Coast, and who make up less than 1 percent of the U.S. population. Still, consumption of tempeh is trending upward, riding the wave of increased awareness. Joe Yonan, food editor for the *Washington Post*, suggested to his readers in 2015 that they include tempeh in their healthy-eating New Year's resolutions and rightly pointed out that tempeh has "more character than tofu will ever possess." Still, tempeh advocates will have to admit that the dream of a tempeh in every pan is yet to be realized. We hope some of our recipes will tempt and inspire you.

Meet the Microbe

Tempeh is brought about by a diverse group of microorganisms, including fungi, yeasts, and lactic acid bacteria, but really it is rhizopus that is the big player here.

Rhizopus is both the common name and the genus name for a group of molds that, like most molds, are found in soil and plant material. Tempeh is most commonly made from one of two rhizopus species: *Rhizopus oryzae* and *R. oligosporus*, both of which are found in Indonesian tempeh. In Indonesia, tempeh makers produce their starter by placing soybeans in fresh unwashed hibiscus leaves, as the fine hairs on the undersurface of the leaves contain the inoculant (the rhizopus fungus, plus a host of other microbes). Layers of leaves and beans are built and then bundled together with rice straw. After a few days, the white mycelium appears and the leaves are opened up and dried for another couple of days until they are covered with black spores. This dried starter is then crushed and sprinkled over soybeans. Traditionally the inoculated beans are then swaddled in banana or sometimes teak leaves, though today they are mostly fermented in plastic bags.

Mildly Alcoholic Rice Pudding, Anyone?

In 1974, we were at opposite sides of the planet. For Christopher, in Missouri, pudding came from a box. If there was anything certain from his childhood, it was that pudding came from Jell-O boxes and it was instant — you could eat it right out of the bowl after mixing. Kirsten was on the island of Ambon in Indonesia enjoying pudding as well. But there was no box.

There was nothing instant about the pudding Kirsten ate. For starters, a sago palm had to be felled first. The pudding she ate was made from the lightly fermented starchy pith of said sago palm (called *papeda*), which is the traditional starch, or core, of the Moluccan Islands. She didn't love (read: hated) the regular, unsweetened version, but she was all about the sweet version made with the local palm sugar. Once in a while she also ate *tape*, or *tapai* (pronounced "tah-pay"), a Javanese dessert that is both sweet and mildly acidic, almost citrusy, and always pushing to become more alcoholic. It can be made with either starch from cassava (*tape telor*) or glutinous rice (*tape ketan*). You see variations on this theme throughout Southeast Asia and East Asia in China. The same group of microbes that makes it is also responsible for *jiu niang*. It is also fermented further to make an alcoholic beverage that is sweet, not unlike sake.

Fast-forward to 2018, when Kirsten went back to Ambon and tasted these "puddings" with an adult palate. She loved papeda — both regular and as a sweet pudding. She tasted cassava *tape* and was told that traditionally, it is often made by men because when *tape* is made by a menstruating woman it will sour.

The consistency of the cassava *tape* varies from pudding to pound cake. We haven't made the pound cake; however, we have made the glutinous rice version. It's supersimple to make once you have the starter (called *tape ragi*), which is a wonderful cooperation of molds (both *Aspergillus oryzae* and *Rhizopus oryzae*), yeasts, and bacteria. You can purchase the starter (with full instructions) from Indonesia very reasonably via the Internet (see the source guide on page 388). This ferment is quickly assembled in a jar (cooked sticky rice with the ragi sprinkled on top), or use the low yogurt setting if you have an electric multiuse pressure cooker. It is ready to go after a few days of fermentation. The result is creamy rice pudding on top and a mildly alcoholic rice wine on the bottom. It's like fruit-on-the-top yogurt for adults. A friendly warning: it is quite addictive.

Modern inoculants are grown on rice or cassava powder. In Western countries, tempeh production has been turned over to microbiologists, who create pure cultures. Back in 1980, when Betty Stechmeyer and her husband, Gordon McBride, began GEM Cultures (see page 16), they produced the only widely distributed tempeh starter: *Rhizopus oligosporus*, specifically NRRL 2710. This strain came from the USDA Northern Regional Research Lab in Illinois. At that time, *R. oryzae* wasn't available in the United States to tempeh makers.

Here's where the science gets interesting. Studies have shown that many types of fungi have been isolated in Indonesian market tempeh, many of which are in the *R. oligosporus* family. Others include, of course, *R. oryzae*, but also *R. stolonifer, R. arrhizus,* and *R. formosaensis*. We have talked to some scientists who hypothesize that a diverse array of cultures (as opposed to the industrialized version that isolates just one) is better for our microbiome. However, some of these fungi have potentially harmful effects. *R. arrhizus,* for example, is a known food spoiler, and these traditional home-grown starters come with the risk of adverse bacterial contaminations, which can include coliform, salmonella, or other pathogenic bacteria. But don't stop reading, because here's the thing: rhizopus protects itself and ultimately us by producing antifungal and antibacterial compounds.

An article published in the *International Journal of Gastronomy and Food Science* on the uses of *R. oryzae* in the kitchen states, "To our best of knowledge, no toxin production by *R. oryzae* has been reported in scientific literature. Actually, the *Rhizopus* species has been used on the one hand, as a detoxifying agent against food toxins . . . and on the other hand, to increase the digestibility of certain legumes."[22] The legumes the researchers are referring to are African yam beans, which, while a great source of low-cost protein, have several barriers to wide consumption, including that they cause high levels of flatulence and diarrhea in many people. Oh, and they take 4 hours of boiling to reach an edible state. The researchers made tempeh with African yam beans, cooking them for only 30 minutes before fermenting them, and the process reduced the chemicals responsible for the negative effects by 97 percent.

Which Rhizopus Do I Choose?

Both *Rhizopus oryzae* and *R. oligosporus* are readily available online (see the sources listed on page 388). Is one better than the other? We have undertaken many side-by-side trials where the only thing that varied was the culture. It was fascinating to see how differently the cultures reacted. Sometimes *R. oryzae* would start faster but then *R. oligosporus* would give us a stronger mat. We didn't come out with a definitive answer, but in our experience, *R. oligosporus* was always stronger in the non-soy ferments. In soy-based ferments, *R. oryzae* was comparable.

Whichever starter you decide to purchase, it is important to make sure that you keep your supply moving and fresh. We have found that the starter cultures weaken over time, even when stored in the freezer. Buy smaller amounts more often. If you suspect your starter is getting weak, double the amount called for.

Meet the Maker

Gunter Pfaff and Betsy Shipley
Betsy's Tempeh

Timing is everything. Gunter Pfaff and Betsy Shipley were 50 years old and living in Michigan when the 1980s recession found Gunter, a documentary filmmaker at Michigan State University, first under-employed and then unemployed. Enter tempeh. In 1980, *Organic Gardening* magazine ran an article titled "Hello, Tempeh Lovers." Soon after Betsy read it, she happened upon a small tempeh starter kit at the East Lansing co-op where she shopped. The kit contained a pound of hulled soybeans, starter culture, and instructions. She bought it as a birthday gift for Gunter.

In order to incubate the tempeh, Gunter and Betsy spread the inoculated beans on a covered tray in the oven, with a lightbulb and a thermostat. The fermentation worked, and they loved the tempeh. They began to make more for themselves. For Betty, a vegetarian, it was a great way to get protein into her diet. For Gunter, it was the beginning of the end of his love of bratwurst. With Betsy's marinades, tempeh would soon replace the flavors and texture Gunter sought in the sausages.

As Betsy and Gunter made more tempeh, the oven proved problematic. The first challenge was that covering the tempeh with a clean cloth allowed air currents to interface with the tempeh more freely, causing uneven mycelium coverage. The second was a matter of quantity. If they made more than one tray, the oven was too warm in the last 12 hours, when the mold was making its own heat, and the batch would overheat. And then there was the small matter of forgetting that the oven is off-limits for 22 hours and turning it on to bake (a problem that many of us who use our ovens as incubators share). By the time the smell reminded them, it was too late

for the tempeh. As an alternative, Gunter rigged up an old refrigerator with a lightbulb.

One day as Gunter was driving along some back roads through Michigan farmland, he saw two cement oceangoing yachts being built. Curious (Michigan is landlocked, after all), he stopped to talk to the builders and learned that they were using ferrocement — a waterproof, mold-proof building material. He was already thinking about the perfect tempeh production facility. He had read *The Book of Tempeh* (1979) and *Tempeh Production* (1980) and now added Stanley Abercrombie's *Ferrocement: Building with Cement* (1978) to his reading list.

Gunter began drawing up plans for a building, diving into the state licensing process, looking for equipment at auctions, and most importantly experimenting like crazy. The modern standard method of incubating tempeh, even in Indonesia, was to place inoculated soybeans into plastic bags perforated with tiny holes and incubate it on racks in warm rooms. Gunter knew there was a better way, despite being told "it can't be done." He just needed to figure it out. University researchers had developed the plastic

bag method in the mid-1960s after trying myriad techniques, including perforated stainless-steel trays with air circulated around them, but the results were poor and the bags won out. Gunter's genius was in seeing the connection between water bath temperature regulation and the stainless-steel trays. Gunter started growing batches of tempeh on stainless-steel trays suspended in temperature-controlled water. There was no waste, the trays were reusable, and the tempeh was much tastier.

Eventually Gunter hand-built an ecological tempeh fermentation facility with a 14-tray fermenting system, and on June 21, 1987, Betsy's Tempeh opened for business. Their first sales were to the very same East Lansing co-op that Betsy had bought the little tempeh kit from seven years earlier. They received immediate feedback that their tempeh was good. Descriptions ranged from "mild" and "buttery" to "firmer" and "more meaty" than the bagged versions, which were a bit slimy and bitter, with a rubber mouthfeel. It was a hit. Folks also wanted to know why it was different.

Gunter believed that his tray method was superior because the fermentation was unrestricted. He also speculated that some of the bitter taste and textural challenges endemic to bag-fermented tempeh were because the tempeh must have access to oxygen and needs to be able to get rid of other gases; in other words, it needs to breathe.

Over the next years, Betsy and Gunter traveled to co-ops, fairs, and trade shows, demonstrating their production method. They wanted to do more than just sell their tempeh; they wanted to see this food adopted as a staple, and Gunter's method brought that flavor profile up.

They dreamed of seeing worker-owned tempeh production co-ops throughout the land.

Everywhere they went, the response to their tempeh was amazing. In 1993 and 1994, Gunter received patents for his "apparatus and method for culturing plant materials as foods." Now, though, Gunter and Betsy were at the point where they had to decide whether they wanted to grow the business. They decided instead to retire. They spent a few years trying to find someone to take over the business, but to no avail. In 1996, Betsy's Tempeh closed.

Retirement was good and included moves to sunny places and plenty of tennis, but Betsy shared that they didn't make tempeh for quite a few years until one day Gunter happened to see some hulled soybeans for sale, brought them home, and made a small incubator out of a picnic cooler. It was now in the early 2000s, and they were in drought-prone Southern California, and it bothered Gunter that every time he made tempeh there was wastewater. He designed a dry incubation system for making tempeh and received a patent for this dry system in 2015, a few months after he passed away. His system is being brought to market by DuPuis Group under the name TempehSure.

Gunter still had a lot of things he wanted to explore, including the question of why his tempeh's mouthfeel was so good. Or understanding the role of the banana leaves in the traditional Indonesian method of incubation. How much "breathing" takes place? How much cooling through transpiration happens, given that the leaves are still green? He and Betsy dreamed of a tempeh institute, where all the beans, grains, and starter cultures could be researched to make all kinds of protein-rich food.

Health Benefits

There is a reason why so many people have been working to make tempeh a staple in the Western diet: its health benefits are impressive. Like natto and miso, it enjoys the functions associated with fermented soybeans — specifically, lowering cholesterol, increasing bone density through improved calcium uptake, reducing menopausal symptoms, and improving muscle recovery through better protein bioavailability. Unlike miso, it is also extremely low in sodium, yet high in fiber and easily digestible. Like the other fermented legumes and grains, tempeh has positive effects in preventing diabetes and minimizing blood sugar spikes. Finally, tempeh is uniquely high in vitamin B_{12} and natural antibiotics. To understand all of these attributes, let's look a little deeper.

The process of making tempeh falls into three phases: the initial soaking fermentation, the cooking, and finally the mold fermentation. Between the first and final stages, vitamins and minerals are enhanced, bioavailable essential amino acids and fatty acids are unleashed, and antinutrients that typically keep the vitamins and minerals locked away from our bodies are drastically reduced. To believe these benefits, we wanted to know more about the science behind them.

The first two stages of making tempeh — the initial soaking, with its spontaneous lactic acid fermentation, and the cooking — are also used to prepare legumes for natto. What makes tempeh unique is the last step of fermentation by mold. *Rhizopus* species of fungus produce three enzymes — lipases, proteases, and amylases — to digest the substrate, be it soybeans, chickpeas, hazelnuts, or whatever. There is nothing magical or foreign about these enzymes. We have them within us, too. Whenever we eat, our pancreas secretes these enzymes into our intestines, and each enzyme goes to work processing carbohydrates, fats, and proteins. In the case of people who have exocrine pancreatic insufficiency, these enzymes aren't sufficiently produced, so not enough nutrients are absorbed by the body, leading to weight loss, abdominal pain, vitamin deficiency, and worse. As part of their treatment, patients are given synthetic enzymes. Tempeh comes not only already loaded with the necessary enzymes but with much of the enzymatic work of digestion already done.

The rhizopus isn't making a magnanimous gesture to feed us; it's feeding itself. Warning: here comes the biology bit. The lipases that our body produces digest the fat in our food — mostly triglycerides, which cannot pass through our intestinal wall to our bloodstream — into fatty acids. The fatty acids are then loaded into our bloodstream, where they are sent to our organs as energy. The rhizopus is doing the same thing: it's secreting lipases to create fatty acids to use as energy for its mold. Similarly, proteases break down proteins into their building blocks — amino acids — for rhizopus's benefit. As this happens, some key vitamins — like iron, vitamin B_{12}, magnesium, and folic acid — that were locked up in protein complexes are freed, making them more available to us when we eat tempeh. Finally, amylase breaks down the starches and carbohydrates in our food into simple sugars so that our body can more easily absorb them into our bloodstream to become energy. In unfermented legumes, some of these complex sugars are not easily broken down by our body and pass all the way through to our colon, where they are digested but produce flatulence. However, with the fermentation process, these complex sugars are removed and the simple sugar glucose is created, which is likely what gives tempeh its sweet-savory taste.

That is the increased nutrition side of the health benefits, but there is another: the reduction of antinutrients. During tempeh production, antinutritional compounds are reduced by at least 65 percent, and in some studies, some of the antinutritional factors were reduced by as much as 90 percent.[23] This is a good place to talk about antioxidants because they are a major benefit of tempeh. First, we need to understand that there are unstable molecules in our bodies, known as free radicals. We know, not a comforting thought, is it? Actually, free radicals are part of normal metabolism and oxidation, but if there are too many of them, they can lead to oxidative damage in some important parts of us, like our DNA. They can also weaken our body's defensive mechanisms and possibly trigger conditions like rheumatoid arthritis, Parkinson's disease, and Alzheimer's. Enter antioxidants, which, as you can guess from their name, fight this oxidation process when it runs amok. Tempeh made with soybeans contains high levels of isoflavones, and a number of studies have concluded that isoflavones and other phytochemicals in tempeh lead to a high level of antioxidant activity.[24] Maker Chad Oliphant (see page 154) believes tempeh also acts as a substrate for our gut flora, providing nourishment for them.

Making Tempeh at Home

Why make your own tempeh? Here are a few reasons.

First, fresh tempeh has more flavor and a better mouthfeel than most of the tempeh you find at the grocery store. That tempeh has either been pasteurized and vacuum-packed or left unpasteurized but immediately frozen. Both processes are done to stop the fermentation process and to preserve the food. Nevertheless, there are some really great commercial products on the market. So even if you never see yourself making it at home, you could still incorporate some tempeh into your diet.

Second, making tempeh at home gives your inner food artist a brand-new and exciting canvas. Your medium is the substrate for the rhizopus, so think up flavorful combinations of legumes, grains, seeds, and nuts to try. The first time we had a steaming bowl of ayocote morado beans, Christopher fell in love with their hearty consistency and vowed to tame them in a tempeh, which he did, eventually. (You will find a recipe using these beans on page 174.)

Finally, it's just so cool. You start out with dry beans in a jar from your pantry and you end up with this tasty bean cake of your own creation. As you will see in this chapter, there are many ways to eat tempeh, and many of the dishes you see on Instagram or websites are made with very plain soybean tempeh from the store. Making your own, customized to what you have in your pantry or what you expect to come through in the dish, takes cooking and enjoying to a whole new level.

Making Tempeh: An Overview

Making tempeh is pretty straightforward, though it requires a bit of attention. Once the mycelium metabolizes the substrate, it creates its own heat, increasing the temperature. If the mycelium gets too hot, it will die, and the substrate will become an ideal medium for contaminating microbes. After about 12 hours of incubation, it is important that you monitor the ferment carefully, at least for your first few attempts, until you understand the vagaries of your incubation space. Once you understand the basic technique and get into a groove, you will find tempeh making fun and delicious.

Making tempeh has five distinct steps: soaking and acidifying, dehulling, cooking, inoculating, and incubating. If you start step 1 in the evening of day 1, you should be able to pull out your beautiful finished tempeh cakes on the afternoon of day 3. For example, start your beans after dinner on Friday night, and you can be serving homemade tempeh tacos for dinner on Sunday afternoon. Though 3 days may seem like a long stretch, don't worry; the microbes will do the work and don't need much help from you.

Soaking and Acidifying

Soaking your tempeh substrate (whether beans, grains, seeds, or nuts) softens it, and as we discussed in chapter 3, it is the first fermentation — it breaks down these foodstuffs into foods that our bodies can more easily assimilate. After the substrate is soaked, it needs to be acidified. Acid isn't traditionally added in Indonesia because lactic acid bacteria naturally acidify the substrate during the soaking period. However, in a temperate climate, an overnight soak may

not be enough to ensure that the acidity is adequate to keep salmonella from growing in the tempeh. Adding vinegar or lactic acid (which can be purchased online) is a simple way to ensure a good tempeh.

Vinegar or lactic acid can be added in various stages of the fermentation — in that first soaking water, in the water that the beans are cooked in, or on the beans themselves right before inoculation. Our profiled tempeh makers each have a preferred spot. Chad Oliphant of Smiling Hara (page 154) suggests adding lactic acid to the soaking water, while Tara Whitsitt (page 168), who was taught by tempeh innovator and whisperer Barry Schwartz (of Barry's Tempeh in Brooklyn), puts the vinegar in the cooking water. Jon Westdahl of Squirrel and Crow (page 141) adds raw apple cider vinegar to the hot drained beans before drying and inoculating. Some people feel that lactic acid does the job better, but most agree that vinegar works well. We have found that regular apple cider vinegar, raw apple cider vinegar, and white distilled vinegar all work equally well. Choose a vinegar with a 5 percent acidity and add 1 tablespoon per batch. Lactic acid is available online and at many brewing or homesteading supply stores. Look for 88 percent food-grade lactic acid and use 1 tablespoon per batch.

Now that we have given you so many options you might be thinking, "What should I do?" We decided to "ask" the flavor. We made countless batches of tempeh trying to figure out if there were any discernable flavor differences between vinegar and lactic acid, or between adding the acidifier at one stage versus another. We found that adding vinegar to the soaking water sometimes gave us a sour tempeh — as in, it tasted like pickles. (This can also happen when the mycelium growth is weak and bacteria move in— see troubleshooting on page 373.) Instead add lactic acid to the soaking water for a more neutral end flavor. At this point it will also neutralize any ammonia made by bacterial enzymes early on, though we have never had a problem (that we know of) with that. Since we prefer to use our own raw apple cider vinegar, we generally add it after cooking and draining, before cooling and drying. In short, we recommend lactic acid during the soaking or vinegar right before inoculation. Feel free to follow the instructions on your starter if it is conflicting.

Dehulling

If you are making whole bean tempeh, dehulling is the next step. The hull needs to come off, or at the very least be broken apart, so that the mycelial threads can get inside to the starches, which they need in order to thrive. Soybeans are easier to dehull than most beans. We find most of the common beans are much more difficult — so difficult that we choose not to dehull them, but chop them coarsely after cooking. You can also choose to use beans that are already dehulled — like split peas or chana dal (split chickpeas) — and skip this step entirely.

Get on your Zen mind-set, because it takes a while. We have found they are significantly easier to hull if we have let them soak for more than 12 hours. Our method is to keep the beans submerged in plenty of water, then pinch or massage them — whatever it takes to free the beans. Then we pluck out the floating hulls and repeat. You don't need to get every single hull off the beans, nor do you need to pluck out every stray hull — the errant hulls will just add a little more fiber. If you can get into a rhythm, it can be relaxing. If you have tried it a few times and find that you are not a Zen master of hulling — and instead you want to scream "what the hull!" — then you can, instead, boil the beans whole, then roughly chop them with a knife or a food processor. This method works really well for most common beans, especially runner and lima types, as they have very starchy interiors and thick skins. As a general rule, as you play with the medium, you will find that the ways in which you treat the substrate will change your outcome.

If you are making grain tempeh, as is the case for beans, the outer layer of whole grains must be broken so that the mycelium can get through. This may mean that the grains need to be cracked, hulled, or pearled. We have also found that a long soak will open up some grains, such as wild rice.

Cooking

A successful tempeh starts with beans that are cooked to the right consistency. You want them to be al dente; beans that have burst are overcooked and often make a poor tempeh. You can either pressure-cook or boil them — it's really a matter of personal preference. Sometimes we prefer to boil them because it allows us to keep an eye on the beans on the stovetop. The one exception is tepary beans, which take forever to cook on the stovetop (3 to 4 hours) versus 45 minutes steamed in an electric pressure cooker. To steam beans, place them in a steamer basket and place a small amount of water in the pressure cooker. When cooking any beans in a pressure cooker, it is important to steam, not boil, them. Place no more than 2 cups of water in the bottom of the pot and place the beans in a collapsible steamer basket. We found that hulled soybeans are done in 9 minutes at high pressure. We have included times for other beans in Some Notes on Other Tempeh Substrates, page 144.

To boil beans and any grains, place them in a pot, then cover with water. Bring to a boil and skim off any flotsam, in the form of hulls and foam, that floats to the top. Again, don't worry about catching all the hulls. The beans and grains should be cooked thoroughly but not mushy. As soon as they are done, drain into a colander so that they begin to cool and stop cooking. Don't rinse. Spread them on a clean baking sheet or casserole dish in order to allow more steam to dissipate (this helps them dry). Stir in the vinegar if you haven't soaked with an acid.

Inoculating

To inoculate your freshly cooked legumes and grains, you need them to be two things they aren't right out of the pot: at body temperature and damp (somewhat dry to the touch but not dried out and definitely not wet). Wet beans (especially soybeans) will have a sheen and once dry will look dull. Don't worry — you will get a feel for the right moisture level. Once the beans and grains have finished cooking, drain them, then transfer them to a tray and spread them out evenly. Usually by the time they are cool (in about 5 to 10 minutes), most of the excess moisture has dissipated, but they will likely need a little more help to finish drying. We have tried three methods of drying cooked beans and grains:

▶ **Dehydrator.** Spread the legumes and grains as thinly as possible on dehydrator trays and dehydrate at 145°F/62°C until the excess exterior water is removed and they are moist but not wet. This can be as quick as 5 to 10 minutes, or up to 20 or 30 minutes.

▶ **Hair dryer.** Carefully stir the legumes and grains on the tray while drying with the hair dryer, being careful not to get too close or you will have them everywhere.

▶ **Towels.** Spoon the legumes and grains onto clean towels and then pat dry. The toughest part of this technique is getting everything to let go of the towels. (And your towels must be quite clean so as not to add impurities.)

During the process of drying off the beans and grains, the temperature will come down but they should still be warm. Before you add the spore powder, make sure they've cooled to body temperature (about 98°F/37°C). Inoculate your cooked legumes and grains with the rhizopus starter and stir to incorporate it evenly throughout. If your substrate has cooled below body temperature, you can still inoculate it — just watch during the first hour to make sure you've brought it up to temperature. You don't want it to be cool for too long.

Incubating

We will cover two incubation techniques: plastic bags in hot air and trays in warm water. Both will bring your tempeh to the same place, which is fully formed mycelium encasing your legumes and grain.

Through much of our recipe testing, we used the bag method in our dehydrator and in our bread proofer with similar results. The key to this technique is spacing the holes in the plastic bag evenly, separating them by about the width of a U.S. quarter, and using something small like the tip of sharp ice pick or darning-size sewing needle to poke the holes. If the holes are too big, you could get premature sporulation around the holes.

When we came across the work of Betsy Shipley and Gunter Pfaff (see page 132), we had to try Gunter's ferment terrarium, so we built one (see the instructions on page 29) and it became our new favorite way of incubating

Tempeh from the Source

KIRSTEN WRITES: The heart of tempeh production is in Java. I was on the island of Ambon in the heart of the Moluccas known for nutmeg and cloves, not tempeh, yet I was determined to talk to a tempeh maker. I was told repeatedly, "Tempeh is not made here, only Java," but I'd seen fresh tempeh in the market, some of which was only half-fermented. The market vendors didn't understand me — they just pointed to the east. From what I knew about transportation in the region, it couldn't be coming from Java. I kept asking.

Finally, a cook in the hotel said he knew where some Javanese tempeh makers had a shop. A few days later he took me there on his day off, on the back of his small motorcycle. We wove through the streets along with all the other small motorcycles. He turned into an alley and I noticed the huge piles of firewood stacked along the walls of buildings. When we pulled into one of these places I realized we were there, and the firewood was the fuel to cook soybeans.

At the entrance of this open air "factory" were 500 kilos of steamed soybeans piled high in large plastic baskets, where they dried in the open air as they cooled. The maker explained how important it is that they dry or the tempeh will go bad. He then showed me the *ragi* (rhizopus spores) and rice flour. The ragi is mixed with the rice flour to disperse it. Most of the tempeh starter cultures that are sold commercially in the United States are already dispersed in rice flour; if yours is not, you can add it to a tablespoon or so of rice flour.

As he emphasized the rice flour, I saw that his inoculated beans had a slightly heavier dusting than we were used to seeing on our tempeh in the United States. When experimenting with some non-soy tempeh, we found that adding just a bit of rice flour (about a teaspoon per pound of substrate) gave the rhizopus some easy food and helped it establish.

tempeh. The key to this technique is to get your proportions and your pan size working together so that your bean and grain layer is about ¾ inch thick. If it is too thick, you might develop anaerobic conditions in the center of the legumes and grains, which will cause poor mycelium growth there.

Incubate your legumes and grains at about 88°F/31°C (85°F/29°C to 90°F/32°C is your workable range) for 18 to 24 hours. After about 12 hours, begin taking the temperature of the substrate with a thermometer. If you have a remote-read thermometer, you can poke it into your substrate and leave it in place. Otherwise check manually. Adjust the settings of your heat source accordingly to keep the temperature in the proper range. Remember that the microbes will generate their own heat, so you will need to adjust your external heating downward as the ferment continues, to the point where you may need to cut off the external heat source. Multiple bags in a small chamber can raise the temperature so much that everything can quickly overheat and go bad.

After 18 to 24 hours, the white spores will begin to noticeably knit everything together, giving them the appearance at first of a dusting of snow, then more like sidewalk pebbles rising up through a clean snowfall. At this point, the tempeh is doing its thing and doesn't need an outside heat source (assuming comfortable room temperature). If you are using a water bath, keep it on. The temperature-controlled water serves to cool tempeh, keeping it at just the right temperature. In a dry incubation space, turn off the heat source and let the mycelium continue to grow for an additional 6 to 12 hours, or until the tempeh has become a firm white mycelium cake. You will see your

soybeans peeking through, but the cake will predominantly be white. The total incubation time can be anywhere from 24 to 36 hours. After this point, tempeh moves into the realm of overripe. If it still hasn't finished, something is likely wrong (see troubleshooting on page 373). When the tempeh is done, refrigerate, freeze, or eat fresh.

Storage and Pasteurization

Fresh tempeh has a limited shelf life due to continued microbial enzymatic action. When stored too long, the tempeh will turn brown or black as the fungus begins to sporulate. It will also develop an ammonia smell. In Indonesia, where refrigeration was not traditionally possible, aged or overripe tempeh is referred to as *tempe bosek*, which you can read more about on page 153. When Chad and Sarah eat tempeh at home, they said they often let it age for 3 to 4 days. Tempeh has many health benefits, but unlike a number of other ferments, it isn't meant to be consumed raw for probiotic effects. So, pasteurization does not detract from its benefits and will give you a longer period in which to use your tempeh fresh. On the other hand, some people believe that pasteurization detracts from the flavor. We ourselves like tempeh when it is fresh. Every day that you wait to eat it can change the flavor. In Indonesia, Kirsten saw tempeh for sale at all stages of development at the market, including recently inoculated bags of beans. Given the tropical temperatures, buying it partly done in the morning will give you fresh tempeh for dinner.

To pasteurize means to heat food for a sufficient period of time to destroy certain microorganisms, and in this case to keep the tempeh from continuing to ripen. To pasteurize your tempeh, simply cover to keep in moisture and bake it for 30 minutes at 180°F/82°C right after it comes out of incubation. If you do not wish to pasteurize it or don't intend to eat it within 1 week of making it, you should freeze it immediately.

Store tempeh in a single layer in an airtight container in the refrigerator. If unpasteurized, it will easily keep for 3 to 4 days, and maybe a few more. If pasteurized, it will stay good in the refrigerator for at least 5 to 7 days, and often close to 2 weeks. Pasteurized or not, it will keep for many months in the freezer. When we pasteurize our tempeh, we usually keep some for the week and then freeze the rest. Thaw frozen tempeh in the refrigerator, directly in a marinade if you like. Once it is thawed, use it within a day and do not refreeze.

Meet the Maker

Jon Westdahl and Julia Bisnett
Squirrel and Crow

Jon Westdahl and Julia Bisnett began experimenting with tempeh in 2014. By 2015 they were making small batches commercially in Portland, Oregon. Unlike most makers, they never made soy tempeh. Instead, they began with everything but soy, and now they regularly use lentils, chickpeas, peanuts, quinoa, peas, pumpkin and sunflower seeds, nixtamalized corn, buckwheat, toasted coconut, and heirloom common beans — some of which are pictured on page 143. They make small batches and still experiment — the running joke is "will it tempeh?"

We asked Jon what advice he has for a home tempeh maker who is looking to see "what will tempeh." He said that it is all in the dehulling and cooking. His advice is to look for any bean with the word "dal" after it — moong dal (mung beans), chana dal (chickpeas), urad dal (black lentils), toor dal (pigeon peas), or masoor dal (red lentil) — because they will be dehulled. On cooking he says, "obviously each bean is different, but once the pot starts boiling they're close to done. Mushy beans, where the starch has ruptured to the outside of the hull, are nearly impossible to get dry enough to grow mold on. There is no space, and it's just a mess."

Choose Your Own Tempeh Adventure

When made with soybeans, tempeh is a complete protein, providing us with all of the essential amino acids, plus all that other good stuff we talked about earlier in the book. However, soybeans are not the only protein-rich substrates that will collaborate with mycelium to build tempeh. In fact, with a few simple guidelines, you and rhizopus can turn all kinds of legumes, grains, seeds, and nuts into tasty tempeh.

Even if you are not a fan of soy, the first few times you make tempeh we encourage you to try soy to set the bar on what you are aiming for. If you don't want to use soy, we have found that dehulled chickpeas (called chana dal at Indian markets) is also a good first tempeh choice. Or choose beans with a similar high-protein profile because these produce tempeh cakes with the best structure. Adzuki beans are a good soy alternative, and tepary beans are even better. Both will give you a good sense of the process and what to expect. Chad Oliphant (see page 154) explained that the hardest part about coming up with a consistent well-formed tempeh cake is getting the protein-to-starch ratio right. The trick is to have a lot of protein and low starch. As Chad and Sarah tested non-soy tempeh recipes,

they found that hemp seed works well for boosting the protein content.

If you are using beans that are starchy, like the runner-type varieties, be sure to monitor them during cooking to prevent them from getting mushy. We found that cooking these types of beans whole, drying them off, and then chopping them coarsely on a cutting board with a chef's knife or a cleaver worked best; you can also give the hot cooked beans about 8 to 10 short pulses in a food processor, but be careful — they will easily become mushy. (If you chop them before cooking, they are guaranteed to be mushy.)

Combining other legumes with soy will also create a high-quality tempeh. But soybeans cook differently than most other beans and are, again, different when hulled (see page 137). We have the best results, even though it is a little more cumbersome, when we cook each type of bean separately and then combine them at inoculation.

All-grain tempeh is also delicious. It has a softer texture than legume tempeh, and we find that the distinctive nutty, mushroomy aroma is usurped by a sweeter yeasty aroma — which shouldn't be a surprise, as we are talking about grains.

Non-soy tempeh made by Squirrel
and Crow (see page 141)

Some Notes on Other Tempeh Substrates

CEREAL GRAINS

Barley worked well when pressure-cooked for 35 minutes. We also found that soaked barley had a similar cooking time to that of common beans and could be added to the beans after the first 5 minutes of boiling. (Just make sure that there is enough water, as the barley may thicken the cooking water.)

Bulgur worked best when cooked al dente, which takes about 5 minutes. No need to soak. Use it in small quantities mixed with larger substrates like beans.

Steel-cut oats can be used in much the same way as bulgur. Whole oat groats should be soaked and cooked for about 10 minutes.

Corn can be cooked without soaking first. Whole corn works well when soaked and then cooked and cracked (whole nixtamalized is our favorite). Or you can sprout it and then cook and crack it. See Tara Whitsitt's recipe for sprouted corn tempeh on page 166.

Whole wheat family members such as kernza, kamut, and farro work well mixed with other grains or legumes. Cook them al dente, just as you would bulgur, barley, or any other whole grain.

Rice (rhizopus loves rice) and other grains work well when cooked al dente. Soak and cook as you would for koji.

Millet, when soaked, practically disintegrates into mush in the first few minutes of cooking, which is great for porridge — not so much for tempeh. We found that toasting it first and then adding 2 cups of water for every 1 cup of millet is perfect. Bring the water to a boil and simmer for 10 minutes, turn off the heat, and allow to sit with the lid on the pot for another 5 minutes.

Sorghum is delicious as a tempeh substrate. Whole sorghum should be soaked for 12 to 20 hours, then simmered on the stovetop for 45 to 50 minutes. Use 2½ to 3 cups of water per 1 cup of sorghum. We have found that often when the sorghum has finished cooking there is still water in the pot. This is okay, just be sure to drain it.

Teff is a little tricky in tempeh because the tiny seeds cook as a thick, almost viscous porridge. We found that it's best to mix in 1 or 2 tablespoons of soaked but uncooked seeds into a substrate of cooked beans and/or grains.

NUTS AND SEEDS

We found that it's best if your substrate is no more than a third nuts by weight, and we prefer closer to a quarter. Our all-nut tempehs had poor mycelium growth. If you are combining the nuts with a dehulled bean, chop the nuts before cooking. Otherwise, you can chop the nuts and beans together after cooking. See our hazelnut tempeh (page 171) and chestnut tempeh (page 172) for ideas.

Seeds are especially wonderful in tempeh. Not only do they add flavor, but many seeds have a high protein and fat content, which helps bring non-soy tempeh into balance. Small seeds, like sesame or hemp, fill in the gaps in tempehs made from larger beans, which helps make a firm cake. Use about 2 or 3 tablespoons per cup of beans or grains. Dry roast on a skillet and add to freshly cooked and drained beans. They will absorb the moisture they need. Make sure larger seeds, like pumpkin and sunflower seeds, are hulled. There is no need to soak them. Boil them with your

CHICKPEA QUINOA TEMPEH

Some Notes on Other Tempeh Substrates, *continued*

legumes and give them a rough chop so that the mycelium can access its food.

Quinoa is tasty in tempeh and pairs nicely with various beans. It also comes in red, black, and white varieties, which can make your tempeh beautiful, too. Quinoa isn't soaked but is instead rinsed thoroughly. Bring water and quinoa to a boil at a 2:1 ratio. Turn down the heat to low and simmer for 10 to 15 minutes.

OTHER LEGUMES

Chickpeas work quite well and, in fact, they are one of the easiest non-soy choices. Dehull them and use the halves as you would soy. Or buy them as chana dal (which is the dehulled halves) in Indian markets; just boil these dehulled halves until al dente. They will be done quickly — about 20 minutes on the stovetop or 8 to 9 minutes on high in a pressure cooker.

Peas and some lentils are often dehulled and make a beautiful tempeh. We boil these for just a few minutes so they aren't overcooked. We haven't had success cooking them in an electric pressure pot.

Common beans can be tricky to dehull. We found they work best when soaked overnight, boiled whole for 25 to 45 minutes on the stovetop or steamed for 16 to 17 minutes on high in an electric pressure cooker, and then roughly chopped. Steaming in a pressure cooker gives most varieties the right texture. You will find that different beans have very different hulls. For example, the small delicate Santa Maria pinquito has such a thin skin that it can be made into tempeh whole (and it is very tasty) without worrying about chopping or dehulling.

Peanuts are actually legumes, even though they are often thought of as nuts. They are a very traditional ingredient in tempeh. Add them to any legume or grain, or make an all-peanut tempeh. They should be soaked for 12 to 16 hours (dehulled if needed) and steamed under low pressure for 15 minutes. Be sure to add a tablespoon of vinegar or lactic acid to the cooked peanuts for extra measure before inoculation. (Peanuts are susceptible to the molds that produce aflatoxins, but the acidification of vinegar will discourage them.) Peanuts will benefit from a few quick pulses in the food processor before inoculation.

Runner beans (and limas) have incredibly thick coats; you'd think these beans benefit from being hulled, but they are nearly impossible to hull and we found that they cook better when fully wrapped in their little coats. In fact, when we did side-by-side trials of prechopped versus whole, not only did the whole come out with a better texture and flavor but they were actually done a good 10 minutes faster. Cooking times vary by variety and can take anywhere from 35 to 60 minutes at a gentle boil with the lid off. After about 30 minutes, watch the beans carefully and check often by pulling one out and breaking it open. Because of the starchy texture, they can go soft and mushy in an instant. When they have split, they are overcooked.

Tepary beans make excellent tempeh. They sometimes rehydrate inconsistently and can require a longer soaking time (20 to 24 hours). These are best steamed under pressure for 45 minutes. Once cooked, their skins are fairly thin and don't need much "roughing up."

OTHER FUN THINGS

Toasted coconut, cocoa nibs, and seaweed are some other possibilities. Keep the percentages of these additions low — they are finishing touches, not the bulk of the substrate. We found this out when we used too much toasted coconut flakes and the tempeh had trouble taking hold.

Once you've made a few rounds of tempeh, you might start to wonder whether tempeh can be preseasoned. In other words, can you can add herbs and spices to the mix before inoculation and incubation? Feel free to play around with this idea. We personally have found that the flavors are lost in the process, and to truly get a nice strong showing of an herb or spice, we had to add a lot more of the spice than would be detectable in the final tempeh. We have found that infusing flavor after production with a little shio koji soak (page 225) and our favorite spices is not only more economical but much tastier.

STEP-BY-STEP

Let's Make Tempeh!

What You Need

- Sanitizing equipment (we often use a spray bottle filled with 190-proof alcohol or boiling water)
- Dry soybeans (or other suitable substrate)
- Colander or strainer
- Large bowl for soaking the beans
- Vinegar
- Spoons
- Food processor for chopping the beans (optional)
- Large pot or electric pressure cooker for boiling or steaming the beans
- Large pan or tray

- Hair dryer for drying off the cooked beans (or a dehydrator, clean towels, or some other drying tool)
- Tempeh starter
- For the plastic bag incubation: 2 quart-sized ziplock bags, a large needle, and a dehydrator, bread proofing box, or other incubation setup (see chapter 2)
- For the water bath incubation: stainless-steel tray (a restaurant warming tray works well) and a ferment terrarium (see page 28) or other water bath setup
- Instant-read food thermometer with probe (a remote-read thermometer is not necessary but nice)

Instructions continue on next page

148

1. Sanitize your tools and work surface before getting started.

2. Following the recipe instructions, rinse the beans, transfer them to a large bowl, and cover by at least 4 inches of water. (Add the lactic acid if using and stir in; if you do this you won't add vinegar to the cooked beans.) Leave on the counter for 8 to 24 hours.

3. Drain the beans, put back into the bowl, and cover with fresh water. Massage the soybeans between your hands to slip the hulls off. The hulls will accumulate in the water on top of the soybeans; skim them off. Alternatively, drain and roughly chop the beans in a food processor. If you do this, you will need to boil the beans in the next step, instead of steaming them, and skim the hulls off the water as the beans boil.

4. Steam the soybeans until they are al dente, following the instructions on page 54. In our experiments, a 10-minute steam in a pressure cooker worked well. Or bring a pot of water to a boil, add the beans, and simmer until al dente, about 45 minutes.

5. Pour the cooked beans into a colander to drain. Place them in a casserole dish and add the vinegar, mixing to disperse the vinegar and release steam. A lot of moisture will dissipate with the steam. Using a hair dryer on high heat, dry the beans until they are damp dry, stirring them gently as you work.

6. Add the tempeh starter and stir with a spoon or clean hands until it is well incorporated. Now you can ferment the beans in either plastic bags in a dehydrator or in a water bath setup.

149

Plastic Bag Method

7. Using a large needle, perforate both sides of the ziplock bags in a grid with holes spaced apart about the width of a U.S. quarter. There is no specific pattern, so you can get creative if you like; just stay fairly uniform.

Instructions continue on next page

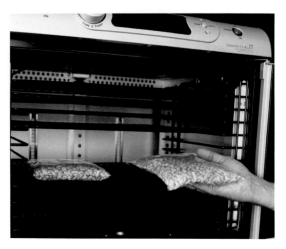

8. Fill each bag about three-quarters full with the beans so that when you lay it flat it is about 1 inch thick (and no more than 1½ inches). Squeeze out the excess air, seal, then lay flat on a cutting board and spread the beans in the bags with your hands so that they are evenly distributed. If either bag is not completely full, fold over the unused portion of the bag.

9. Place the bags in the center of a dehydrator and set the temperature to 88°F/31°C and the timer to 24 hours. Begin checking the temperature after 12 hours, and reduce the dehdyrator heat as needed to keep the tempeh at 88°F/31°C.

10. After 18 to 24 hours, white spores will begin to knit everything together. Keep the tempeh in the dehydrator, with the heat turned off, for another 6 to 12 hours.

11. When the tempeh is done, it will be a firm white cake.

Instructions continue on page 152

Don't Want to Bag It?

Single-use plastic bags are not the most ecological choice for making tempeh. If you are making a lot of tempeh, they can feel like a lot of waste. Don't despair — there are many other choices out there.

Banana leaves (available frozen from some Asian markets) are the original tempeh wrappers — they breathe and hold in moisture naturally. Simply thaw out the package and cut off about 10 to 12 inches to form a rectangle that is close to square. Place 1½ to 2 cups of substrate in the center of the leaf. You want enough to make a decent cake, but not so much as to overstuff and split the leaves. Wrap carefully. Often there is enough overlap that just placing the bundle on the flaps is enough to keep it closed. If not, cut off a small strip (with the grain) of the leaf and use it as a tie. Banana leaves work best in a proofing box, where you can keep a bit of water in the box so the leaves don't dry out. If you are using a dehydrator, wrap the leaves in moist, clean cotton tea towels (rewet as they dry).

Silicone soap molds are just plain fun. The mycelium mold grows into the nooks and crannies of the design and gives you a fun shape. The round flowers are perfect for slider-sized "burgers." Simply fill the mold with your tempeh substrate and press in place. If using a water bath, put the mold into a casserole dish to float on the water (no need to cover with plastic). If using a proofing box or dehydrator, place a sheet of BPA-free plastic wrap on the top and perforate over the areas with the tempeh mixture.

You can also make a reusable fermenting container out of a plastic food storage container. Depending on your batch size, you will likely need more than one. You can make nice-sized blocks with shallow sandwich containers or burger shapes with small round containers. You will want to poke holes (no bigger than the diameter of a toothpick) on the top and bottom of the tub, spaced about 1 inch apart. The best tool, if you have it, is a tiny-diameter drill bit. You can use a small-diameter awl, too; just be sure to poke the holes from the inside out so that the edges of the pushed plastic don't catch on the tempeh.

151

Water Bath Method

12. Spread the beans evenly in a stainless-steel pan and press down with the back of a spoon. They should form a layer about ¾ inch thick.

13. Bring the temperature of the water in your water bath incubator to 88°F/31°C. When the water is at temperature, lower the tray into the water until it is floating.

14. After 20 hours the white spores will have begun to knit everything together. Let the tempeh remain in the incubator for another 2 to 10 hours until fully formed.

15. When the tempeh is done, it will be a firm white cake.

Almost Done . . . Whoa, What Happened? Overripe Tempeh

Overripe tempeh (*tempe bosek*) is common in Javanese cuisine. With a warm climate and no refrigeration, they don't set out to make overripe tempeh — it just happens. During fermentation, the microbes are constantly jockeying for dominance. In the late stages of fermentation, the mold weakens and is supplanted by bacteria fermentation. This, of course, explains why overripe tempeh is a little more sour and bitter than its younger counterpart. Studies have shown that overripe tempeh has a higher glutamic acid content, which is why it has more umami. The amino acids degrade, and we get the unique pungent odor — decidedly ammonia headed toward rancid.

Because all tempeh is headed toward being overripe, you don't need to do much to make it. That said, it is nice to manage the process so that you don't head into the rancid phase. Unlike miso, wine, sauerkraut, or cheese, where flavor continues to develop during aging, overripe tempeh quickly goes from edible to inedible. To make overripe tempeh, you can keep your freshly harvested tempeh at room temperature for another 2 or 3 days. The cake will soften and darken with plenty of black spots. In *The Book of Tempeh*, the authors suggest a second option: Remove the perforated plastic bag and sprinkle the surface of the tempeh block with salt (this will kill the fungus, but not the enzymes or bacteria). If in a casserole pan, sprinkle salt on both sides of the tempeh. Return the tempeh to its bag or pan and keep at 80°F/ 27°C to 85°F/29°C for about 24 hours. Use in any recipe that calls for overripe tempeh.

Chad Oliphant and Sarah Yancey
Smiling Hara

It was a tumultuous spring day when Kirsten drove out to the Smiling Hara tempeh production facility in the Blue Ridge Mountains outside Asheville, North Carolina. The verdant green was punctuated by an angry gray sky and wind that blew small branches onto the car as she drove. It was nothing short of beautiful.

Kirsten was welcomed immediately by Chad Oliphant and Sarah Yancey, founders of Smiling Hara and makers of Hempeh, a hemp-fortified tempeh, made with the traditional soy or peanuts. Chad, who had traveled to Java to study, explained that in Indonesia, tempeh is produced in communities, sometimes within larger cities or towns. Everyone in these communities plays a part, whether their operation is from bean to cake or they simply do one thing, like cooking the beans or thinly slicing long circular logs of finished tempeh (with machetes) and deep-frying them to make snack chips called *kripik tempe*. Every home produces tempeh, often in the open air, which is smoky from the fires that cook the beans and chips. Many families have been producing tempeh for five or six generations. Chad came home with a renewed respect for the craft and a grounding in its roots.

Sarah cooked Hempeh for Kirsten to taste. She said, "Tempeh has the qualities of mushrooms because it's mycelium that we're growing on beans. These microscopic mushrooms are so phenomenal for your health because they have the ability to find cancer-causing agents in your body that they absorb and ingest. It's superfood, a super-protein." Chad agreed and added, "Not all fiber is equal. The mycelium hasn't been looked at enough and what it does for the microbiome; there is more going on than we fully understand." Sarah explained Hempeh and the road that brought their product to where it is now.

In 2008, Sarah came home with a block of tempeh. She'd recently been laid off, and that meal sparked a conversation. Chad, who had had fresh tempeh when he'd studied at the Kushi Institute in Massachusetts years earlier, told Sarah that it was delicious when fresh and could be made with many types of beans. They soon had a project going. They soaked and hulled chickpeas and rigged up a shoebox to incubate. It wasn't easy, but the tempeh was delicious. "We started eating it all the time and felt so good," Sarah told Kirsten. With Chad's enthusiasm and Sarah's time and ability to bring the idea to fruition, the business was born in 2009 as Smiling Hara.

They started making 20-pound batches in a community commercial kitchen as they learned. "In the beginning, we endured 300-pound losses at a time. The challenge at that point was keeping at it. We had to hold tight to our vision of success and persistence," Sarah said.

Initially, they sold traditional frozen unpasteurized raw soy tempeh. Early on, they discovered two things: While soy-free tempeh was in high demand, it was difficult to find. And their customers, who loved the product, would often take it home and forget about it in the freezer. To meet the needs of folks who want a vegetarian protein that's not derived from soybeans, they created two additional products: Black Bean Tempeh and Black-Eyed Pea Tempeh. They also started to imagine a different product, one that would be simple — flavored tempeh that customers could just heat and eat. Their Hempeh now comes in three flavors: Asian Miso Ginger, Raspberry Habanero BBQ, and Smoked Salt and Pepper Steak.

Chad's passion for improving, adapting, and adding depth to both the finished product and the tempeh-making process are evident as you walk through the facility. When Kirsten was there, he was experimenting with growing tempeh in long round rolls (tempeh is traditionally formed into squares or rectangles). These logs can be sliced in any thickness, from "burger patties" to superthin wafers that can be deep-fried into chips. Tempeh chips are popular in Indonesia, probably for two reasons: First, who doesn't love crispy deep-fried snacks? We believe the world's people can agree on that. The second is more practical. In the hot climate of Java, where there is often little or no refrigeration, deep-frying increases the shelf life of the tempeh from less than 1 week to 2 to 4 weeks.

Chad is also trying to tackle a problem with making tempeh, which is that when tempeh is in full growth mode it gives off a lot of heat, and the tempeh room can easily get too warm for good tempeh growth — it's like each little bean cake has a fever. (You will see when you make your own.) In addition, at this stage the mycelium is rapidly consuming oxygen and giving off

CO$_2$. When CO$_2$ levels get too high, the mycelium can choke itself out. Most tempeh makers would be content to use vents and fans to send the air outside, but Chad is trying permaculture techniques. He's trying to capture the CO$_2$ in fish ponds that, in addition to growing fish to eat, grow plants to keep the air and water clean.

It is clear that both Chad and Sarah are committed to being as ecologically conscious as they can while making this food. Chad said, "In Indonesia, tempeh is the people's food. I want to bring tempeh into the mainstream, not as a meat substitute, but for what it is: a healthy, low-cost protein — the people's food. Tempeh is one of the most important foods to feed a growing global population, and Indonesians should be proud for the contribution they have made to the world."

Basic Soybean Tempeh

YIELD: 2 CAKES, 1 POUND EACH

This is the most basic traditional style of tempeh. While we encourage you to play with different beans and grains, don't look past plain soy tempeh yet. It is delicious and has a distinctive nutty taste and meaty texture; we recommend you start here or with the variation below that steps it up with the addition of barley or another grain. Tempeh can be a little finicky, especially with alternative substrates, so it is a good idea to practice with the easier ones first.

In Indonesia, sometimes they make tempeh from beans laid out in a single layer, which yields a higher mycelium-to-bean ratio. You don't need to go to the width of a single bean, but you can make this tempeh thinner to be used as patties for myriad sandwiches.

PROCESS **tempeh (page 147)**	FERMENTATION **24–36 hours**

3 cups (500 g) soybeans

2 tablespoons (28 g) vinegar

1 teaspoon (1 g) tempeh starter

Note: Confirm the quantity of tempeh starter against what is recommended by the manufacturer of your starter.

1. Rinse the beans by placing them in a bowl and running water into it.

2. Fill the bowl with enough water to cover the beans by 4 inches. Let sit at room temperature for 8 to 24 hours. After 12 hours, change out the water.

3. Drain the beans, then pour them back into the bowl and cover with fresh water. Massage the soybeans between your hands to slip off the hulls. As these transparent bean coats accumulate in the water on top of the soybeans, tip them out, and keep massaging until you have removed nearly all of the hulls or you have just had it — whichever comes first. Alternatively, roughly chop the rinsed beans in a food processor to break them up.

4. Steam for 10 minutes in a pressure cooker. If you have chosen to chop the beans, put the chopped beans in a pot and cover by about 3 inches of water. Bring the beans to a low boil and skim off the foam and floating hulls. Cook for about 45 minutes, or until the beans are al dente. You may need to skim off the hulls a few more times. Don't worry if you don't get them all.

5. Pour the cooked beans into a colander to drain. Then put them into a casserole pan and add the vinegar, mixing to disperse the vinegar and release steam. A lot of moisture will dissipate with the steam. Using a hair dryer on high heat, dry the beans until they are damp dry, stirring them gently as you work. Alternately you can use a dehydrator or clean towels (page 138).

6. Add the tempeh starter and stir with a spoon or clean hands until well incorporated.

Recipe continues on next page

Basic Soybean Tempeh, *continued*

7. Transfer the soybeans to a stainless-steel pan (if incubating in a hot water bath) or two perforated quart-sized ziplock bags (if incubating with the plastic bag method). If using bags, divide the beans evenly between the two bags, squeeze out the excess air, seal, then lay flat on a cutting board and spread the beans in the bags with your hands so that they are evenly distributed. If either bag is not completely full, fold over the unused portion of the bag.

8. Incubate at 88°F/31°C (85°F/29°C to 90°F/32°C is your workable range). After about 12 hours, begin taking the temperature of the beans with a thermometer. When you notice the tempeh's internal temperature rising, the microbes are starting to generate their own heat. At this point you may need to adjust the heat source down. If the mycelium sits at a temperature of 105°F/41°C or above for over an hour, it will die, so it's important to monitor the temperature.

9. After 18 to 24 hours, the white spores will begin to noticeably knit everything together, giving them the appearance at first of a dusting of snow, then more like sidewalk pebbles rising up through a clean snowfall. At this point, the tempeh is doing its thing without any outside heat source. Once you see this, turn off the heat source if using a dehydrator or proofing box (not if using a water bath) and let the mycelium continue to grow for an additional 6 to 12 hours, until the tempeh has become a firm white mycelium cake. You may see your soybeans peeking through, but it will predominantly be white. You want to make sure it is holding together nicely. It can continue to ferment until you see no beans. However, watch the tempeh because at a certain point, especially if it is warm, it will begin to sporulate, at which point it is definitely time to stop fermentation. If this happens, you will begin to see black patches of forming spores. (If you see the patches, don't worry, it is fine — it's the beginning of overripe tempeh.)

10. If pasteurizing, preheat the oven to 180°F/82°C and bake the tempeh for 30 minutes. Unpasteurized tempeh will keep in the refrigerator for 2 to 3 days. Pasteurized tempeh will keep in the refrigerator for at least 5 days and up to 2 weeks.

Variations begin on page 160

CHICKPEA QUINOA
TEMPEH

BASIC SOYBEAN
TEMPEH, page 157

BLACK SOYBEAN
TEMPEH, page 160

YELLOW AND
GREEN PEA
TEMPEH

CHESTNUT
TEMPEH, page 172

LENTIL TEMPEH

BLACK-EYED
PEA AND PECAN
TEMPEH

BLACK ONCOM,
page 163

Variation: Black Soybean Tempeh

Black soybean tempeh has a mild taste and is quite beautiful. When the beans soak, their pistachio-green inside is revealed. Surprisingly, some of this green is retained through cooking. When the finished tempeh is cut, it has a mottled green appearance, almost like a granite countertop.

If you've made natto with black soybeans, you might remember a shorter soaking time due the possibility of sprouting. While the beans may begin to sprout for tempeh, this isn't an issue. The biggest difference in making tempeh with black soybeans is that hulling is quite difficult and can make the process feel long and frustrating. So don't bother. Instead, after soaking the beans, place them in a food processor and roughly chop them so that most of the beans are in pieces, about one-third of their original size. It will be easiest to do this in batches. Don't worry about having a few whole beans or unevenness. Bring the beans to a boil and boil them for about 45 minutes, or until al dente. The loosened hulls will float to the top; scoop them out with a slotted spoon or mesh strainer every so often. Don't worry about getting them all. Proceed to step 5 in the Basic Soybean Tempeh recipe (page 157).

Variation: Barley Soybean Tempeh

Adding pearled barley to the Basic Soybean Tempeh recipe (page 157) not only adds a nuttiness to the flavor but also makes for a firmer cake because the smaller barley fills the gaps between the soybeans. Think of soybean tempeh as filling a gallon jar with golf balls — from afar it looks pretty packed but up close you see the gaps between the golf balls. Fill that same jar with golf balls and marbles and you get a tighter pack as the marbles find the spaces in between. It's the same with barley and soybeans.

- 3¾ cups (625 g) soybeans
- ⅔ cup (110 g) pearled barley
- 2 tablespoons (30 mL) vinegar
- 1 teaspoon (1 g) tempeh starter

Note: Confirm the quantity of tempeh starter against what is recommended by the manufacturer of your starter.

Follow the Basic Soybean Tempeh instructions (page 157) with the following exception: If you are steaming your soybeans, add the barley to the soybeans before steaming. If you are boiling your soybeans, add the barley after the first 5 minutes. Make sure there is plenty of water, as the barley will thicken the liquid a bit.

Variation: Chickpea (or Any Dal) Tempeh

When we want to make a quick tempeh and not fuss with creativity, dehulling, or even long cooking and soaking times, we use chana dal (split and hulled chickpeas). Of course, that is not the only reason; the flavor is great. It is a mild and easy tempeh to pair with just about anything.

Soak the dal beans for 8 hours or overnight. Place in a pot with fresh waster and bring to a boil, then turn down to a simmer and check them frequently. Chana dal is ready in about 15 to 25 minutes. Some of the smaller dal beans are ready very soon after the boil. The chana dal beans can be steamed in an electric pressure cooker for 8 minutes on high pressure (the smaller dal beans are done too quickly when made in the pressure cooker). Follow the same instructions for split peas. Once cooked, add the vinegar and follow the procedure for Basic Soybean Tempeh on page 157.

- 3 cups (500 g) chana dal (split, hulled chickpeas)
- 2 tablespoons (30 mL) vinegar
- 1 teaspoon (1 g) tempeh starter

Note: Confirm the quantity of tempeh starter against what is recommended by the manufacturer of your starter.

Variation: Horse Gram Tempeh

This tempeh is made with the beans left over from making Pone Yay Gyi (page 95), the sour bean paste from Myanmar. Interestingly, these beans are still mostly intact even after many hours of cooking and being squeezed on a sieve. They make an excellent tempeh with an extremely meaty texture.

Because this tempeh is a by-product, we don't make it much, but when we do, it is a favorite for tempeh burgers. If you decide to make this tempeh straight from horse gram beans and don't make the paste first, be sure to cook the beans until they are soft, which will take about 4 hours if boiling or 45 minutes if pressure cooking. Contrary to most of our tempeh recipes, you don't want these beans to be al dente — the flavor will be good, but the texture won't be. Be sure to use *Rhizopus oligosporus* starter with this bean, as it makes a much sturdier tempeh than *R. oryzae.*

- 1 pound (450 g) horse gram beans, from Pone Yay Gyi (page 95) or soaked and cooked until soft
- 1 tablespoon (14 g) vinegar
- 1 teaspoon (1 g) *R. oligosporus* tempeh starter

Note: Confirm the quantity of tempeh starter against what is recommended by the manufacturer of your starter.

SCIENCE SAYS . . .

Getting to Know Saponins

When you run water through raw soybeans, a soapy froth comes forth to float upon the water. That's the first hint of the presence of saponins, glucosides found in plants that produce a soapy lather. (They're the reason behind the name of plants such as soapwort and soapbark.) Between 1 and 6 percent of the dry weight of raw soybeans is saponins, depending upon the variety. Saponins are also in things like quinoa, olives, asparagus, and ginseng. Some types of saponins are toxic to insects and cold-blooded animals, as part of the plant's defensive mechanisms, but thankfully, soybeans do not contain these toxic types. Soybeans have two types of saponins, and they are actually quite healthful to humans. The first type of saponin has an astringent taste and is found mostly in the germ. The second type has the healthful properties and is also found in the germ as well as in the cotyledons — the baby plant in the center of the bean.

Soybean-based saponins may help to lower our levels of cholesterol and reduce our risk of developing colon cancer.

Black Oncom
(Fermented Bean Cake)

YIELD: 1 ONCOM CAKE, 17 OUNCES (PLUS 3 PINTS SOY MILK)

Oncom (pronounced "on-chom") is another traditional Indonesian ferment. It is made by growing mold on peanut, coconut, or soy press cake (the by-product of making oils, soy milk, or tofu). There are two types of oncom: red and black. Red oncom is fermented with an entirely different microbe (*Neurospora intermedia* var. *oncomensis*) that grows a beautiful orange mold. As of this writing, we have not found a source for red oncom culture in the United States. Black oncom is very similar to tempeh — so similar, in fact, that the difference is in the medium, not the mold or the process. All black oncom is made with either *Rhizopus oligosporus* or *R. oryzae*, or both, so you don't need to buy separate oncom starter. Contrary to its name, black oncom is not black but white.

Oncom is quite mild, has a smoother texture than tempeh, and easily soaks up flavor. For these reasons, we use it to make our favorite fermented meat-free sausages — chorizo (page 338) and maple breakfast sausage (page 336). The mouthfeel is more like sausage than tempeh.

If you already make soy milk or tofu, oncom is a wonderful way to use the pressed beans that are the by-product, known as the okara; you'll need about 5 cups (wet weight). If you make oncom, you will have soy milk as a by-product. We don't drink soy milk but we don't want to waste it, so we include a recipe for soy yogurt on page 165.

PROCESS **tempeh (page 147)**	FERMENTATION **22–36 hours**

3 cups (500 g) soybeans or okara

2 tablespoons (30 mL) vinegar

1 teaspoon (1 g) tempeh starter

Note: Confirm the quantity of tempeh starter against what is recommended by the manufacturer of your starter.

1. Soak the soybeans for 12 hours or overnight.

2. Drain and rinse the soybeans. Transfer the soybeans to a food processor and process until they are finely ground. You may need to do this in two batches. The beans should be moist enough from the soaking to grind up cleanly; however, if they become too gummy, you may add a bit of water to facilitate the grinding.

3. Combine the ground soybeans in a large pot with 8 cups of water. Bring to a boil over medium-high heat and watch it carefully, as it will foam up quite dramatically. Immediately reduce the heat to low and simmer for 20 minutes, stirring frequently to prevent the bean mash from sticking to the bottom of the pot. The milky mixture will stop foaming as it cooks. As the process nears completion, the liquid will look more opaque and the ground beans will look grainy as the milk separates from the fibrous part of the soy.

4. Preheat the oven to 250°F/120°C.

Recipe continues on next page

5. Place a piece of tightly woven cheesecloth or butter muslin in a colander over a bowl. Strain the beans through the cheesecloth and allow to cool. When the beans are cool enough to touch, squeeze the pulp to release the rest of the milk. Place the warm milk in a 2-quart jar; you can use it to make yogurt, but if you're not going to ferment it or drink it right away, put it in the refrigerator, where it will keep for up to 3 days.

6. Spread the pulp over a large baking sheet. The sheet should be large enough that the oncom is just 2 to 3 inches thick. Place the sheet in the oven and bake for 20 minutes, stirring at the halfway mark.

7. Spread a few paper towels or a clean cotton towel on a baking tray. Spread the hot soy mash over the towels and mix to dissipate the steam. Let rest until cool to the touch. It is important that the beans be rather dry, as extra moisture can lead to spoilage.

8. Transfer the beans to a clean bowl and thoroughly mix in the vinegar. Add the starter and mix to fully disperse the spores.

9. Transfer the beans to a stainless-steel pan (if incubating in a hot water bath) or two perforated quart-sized ziplock bags (if incubating with the plastic bag method). If using a pan, spread the beans in an even layer about 1 inch thick. If using bags, divide the beans evenly between the two bags, squeeze out the excess air, seal, then lay flat on a cutting board and spread the beans in the bags with your hands into a layer about 1 inch thick.

10. Incubate at 88°F/31°C (85°F/ 29°C to 90°F/32°C is your workable range). After about 12 hours, begin taking the temperature of the oncom with a thermometer. When you notice the oncom's internal temperature rising, the microbes are starting to generate their own heat. At this point, you may need to adjust the heat source down. If the mycelium sits at a temperature of 105°F/41°C or above for over an hour, it will die, so it's important to monitor the temperature.

11. After 18 to 24 hours, the white spores will begin to noticeably knit everything together, giving them the appearance at first of a dusting of snow. At this point, the oncom is doing its thing without any outside heat source. Once you see this, turn off the heat source if using a dehydrator or proofing box (not if using a water bath) and let the mycelium continue to grow for an additional 6 to 12 hours, until the oncom has become a firm white mycelium cake. It will predominantly be white. You want to make sure it is holding together nicely. Watch the oncom because at a certain point, especially if it is warm, it will begin to sporulate, which is definitely time to stop fermentation. If this happens, you will begin to see black patches of forming spores. If you see the patches, don't worry it is fine.

12. If pasteurizing, preheat the oven to 180°F/ 82°C and bake for 30 minutes. Unpasteurized oncom will keep in the refrigerator for 3 to 5 days. Pasteurized oncom will keep in the refrigerator for at least 7 days and up to 2 weeks.

Fermented Soy Yogurt

YIELD: ABOUT 2½ QUARTS YOGURT

We developed this recipe as a way to use leftover soy milk from making oncom (page 163). To get the consistency and texture right, we tried everything from agar to pectin and found that tapioca flour (also called tapioca starch) and some chia seeds worked the best. Don't cut out the coconut sugar — it feeds the microbes. The fat from the coconut milk provides the richness of flavor the soy milk is missing.

FERMENTATION **8 hours**

---

2 quarts (1.9 L) soy milk

2 cups (473 mL) full-fat coconut milk

2 tablespoons (28 g) coconut sugar or cane sugar

6 tablespoons (84 g) tapioca flour

1–2 packets (5 g) yogurt culture (see note)

3 tablespoons (32 g) chia seeds

Maple syrup (optional)

Note: *Confirm the quantity of yogurt culture against what is recommended by the manufacturer of your culture.*

---

1. Combine the soy milk, coconut milk, and sugar in a large pot and warm over medium heat until it is 150°F/65°C.

2. Pour 1 cup of the hot milk into a bowl and whisk in the tapioca flour until smooth. Whisk the mixture back into the pot. Continue to heat the soy milk mixture, stirring frequently, until the temperature is 180°F/82°C.

3. Remove the pot from the heat. Let the milk cool until it is 115°F/46°C or the temperature suggested on your starter culture.

4. Sprinkle the starter over the surface of the milk and allow it to sit for a moment to hydrate, then whisk to combine.

5. Pour the yogurt into three clean quart jars and place in a dehydrator, proofing box, or yogurt maker for 8 hours at 105°F/40°C to 112°F/44°C. Alternatively, place the yogurt in an electric pressure cooker on the yogurt setting for 8 hours.

6. Stir the chia seeds into the yogurt. Add the maple syrup, if using, and stir, then refrigerate immediately. The yogurt will continue to thicken in the refrigerator, reaching its full thickness in 8 to 12 hours. The yogurt will keep in the refrigerator for up to 2 weeks.

Sprouted Corn Oregon Tempeh

From Tara Whitsitt, author of *Fermentation on Wheels*

YIELD: 30 OUNCES OR 3 CAKES, 10 OUNCES EACH

TARA WRITES: I am in love with the versatility of this medium. Even though I grew up vegetarian, tempeh was surprisingly not on my must-haves grocery list. It took Barry Schwartz (of Barry's Tempeh in Brooklyn) to truly introduce me to tempeh. He once gifted me three 12-ounce tempeh blocks, each different. I was taken with the aroma and the fact that each block was unique — one was made with adzuki bean and brown rice, another with quinoa and buckwheat, and the final with navy beans and brown rice. My tempeh obsession began soon after.

My obsession would not have gone so deep if it had not been for the flavor force of *Rhizopus oligosporous* and how far you could bend the "rules" for making tempeh. I've experimented with roasted walnuts, mung beans, sprouted corn, sunflower seeds, and much more. Store-bought tempeh is usually pasteurized, whereas Barry's tempeh was always served fresh (or he would freeze it). With fresh tempeh, you can actually taste the complexity and take in the aroma of the fungus, which enhances the flavor dramatically.

Scarlet runner beans, pintos, and corn were inspired by local farmers of the Willamette Valley, where I am based when not driving around the country teaching fermentation. This recipe has proven delicious and nourishing.

PROCESS **tempeh (page 147)**	FERMENTATION **24–36 hours**

4 ounces (115 g) dry corn

8 ounces (225 g) dry pinto beans

6 ounces (170 g) dry scarlet runner beans

¼ cup (60 mL) vinegar

½–1 teaspoon (1 g) *Rhizopus oligosporus* or *R. oryzae* tempeh starter

Note: Confirm the quantity of tempeh starter against what is recommended by the manufacturer of your starter.

1. Soak the corn for 36 to 48 hours or until you see it beginning to sprout, rinsing it and replacing the soaking water with fresh water every 12 hours. Soak the pinto beans and scarlet runner beans in separate bowls for 8 to 12 hours.

2. After soaking, rinse the beans. Keeping the two kinds of beans separate, process them in a food processor to break them down a bit, until they are roughly a quarter of their original size.

3. Transfer the beans to separate pots. Add to each pot enough water to cover the beans by 3 inches, along with 2 tablespoons vinegar. Bring the beans to a boil and boil until al dente. Pinto beans will reach al dente after roughly 30 to 45 minutes, while the scarlet runners will need anywhere from 45 to 60 minutes. The beans should be soft enough to comfortably bite through but not mushy. As the beans cook, you will notice the hulls float to the surface. Scoop them out with a mesh strainer. You won't get all of them — that's fine.

4. Meanwhile, drain the sprouted corn and transfer to a pot with enough water to cover by 2 inches. Bring to a boil and boil for 15 to 20 minutes, until slightly softened.

5. Once the beans and corn are cooked, drain them and combine them in a large bowl.

6. Dry the beans and corn with a hair dryer while stirring vigorously until they are no longer wet but just barely damp. This takes 10 to 15 minutes.

7. Check the temperature of your bean and corn mixture. Once it falls below 90°F/32°C, add the starter, and stir well to distribute the culture evenly.

8. Lay out two ziplock bags on your counter and poke small holes with a fork, ice pick, or thick needle throughout each of them. Don't be shy — these are breathing holes for the beans as they come to life through the power of fungus. Start with the sides and the corners, then work your way along the entire bag.

9. Spoon equal amounts of beans and corn into each bag (I like to use a scale for accuracy). Seal the bags and gently flatten each with your hands, so the bean and corn mixture firmly fills the bag.

10. Place the bags in an incubator (see chapter 2). The temperature should range anywhere from 85°F/29°C to 90°F/32°C. It will take longer to ferment on the lower end of the spectrum and less time on the upper end. After about 12 hours, begin taking the temperature of the bean and corn mixture with a thermometer. When you notice the internal temperature rising, the microbes are starting to generate their own heat. At this point you need to adjust the heat source downward, often to the point that you simply shut it off. If the mycelium sits at a temperature of 105°F/41°C or above for over an hour, it will die, so it's important to monitor the temperature.

11. The mycelium growth creates a unique soft white field with the slightest fuzz. There may be black spots, too — that's sporulation, which is great. When the tempeh looks done, take it out of the incubator. I recommend digging in right away. There is absolutely nothing like tempeh fresh out of the incubator.

Meet the Maker
Tara Whitsitt

It is funny how paths can nearly cross before they actually intersect. The same fall that we had turned in our manuscript for our first book, *Fermented Vegetables,* Tara Whitsitt passed through our out-of-the-way rural Oregon valley on the first stop of what would become her epic journey across the United States in a converted school bus. She gave her first "Fermentation on Wheels" workshop on a farm directly on the other side of the ridge behind our home, only a mile or so as the crow flies but nearly an hour's drive by car.

We missed that workshop, and despite having many of the same friends in the national fermentation community, we didn't meet until three years later. Tara was working on her book, *Fermentation on Wheels,* the same name as her grassroots project to bring fermentation education to communities of all ages and backgrounds. A mutual friend, Cheryl Paswater, who was visiting the Northwest, invited Kirsten and her mother to Tara's home. That day they shared a tempeh meal that Tara made for them and other fermentation friends. It was the first time they'd tasted corn in tempeh. It was delicious. It was a beautiful sunny day and they'd explored Eugene, Oregon, Tara's home base. Kirsten's mother was so excited to be included in the day's events and to get a sense of what Kirsten did. She was so proud. It was her last summer, and it is one of Kirsten's last memories of her in her full exuberant self. We were happy when Tara agreed to share a tempeh recipe that includes sprouted corn for this book. About the recipe she said, "Working with fresh, local foods has always been important to me. Food speaks to us (and our stomachs) through its stories — from the soil microbes that nourish the seed to the hands that care for the plant once it emerges. I'm enamored of fermentation because of all the hidden life forces that create complex flavor and resilient communities, and I love that it encourages healthy, diverse ecosystems in our foodways. I can't think of a better way to celebrate this than to work with local, sustainably grown ingredients that represent my region's microflora."

Tara travels with stories, ferments, starter cultures, and her book, which came out in the fall of 2017. In many ways, she is like a starter culture herself as she heads into communities to inoculate them with a bit of fermentation knowledge and a few billion microbes as they taste her creations. This work makes Tara one of the most knowledgeable, passionate, and brave people we know. Just imagine for a minute driving in a converted school bus from community to community inviting strangers into your fermentation lab, which is also your home. For many of those strangers, it is beautiful and life-changing experience. Tara's passion around fermentation is twofold: it is the fermented food itself and also the beauty of seeing human communities mimic the cooperation of microbial communities. For her work she is rewarded with an incredible network of community throughout the country.

Ancient Grains Tempeh

Change is the thing. We hear it everywhere — you must evolve, or you die. Well, if you want to be in the ancient grains club, you have to go against the grain and not change for a very long time. What defines an "ancient grain" is a bit nebulous, but basically it is a grain that hasn't changed in the past few hundred years. We wanted to create a tempeh that highlighted the officially recognized ancient grains and, more importantly, was tasty. This tempeh is delicious, plus you can enjoy all the health benefits ascribed to these grains.

Making this tempeh is a bit of dance, given that each grain needs a different treatment, but it is worth it. Use this recipe as a springboard for trying all kinds of grains — with or without legumes.

PROCESS **tempeh (page 147)**	FERMENTATION **22–36 hours**

½ cup (100 g) sorghum

2 tablespoons (25 g) whole teff grain

½ cup (100 g) millet grain

¼ cup (50 g) quinoa

2 tablespoons (30 mL) vinegar

1 teaspoon (1 g) tempeh starter

Note: Confirm the quantity of tempeh starter against what is recommended by the manufacturer of your starter.

1. Rinse the sorghum and teff separately, until the rinse water runs clear. Put each grain in its own bowl and add enough water to cover by 2 inches. Let sit at room temperature for 8 to 12 hours. Note that you will not be soaking the millet or quinoa.

2. Thoroughly rinse the quinoa and set aside.

3. Drain the teff through a fine sieve and set aside to give it time to dry. Drain the sorghum and place in a pot with 1½ cups of fresh water. Bring the pot to a boil over high heat, then reduce the heat to low and simmer, covered, for 45 minutes.

4. In the meantime, place the dry millet in a pot and toast on medium-high heat until it smells a bit nutty and is beginning to brown, about 5 minutes. Remove the pot from the heat and add the rinsed quinoa and 1½ cups of water. Bring the mixture to a boil over high heat, then reduce the heat to low and simmer, covered, for 10 minutes. Turn off the heat. Stir the grains, replace the lid, and allow to rest until the sorghum is ready, at least 5 minutes.

5. Drain the sorghum if needed, then transfer to a tray and spread evenly. Add the millet-quinoa mixture and the teff. Add the vinegar and stir to incorporate evenly. Dry the mixture either with a hair dryer or in a dehydrator (see page 138) until the mixture is room temperature and damp dry.

6. Add the starter culture and mix well to fully incorporate.

Recipe continues on next page

Ancient Grains Tempeh, *continued*

7. Transfer the grains to a stainless-steel pan (if incubating in a hot water bath) or two perforated quart-sized ziplock bags (if incubating with the plastic bag method). If using bags, divide the grains evenly between the two bags, squeeze out the excess air, seal, then lay flat on a cutting board and spread the grains in the bags with your hands so that they are evenly distributed.

8. Incubate at 88°F/31°C (85°F/29°C to 90°F/32°C is your workable range). After about 12 hours, begin taking the temperature of the beans with a thermometer. When you notice the tempeh's internal temperature rising, the microbes are starting to generate their own heat. At this point you may need to adjust the heat source down. If the mycelium sits at a temperature of 105°F/41°C or above for over an hour, it will die, so it's important to monitor the temperature.

9. After 18 to 24 hours, the white spores will begin to noticeably knit everything together, giving them the appearance at first of a dusting of snow. At this point, the tempeh is doing its thing without any outside heat source. Once you see this, turn off the heat source (for dehydrator or proofing box, not for water bath) and let the mycelium continue to grow for an additional 6 to 12 hours, until the tempeh has become a firm white mycelium cake. You want to make sure it is holding together nicely. It can continue to ferment until you no longer see any grains, but monitor it carefully to prevent sporulation. Overripe tempeh is still edible (see page 153).

10. If pasteurizing, preheat the oven to 180°F/82°C and bake the tempeh for 30 minutes. Unpasteurized tempeh will keep in the refrigerator for 2 to 3 days. Pasteurized tempeh will keep in the refrigerator for at least 5 days and up to 2 weeks.

Hazelnut–Cocoa Nib Tempeh

YIELD: 1 LARGE CAKE, 20 OUNCES

Peanut tempeh is a thing and has been for a long time. We wanted to make a nut tempeh that was a bit different and also gave a nod to our stomping grounds in the Pacific Northwest, and that led us to hazelnuts. After all, 99 percent of the hazelnuts grown in the United States come from the Willamette Valley of Oregon. Hazelnuts made us think of chocolate (we're Nutella fans), and now you know why we put these two things together in a tempeh.

The first time we made this it was perfect — we'll just call that beginner's luck. The next (way too many) times it didn't work. But that first tempeh was so delicious that we didn't give up. For over a year we tried to re-create that first batch, then we began to think we dreamed it. As we experimented with other nuts, we discovered that 100 percent nut tempehs are hit or miss (you can read more about nuts on page 144). This recipe is far from the original vision, but we think it's better and more interesting (and repeatable . . .). The addition of grated coconut pulls it all together and gives it a wonderful texture. We love this steamed in shio koji, then fried in generous amounts of ghee with deep-fried sage. We know that is a lot of flavors, but it works.

PROCESS **tempeh (page 147)**	FERMENTATION **24–36 hours**

1⅓ cups (285 g) chana dal (split, hulled chickpeas)

½ cup, plus 1 tablespoon (85 g) raw hazelnuts

2–3 tablespoons (20 g) cocoa nibs

2 tablespoons (10 g) grated (dried) coconut

2 tablespoons (30 mL) vinegar

1 teaspoon (1 g) tempeh starter

Note: Confirm the quantity of tempeh starter against what is recommended by the manufacturer of your starter.

1. Place the chana dal in a bowl and cover by at least 2 inches of water. Let sit at room temperature for 8 to 12 hours.

2. Roughly chop the hazelnuts or pulse them in a food processor until they are about a quarter of their original size.

3. Drain the chana dal, then place in a pot with fresh water. Bring the beans to a low boil over medium heat and skim off the foam. Add the chopped hazelnuts and cocoa nibs and cook until the beans are al dente, about 15 to 25 minutes.

4. Pour the cooked mixture into a colander to drain, then place in a casserole dish. Add the coconut and vinegar and mix, stirring to incorporate and disperse steam. A lot of moisture will dissipate with the steam. Using a hair dryer on high heat, dry the mixture until it is damp dry, stirring gently as you work. Alternatively, you can use a dehydrator (see page 138).

Recipe continues on next page

5. Transfer the mixture to a stainless-steel pan if incubating in a hot water bath, or into two perforated quart-sized ziplock bags if incubating with the plastic bag method. If using bags, divide the mixture evenly between the two bags. Squeeze out the excess air, seal, then lay flat on a cutting board and spread the mixture in the bags with your hands so that they are evenly distributed.

6. Incubate at 88°F/31°C (85°F/29°C to 90°F/32°C is your workable range). After about 12 hours, begin taking the temperature of the beans with a thermometer. When you notice the tempeh's internal temperature rising, you may need to adjust the heat source down. If the mycelium sits at a temperature of 105°F/41°C or above for over an hour, it will die, so it's important to monitor the temperature.

7. After 18 to 24 hours, the white spores will begin to noticeably knit everything together, giving them the appearance at first of a dusting of snow, then more like sidewalk pebbles rising up through a clean snowfall. At this point, the tempeh is doing its thing without any outside heat source. Once you see this, turn off the heat source (for dehydrator or proofing box, not for water bath) and let the mycelium continue to grow for an additional 6 to 12 hours, until the tempeh has become a firm white mycelium cake. You may see your chickpeas peeking through, but it will predominantly be white. You want to make sure it is holding together nicely. It can continue to ferment until you see no chickpeas. However, watch the tempeh because at a certain point, especially if it is warm, it will begin to sporulate, forming black patches. Overripe tempeh is still edible (see page 153).

8. If pasteurizing, preheat the oven to 180°F/ 82°C and bake the tempeh for 30 minutes.

Unpasteurized tempeh will keep in the refrigerator for 2 to 3 days. Pasteurized tempeh will keep in the refrigerator for at least 5 days and up to 2 weeks.

Variation: Chestnut Tempeh

We once had a yard with a huge American chestnut tree. It was planted by settlers and because it was isolated in the Willamette Valley of Oregon, it was blight-free. We developed a love not only for the tree but for the chestnuts themselves. We could go on about this love affair, but instead we will just tell you that this tempeh is delicious. The chestnuts act as an accent, but if you want a more chestnut-forward tempeh, you can double the amount of chestnuts and include just 1 cup of beans.

We like this recipe with both traditional soybeans and chickpeas. The difference is that the roasted chestnuts don't need to be cooked. Right before you are ready to strain the beans, add the chestnut pieces to the cooking water of the beans. Let them sit in the water for 1 minute to warm them and then strain. Continue with the instructions for making Hazelnut–Cocoa Nib Tempeh on page 171. If using soybeans, be sure to follow the instructions for hulling in the Basic Soybean Tempeh recipe on page 157.

- 1⅓ cups (285 g) chana dal (split, hulled chickpeas) or soybeans
- ½ cup (100 g) peeled and roasted chestnuts, chopped into raisin-sized pieces
- 2 tablespoons (28 g) vinegar
- 1 teaspoon (1 g) tempeh starter

Note: *Confirm the quantity of tempeh starter against what is recommended by the manufacturer of your starter.*

CHESTNUT TEMPEH

Ayocote Bean & Hominy Tempeh

YIELD: 2 CAKES, 18 OUNCES EACH

When we tasted the full flavor and meaty texture of ayocote morado beans, we knew we had to make tempeh with these enchanting, huge purple beans. Cook these beans whole, dry them off, and then chop them coarsely. Their flavor and texture blend perfectly with hominy. This recipe uses canned hominy, but be sure to get the kind that contains lime, as that indicates it has been nixtamalized. Feel free to use your own nixtamalized corn. We have used this tempeh a number of ways, from filling in tacos to stir-fries.

PROCESS **tempeh (page 147)**	FERMENTATION **22–36 hours**

1 cup (170 g) ayocote morado or ayocote negro or other runner bean

1 cup (170 g) canned hominy or cooked nixtamalized corn

2 tablespoons (30 mL) vinegar

1 teaspoon (1 g) tempeh starter

Note: Confirm the quantity of tempeh starter against what is recommended by the manufacturer of your starter.

1. Place the beans in a bowl and cover by at least 2 inches of water. Let sit at room temperature for 12 to 24 hours.

2. Drain the beans, then place in a pot and cover with fresh water. Bring the beans to a low boil and skim off the foam. Cook for 40 minutes, then add the hominy and cook until the beans are al dente, about 5 minutes longer. Pour into a colander to drain.

3. Roughly chop the beans and hominy when cool enough to handle, or place in the bowl of a food processor and pulse for 8 to 10 times, or until the beans and hominy have been broken up evenly. Pour the mixture into a casserole dish and add the vinegar. Mix to disperse and release steam. A lot of moisture will dissipate with the steam. Using a hair dryer on high heat, dry the mixture until it is damp dry, stirring gently as you work. Alternately, you can use a dehydrator (page 138).

4. Transfer the mixture to a stainless-steel pan if incubating in a hot water bath or two perforated quart-sized ziplock bags if incubating with the plastic bag method. If using bags, divide the mixture evenly between the two bags, squeeze out the excess air, seal, then lay flat on a cutting board and spread the mixture in the bags with your hands so that they are evenly distributed.

5. Incubate at 88°F/31°C (85°F/29°C to 90°F/32°C is your workable range). After about 12 hours, begin taking the temperature of the beans with a thermometer. When you notice the tempeh's internal temperature rising, you may need to adjust the heat source down. If the mycelium sits at a temperature of 105°F/41°C or above for over an hour, it will die, so it's important to monitor the temperature.

6. After 18 to 24 hours, the white spores will begin to noticeably knit everything together, giving them the appearance at first of a dusting of snow, then more like sidewalk pebbles rising up through a clean snowfall. At this point, the tempeh is doing its thing without any outside heat source. Once you see this, turn off the heat source (for dehydrator or proofing box, not for water bath) and let the mycelium continue to grow for an additional 6 to 12 hours, until the tempeh has become a firm white mycelium cake. You may see your beans peeking through, but it will predominantly be white. You want to make sure it is holding together nicely. It can continue to ferment until you no longer see any beans, but monitor it carefully to prevent sporulation. Overripe tempeh is still edible (see page 153).

7. If pasteurizing, preheat the oven to 180°F/ 82°C and bake the tempeh for 30 minutes. Unpasteurized tempeh will keep in the refrigerator for 2 to 3 days. Pasteurized tempeh will keep in the refrigerator for at least 5 days and up to 2 weeks.

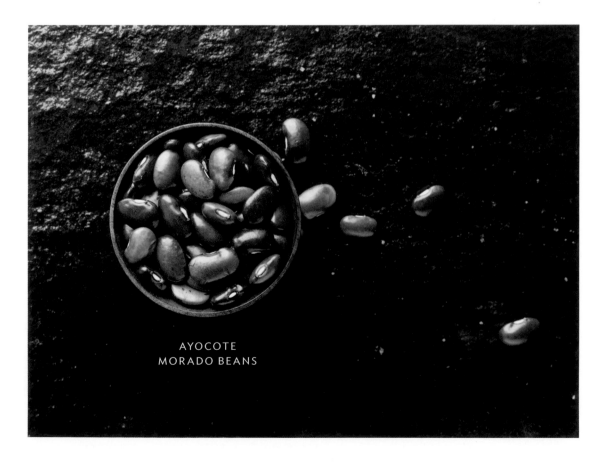

AYOCOTE
MORADO BEANS

Hominy–Pumpkin Seed Tempeh

We see this as a grits-meets-tempeh recipe, and as such we have fried it in bacon fat and eaten it with a mess of greens. If you want the same flavor without the bacon fat, marinade the tempeh as you would Smoky Bacon-ish Tempeh (page 335). Like the previous recipe, this uses canned hominy, but this can be easily substituted for cooked homemade nixtamalized corn.

1½ cups (205 g) canned hominy or cooked nixtamalized corn

½ cup (65 g) hulled pumpkin seeds

2 tablespoons (30 mL) vinegar

1 teaspoon (1 g) tempeh starter

Note: *Confirm the quantity of tempeh starter against what is recommended by the manufacturer of your starter.*

1. Drain the hominy and place in a pot. Add the pumpkin seeds and enough fresh water to cover by an inch or so. Bring to a boil over high heat, then reduce the heat to low and simmer for 10 minutes. Drain in a colander, then roughly chop.

2. Pour the mixture into a casserole dish and add the vinegar. Mix to disperse and release steam. A lot of moisture will dissipate with the steam. Using a hair dryer on high heat, dry the mixture until it is damp dry, stirring gently as you work. Alternately, you can use a dehydrator (page 138).

3. Transfer the mixture to a stainless-steel pan if incubating in a hot water bath or two perforated quart-sized ziplock bags if incubating with the plastic bag method. If using bags, divide the mixture evenly between the two bags, squeeze out the excess air, seal, then lay flat on a cutting board and spread the mixture in the bags with your hands so that they are evenly distributed.

4. Incubate at 88°F/31°C (85°F/29°C to 90°F/32°C is your workable range). After about 12 hours, begin taking the temperature of the beans with a thermometer. When you notice the tempeh's internal temperature rising, you may need to adjust the heat source down. If the mycelium sits at a temperature of 105°F/41°C or above for over an hour, it will die, so it's important to monitor the temperature.

5. After 18 to 24 hours, the white spores will begin to noticeably knit everything together, giving them the appearance at first of a dusting of snow, then more like sidewalk pebbles rising up through a clean snowfall. At this point, the tempeh is doing its thing without any outside heat source. Once you see this, turn off the heat source (for dehydrator or proofing box, not for water bath) and let the mycelium continue to grow for an additional 6 to 12 hours, until the tempeh has become a firm white mycelium cake. You may see your hominy peeking through, but it will predominantly be white. You want to make sure it is holding together nicely. It can continue to ferment until you no longer see any hominy, but monitor it carefully to prevent sporulation. Overripe tempeh is still edible (see page 153).

6. If pasteurizing, preheat the oven to 180°F/82°C and bake the tempeh for 30 minutes. Unpasteurized tempeh will keep in the refrigerator for 2 to 3 days. Pasteurized tempeh will keep in the refrigerator for at least 5 days and up to 2 weeks.

Eating Tempeh

In general, tempeh isn't eaten raw. It is usually steamed and often deep-fried. Eating fresh raw tempeh is a little like eating raw mushrooms, as far as texture and flavor go. If you use raw tempeh in other dishes, any oils in the dish can create an anaerobic environment and encapsulate the (still metabolizing) compounds, causing off or bitter flavors. Sometimes there are very slight bitter flavors in purchased tempeh as well. These flavors are easily dissipated through steaming or boiling in a bit of water with soy sauce. This is also a good way to thaw frozen tempeh quickly when you are ready to cook it for any recipe.

Don't be afraid to deep-fry tempeh. The great majority of tempeh is traditionally consumed deep-fried because it is tasty and practical. We precook or marinate tempeh (page 335) in a bit of water and some kind of amino sauce to give it flavor before frying. Tempeh should be fried hot and fast so that it is crispy, not soggy. Be sure to use good high-heat oils for frying, like avocado, peanut, coconut, or sesame oil.

Before diving in, we want to make sure you understand the vocabulary. Koji's scientific name is *Aspergillus oryzae*. *Koji* is the Japanese word that is now widely used throughout much of the world, in both scientific and popular works on fermentation. *Koji*, *kome-koji*, *koji rice*, *rice koji*, and *malted rice* all refer to koji that has been grown on rice. Koji is also grown on grains (most commonly wheat and barley), legumes, and other starchy things like potatoes or seeds. Traditionally these all have different names that specify what the koji has been grown on, like *barley koji* and *soybean koji*. The starch is dusted with koji spores and incubated, under the ideal temperature and humidity, to grow koji mold. The actual mold, regardless of the substrate it is grown on, is called *koji-kin* in Japan; we will refer to the mold as koji. The koji spores are the starter, or the *tane koji*.

You don't need to be a koji maker to make miso, a tasty paste, amazake, or any of the other koji-based ferments in this book. You can simply buy premade rice koji or barley koji (see the sources on page 388). And this is perfectly legit; in Japan you can find generational koji houses that have been making kojis for centuries. Whether you buy premade koji or make koji, you will harness the power of transformation. We do, of course, encourage you to make your own. If you get excited by the possibility and magic of the microbial world, you will be amazed. If you are practical, then making your own is significantly more economical. And if you are just curious, the perfume of fresh koji is worth the effort.

Part of the Magic

This thing we call fermentation is humbling. We are not in control. The microbes are running the show. Even when we think we are following the correct procedures, or using the same microbes, small differences that we can't perceive show themselves in our ferments.

For example, chef Josh Fratoni, who was part of Sean Brock's fermentation program at McCrady's restaurant, shared that on the same day in Tokyo he bought koji from two traditional koji houses that have made koji in the same neighborhood for hundreds of years. He took each portion of koji and used them to make the same miso recipe. The resultant miso from one shop's koji had amazing flavor characteristics, including a strong floral aroma. The miso from the other koji was much more efficient with a resulting deep umami. In this case, both positive but very different.

All this is to say that sometimes despite all our efforts to manage every detail of the process, the microbes do their own thing, and the result doesn't look or taste like we'd envisioned. This can be frustrating, but we'd like to offer that it is the beauty and the magic.

CHAPTER 7

KOJI

Koji can seem like one of those things that all the cool kids are talking about while the rest of the kids are still trying to figure out what it is but are afraid to ask. It's a fair assessment, because there is nothing else like it. Koji is basically a culinary mold, a type of (very yummy) fungus that is the starting point for freeing up flavor within foods. We can't point to another familiar food and say, "Well, it's like . . ." Koji is the first stage in traditional two-part fermentations such as sake, miso, or shoyu. In contemporary kitchens, koji is used more as a means to transform than as an ingredient. Some chefs are proclaiming it the next great game changer after salt; that is a huge statement.

While koji is edible on its own — it is delightfully sweet, in fact, with an intoxicating scent that calls to your taste buds — it is rarely eaten straight. While we will fully admit to nibbling bits of koji right out of the tray as soon as it done (we just can't help it), it is not considered a stand-alone food, so there isn't a lot of research on its health benefits. Instead, you can look to the benefits of the products you make with koji, like miso, soy sauce, amazake, and shio koji.

For those of us who experiment with harnessing the flavors created by microbes, spending time with koji can be very fulfilling. Kirsten spent a long weekend in the kitchen with Jeremy Umansky (page 184) learning about koji, and it was a special time indeed. Her brain was packed with knowledge and ideas, but the biggest takeaway was how easy (and fun) making and working with koji is. This should be your takeaway, too, after you finish reading this chapter — if we do our job right.

Koji is the engine that drives the fermentation.

BETTY STECHMEYER, FOUNDER, GEM CULTURES

Fermented Soy Yogurt

YIELD: ABOUT 2½ QUARTS YOGURT

We developed this recipe as a way to use leftover soy milk from making oncom (page 163). To get the consistency and texture right, we tried everything from agar to pectin and found that tapioca flour (also called tapioca starch) and some chia seeds worked the best. Don't cut out the coconut sugar — it feeds the microbes. The fat from the coconut milk provides the richness of flavor the soy milk is missing.

FERMENTATION **8 hours**

---

2 quarts (1.9 L) soy milk

2 cups (473 mL) full-fat coconut milk

2 tablespoons (28 g) coconut sugar or cane sugar

6 tablespoons (84 g) tapioca flour

1–2 packets (5 g) yogurt culture (see note)

3 tablespoons (32 g) chia seeds

Maple syrup (optional)

Note: Confirm the quantity of yogurt culture against what is recommended by the manufacturer of your culture.

---

1. Combine the soy milk, coconut milk, and sugar in a large pot and warm over medium heat until it is 150°F/65°C.

2. Pour 1 cup of the hot milk into a bowl and whisk in the tapioca flour until smooth. Whisk the mixture back into the pot. Continue to heat the soy milk mixture, stirring frequently, until the temperature is 180°F/82°C.

3. Remove the pot from the heat. Let the milk cool until it is 115°F/46°C or the temperature suggested on your starter culture.

4. Sprinkle the starter over the surface of the milk and allow it to sit for a moment to hydrate, then whisk to combine.

5. Pour the yogurt into three clean quart jars and place in a dehydrator, proofing box, or yogurt maker for 8 hours at 105°F/40°C to 112°F/44°C. Alternatively, place the yogurt in an electric pressure cooker on the yogurt setting for 8 hours.

6. Stir the chia seeds into the yogurt. Add the maple syrup, if using, and stir, then refrigerate immediately. The yogurt will continue to thicken in the refrigerator, reaching its full thickness in 8 to 12 hours. The yogurt will keep in the refrigerator for up to 2 weeks.

Sprouted Corn Oregon Tempeh

From Tara Whitsitt, author of *Fermentation on Wheels*

TARA WRITES: I am in love with the versatility of this medium. Even though I grew up vegetarian, tempeh was surprisingly not on my must-haves grocery list. It took Barry Schwartz (of Barry's Tempeh in Brooklyn) to truly introduce me to tempeh. He once gifted me three 12-ounce tempeh blocks, each different. I was taken with the aroma and the fact that each block was unique — one was made with adzuki bean and brown rice, another with quinoa and buckwheat, and the final with navy beans and brown rice. My tempeh obsession began soon after.

My obsession would not have gone so deep if it had not been for the flavor force of *Rhizopus oligosporous* and how far you could bend the "rules" for making tempeh. I've experimented with roasted walnuts, mung beans, sprouted corn, sunflower seeds, and much more. Store-bought tempeh is usually pasteurized, whereas Barry's tempeh was always served fresh (or he would freeze it). With fresh tempeh, you can actually taste the complexity and take in the aroma of the fungus, which enhances the flavor dramatically.

Scarlet runner beans, pintos, and corn were inspired by local farmers of the Willamette Valley, where I am based when not driving around the country teaching fermentation. This recipe has proven delicious and nourishing.

PROCESS **tempeh (page 147)**	FERMENTATION **24–36 hours**

4 ounces (115 g) dry corn

8 ounces (225 g) dry pinto beans

6 ounces (170 g) dry scarlet runner beans

¼ cup (60 mL) vinegar

½–1 teaspoon (1 g) *Rhizopus oligosporus* or *R. oryzae* tempeh starter

Note: Confirm the quantity of tempeh starter against what is recommended by the manufacturer of your starter.

1. Soak the corn for 36 to 48 hours or until you see it beginning to sprout, rinsing it and replacing the soaking water with fresh water every 12 hours. Soak the pinto beans and scarlet runner beans in separate bowls for 8 to 12 hours.

2. After soaking, rinse the beans. Keeping the two kinds of beans separate, process them in a food processor to break them down a bit, until they are roughly a quarter of their original size.

3. Transfer the beans to separate pots. Add to each pot enough water to cover the beans by 3 inches, along with 2 tablespoons vinegar. Bring the beans to a boil and boil until al dente. Pinto beans will reach al dente after roughly 30 to 45 minutes, while the scarlet runners will need anywhere from 45 to 60 minutes. The beans should be soft enough to comfortably bite through but not mushy. As the beans cook, you will notice the hulls float to the surface. Scoop them out with a mesh strainer. You won't get all of them — that's fine.

Meet the Microbe

Koji, like mushrooms and mold, is a fungus. More specifically, it is a filamentous fungus, a surface mold whose mycelium creates a network of filaments forming (often circular) colonies that metabolize the substrate and produce enzymes. This type of fungus does not form fruiting bodies (mushrooms) but is cottony or woolly. And perhaps most importantly, it is extremely safe and has been used in food fermentation for centuries. We mention this because of *Aspergillus oryzae*'s nearly identical genetics to *A. flavus*, the mold that produces aflatoxins. Not to worry. Researchers who sequenced *A. oryzae*'s DNA found that the genes for making aflatoxins are so heavily mutated that *A. oryzae* is genetically incapable of producing aflatoxins.

In fact, humans and koji have been collaborating for thousands of years. Koji's history and use are deeply rooted in both Japan and China. The first mention of using mold to transform food was made in 300 BCE, in the time of the Zhou dynasty. Its discovery blew open the nutrition and flavor potential of soybeans and grains. This history is vast and fascinating (well, at least for a select group of self-described fermentation nerds). We won't go into it, but researcher extraordinaire (we know, we've been to his library) William Shurtleff has, with Akiko Aoyagi, put together an extensive history, titled *History of Koji: Grains and/or Soybeans Enrobed with a Mold Culture (300 BCE to 2012)*. While it is 660-plus pages and pretty dense, the gist is that koji is amazing. It has been transforming foods for human consumption for more than two thousand years.

Let's look at what koji does when it takes up residence on a surface. First it produces and lays down a complete set of enzymes so that it can get to work breaking down compounds (it has over 50 different enzymes). It does all this to feed itself and in turn help us feed ourselves. To understand what is happening, we bump into differences of language used to describe fermentation discovery and microbiology. The words we use in a kitchen don't quite match the words used in the lab. Until recently, 99 percent of us described the process simply as fermentation. However, strictly speaking, koji's process isn't fermentation. But for our intents and purposes, the word works. If your eyes start to spin with science, skip the next paragraph; you truly don't need to know any of it to ferment anything you want with koji. This is for those of you who love diving deep.

Koji's process is sometimes described as autolysis— a fancy way to say enzymatic breakdown from within a cell (self-digestion, if you will). Autolysis happens when meat ages and bread is baked. In the case of bread, when the flour and water are mixed together and allowed to rest, naturally occurring enzymes break down the gluten by splitting up the proteins (called by its French name, *autolyse*). But koji is a fungus, so strictly speaking, it doesn't perform autolysis because the fungi are performing extracellular digestion, which has to do with which side of the cell wall the enzymes are hanging out on, doing their thing. This is where people can get hung up on trying to fit the process into an exact lab definition because once the mold is killed, as happens in koji ferments, the enzymes are still active in and outside of cells. Ultimately, a fermented food made with koji is a mix of autolysis, extracellular digestion, and alcoholic, lactic, and acetic fermentation.

The important thing to know is that due to koji's breaking-things-down superpower, it is used to ferment things that are otherwise difficult to ferment — notably grains and legumes. But as you will read, koji pioneers are using this humble mold for a lot more. As koji goes through its life cycle, its main interest is in creating food for itself. After all, it is a live organism with hopes and dreams; okay, maybe not hopes and dreams, but it does want to live, and in order to do so it wants to feed itself. Given the right conditions and food, it will grow on anything. *Aspergillus oryzae* even breaks down biodegradable plastic, specifically polybutylene succinate, which you can find in packaging film, bags, and film for mulching. It's not the first time humans have asked this microbe to do more than process our food. It has been used for the past 30 years in laundry formulations due to its production of lipase enzymes, which break down fats.

This fungus breaks down the bigger starch, protein, and fat molecules into much simpler sugars, amino acids, and fatty acids. *A. oryzae* is particularly prized for at least four reasons. First, it grows rapidly on freshly steamed rice, easily outcompeting other microorganisms. Second, it produces strong carbohydrate-consuming enzymes. Third, and you will understand this as soon as you smell your first tray of homemade koji, it produces flavor compounds that are very pleasant to humans. Finally, unlike some other fungi, it doesn't discolor the rice, which is important because rice is prized for the level of "whiteness" that it can obtain through the polishing process.

There is a long history of humans nurturing and selecting for traits in koji mold to break down starches. *A. oryzae* has more alpha-amylase genes than other related fungi, which is important because the alpha-amylase helps convert starches to molecularly simpler forms of sugar (starch hydrolysis), and we want that in order to produce alcohol. In fact, the production of sake has driven our selection of strains of *A. oryzae* over the centuries.

The smaller molecules of sugar, amino acids, and fatty acids are perfect for fermentation microbes. The microbes we associate most with fermentation are yeast (think alcoholic ferments), acetobactors (think vinegars), and lactobacilli (think cheese, salami, kraut, pickles, and kimchi). You can think of koji as breaking up the rocks and smoothing the surface, in a sense laying the groundwork for many traditional ferments — sake, mirin, soy sauce, amazake, and miso being the most common.

HAIGA RICE KOJI

Amylase Alcohol around the World

To produce alcohol, yeasts need sugar, which is pretty easy to find in sugar-rich foods like fruit. The sugars in starchy cereals — corn, wheat, barley, and rice — are harder for the yeasts to get at; they need to be unlocked. You need an enzyme like alpha-amylase to do this work for the yeasts, so that the yeasts can work for you (assuming you are the one desiring the alcohol in the first place). Historically, depending upon when and where you lived, you went about this in one of three ways:

▸ In Asia generally, you turned to mold, specifically koji, which secretes these enzymes, though malting also has a long history in these areas.

▸ In Europe and North America, you turned to malt (germinated barley), which contains these enzymes. This began earlier, in ancient Egypt and Mesopotamia, with beer.

▸ In the Americas, you turned inward to your own saliva, which contains amylases to break down starch into sugar. By chewing and then spitting out the grains, you inoculated them with the proper enzymes and then they did the rest. (There are also examples of chewing in Africa and Asia.)

The Man Who Brought America Cherry Trees, Adrenaline, and Quick Whiskey

In 1890, an enzyme from *A. oryzae* was patented by an amazing Japanese-American inventor, scientist, entrepreneur, and industrialist named Jokichi Takamine. Jokichi called his patented enzyme taka-diastase, and when applied to the whiskey process, it cut the fermentation time of the grains from 6 months to 48 hours. He established America's first fermentation company, Takamine Ferment Company, and quickly partnered with "the whiskey trust," which was a big player in the American whiskey industry. Unfortunately, just as the company was ready to start production, the whiskey trust folded and the new owners were not interested in continuing. Jokichi discovered that the enzyme could also be applied to a common stomach condition called dyspepsia, or indigestion. He marketed it as "Dr. Takamine's Taka-Diastase."

Jokichi Takamine is perhaps best remembered for his lab's isolation of the crystalline version of adrenaline. He also organized the donation of thousands of cherry trees that you can enjoy in Washington, D.C.

Jeremy Umansky

This chapter would have been very different without the help of chef Jeremy Umansky, who taught Kirsten and supported our koji journey. Koji is having a renaissance. After thousands of years of domesticity, some folks around the world are thinking beyond the traditional uses for koji and asking, *What if . . . ?* Jeremy is in the forefront of this group. When we reached out to him, he generously invited Kirsten to stay with his family in Cleveland while he freely shared his amazing food, enthusiasm, knowledge, and techniques.

When Kirsten was there, Jeremy and his business partners Allie La Valle-Umansky and chef Kenny Scott had just signed a lease on a rad space — an 1850s firehouse that would house Larder: A Curated Delicatessen & Bakery. Larder is a from-scratch eastern European deli and bakery featuring koji-cured meats, wild-foraged ingredients, and local fermented foods. It is one of the first of what we think will be a wave of new microbe-smart kitchens.

Jeremy is a big-picture thinker. He feels that food binds us, and he is always wondering how food can best be enjoyed. He believes that you cannot be mad at someone when you are sharing a good meal with them. He started working with koji as a chef at Trentina in Cleveland. Like everyone else, he started with rice, but soon he was looking around the kitchen wondering what else he could grow koji on, trying pasta and other things. Because koji lays an enzymatic framework as its mycelium digs into the surface of anything it lands on, Jeremy started to speculate about meat. Specifically, curing meat. His first foray into growing koji was on scallops. Scallops spoil easily "so I knew that if I could hold it at 80 degrees for 36 hours and not have it spoil, I might be on to

something cool," Jeremy said. (You can see him explain this in his TED Talk, "Adventure Time in the Koji Kingdom" from TEDxCLE 2015.)

It worked. The scallops were fuzzy, with a sweet smell, and had no signs of spoilage. He cooked them and they were tasty. Over and over again, he found that once the koji was well colonized, it released its enzymes into the meat, tenderizing the cuts. But it doesn't end there. Jeremy hopes that this knowledge can be used to make lower-quality meat better and more affordable, even on a scale that could help relieve the so-called food deserts spread across this country.

Jeremy demonstrated this for Kirsten with three equal cuts of steak. The first steak was dredged in a koji-flour mixture, with a generous helping of flour so that the koji could gain a strong foothold on the surface of the meat. Then he put the meat in an incubation chamber at 80°F/27°C — yes, that balmy condition that you have spent your whole life keeping meat out of. Everything about it feels wrong, and yet, everything about it works. Honestly, even though we understand the science, we are surprised every time by the success.

A day later, Jeremy tucked the second steak into a vacuum-sealed bag with amazake (read more about this tasty koji-rice pudding on page 129). He dropped the bag into a 138°F/59°C water bath warmed by an immersion circulator and kept it there for a few hours.

The last steak was our "control." Jeremy left it in the refrigerator until he was ready to cook, as per usual.

When the koji had fully populated the first steak, Jeremy cooked all three simultaneously. Each steak had an entirely different flavor profile. The control steak was a well-prepared, tasty steak. The amazake steak had more depth of flavor and was more tender than the first. The koji steak had a multifaceted flavor profile; it could be sliced with a butter knife and had a subtly sweet crust formed by the koji, a bit like a chicken-fried steak.

Jeremy is a pioneer in bringing the culinary traditions of East and West together by marrying the art of curing charcuterie with the art of koji. From this union, we get koji charcuterie, in which koji is grown directly on the surface of the meat to improve its preservation. Jeremy and his partners have worked with a safety consultant to develop hazard analysis and critical control point (HACCP) plans that ensure the safety of their koji products. They use only whole cuts of meat (using koji on sausages would require an experienced maker with a devoted professional setup, as it requires top-tier environmental controls and an incredible amount of vigilance), and the meat is given a salt cure as one would give any cured meat. Then the meat is coated with koji spores mixed in flour. In a few days, when the mold has covered the surface, the cured meat is then hung. This step floods the meat's surface with enzymes, accelerating the process such that the curing time is cut by a third or more. (In charcuterie, the curing time, or doneness, is determined by weight. The meat is weighed at hanging and must lose 40 to 50 percent of its weight before it is considered done.) Lucky for her, Jeremy had made a koji coppa in anticipation of Kirsten's visit. It was nothing short of buttery deliciousness.

We are not qualified to teach you koji charcuterie, but Jeremy has some open-sourced work that is available through his website, https://larderdb.com. On pages 237 and 238, he also shares two recipes for growing koji on pork. If you are interested in charcuterie, we recommend *Pure Charcuterie* by Meredith Leigh. Her beautifully written book delivers a very approachable introduction to the craft, and she has a section on koji charcuterie.

When we eat, our tongues only register certain combinations of molecules, amino acids, esters, fatty acids, and sugars. Koji takes these complex things and breaks them down into a mire of other things. What we normally can't perceive, koji opens it all up.

JEREMY UMANSKY

Types of Koji

Aspergillus has been used in fermentation for more than a thousand years. There are dozens of different "seeds," or starter cultures, which the Japanese call *tane,* to create specific types of koji, called *tane koji.* The most readily available koji starters are those for light rice, red rice, barley, soybean, and shoyu. While the spores of the different strains are all *A. oryzae* (same genus, same species), they have different enzymatic abilities, which have been selected and developed for hundreds of years by artisans who use them to make miso, vinegar, shoyu, miso, pickles, and the like.

For example, a soybean koji spore has very high protease activity and as such, it specializes in breaking down the protein in soybeans. However, it's not nearly as good at breaking down carbohydrates. At the other end of the spectrum, light rice starter has high amylase capabilities, so it eats those carbohydrates, but it doesn't have as much enzyme strength for the proteins and fats. Despite this, it is a good choice for a broad spectrum of applications. It is a good starter koji. The red rice and barley starters have more middle-of-the-road enzymatic capabilities.

Any serious makers might want to get a few different types of koji and experiment to find the one that gives the best results for their specific applications. For example, we know a maker who prefers to use red rice koji starter for barley koji instead of the barley starter, because they've experimented and the barley koji gives results they prefer.

One last consideration is that some spores are sold dispersed in another medium, like rice flour, and some are sold pure. Be sure to compare the instructions for your starter with the recipe and adjust accordingly. Aspergillus is considered one of the most successful groups of mold. We don't cover other culinarily helpful members like *A. glaucus* or *A. niger.*

Red Rice Koji

There are two basic types of mold used in Asia for fermented food: *Aspergillus oryzae* (which you will know well by the end of this chapter) and *Monascus purpureus,* often also referred to as red koji, red rice koji, or red yeast rice. Red yeast rice is commonly sold as a natural food supplement to lower cholesterol. *M. purpureus* is no stranger to medicine; it has been used for centuries in traditional Chinese medicine for promoting digestion and invigorating circulation, and it has been proven effective by modern medicine. When grown on rice, this fungus produces a deep red color and is mostly used as a natural food coloring. A recipe dating back to 965 CE called for a roast of lamb to be simmered with red koji. It is common in fish sauce, fish paste, rice wine, and red fermented tofu (*furu*). For the red furu, molded tofu squares are salted and soaked in a brine containing wine, rice inoculated with red koji (*M. purpureus*), and spices and then aged for 1 to 6 months.

We will not be using *M. purpureus* in any recipes in this book, and it should not be confused with the red rice koji that is a variety of *Aspergillus oryzae* mold used in making red miso (see page 267).

Growing Your Own Aspergillus Spores

Sometimes you may wake up in the morning and find that your koji has fully sporulated overnight — instead of a few patches, the whole affair is furry olive green. At this point, in most cases, you will want to make a new batch. Fully sporulated koji will taint the flavor of any ferment you use it in. (See the troubleshooting note on page 377.)

You can look at this event as an opportunity to harvest your own spores. Simply break up the koji a bit and dehydrate it, without any covering, at around 113°F/45°C. Once fully dry, store it in a cool, dark, dry spot. When you are ready to use it, grind it up in a food processor or blender with the lid on. Let the fine particles settle for about an hour before opening the lid and wear a dust mask (see the sidebar Be Spore Aware on page 193). These are not pure spores like you would buy from the lab, since they are dispersed in whatever substrate you used for the koji, and they don't need any flour added to disperse them — they are ready to use. You will, however, need to use twice as much as a pure spore culture, as they are weaker.

Of course, you can also grow some rice or barley koji out to this point on purpose, to propagate your own culture. You will want to control the sanitation of this process as carefully as you can to avoid tainting the koji with contaminating organisms, as their undesirable effects will multiply. If you continue to propagate your own spores, you will find that with every generation the spores weaken, as it is impossible to keep a perfect strain outside of very sophisticated lab conditions.

Betty Stechmeyer, who grew tempeh spores for many years, shared with us some tips for working with microbes, whether transferring a kombucha scoby, inoculating tempeh, or growing koji, and especially on days you are harvesting spores: keep your area clean, use fresh towels, wash your hands . . . So far, pretty intuitive, right? Wait, because here's the pro tip: work with your cultures in the morning before you open a window. Betty points out that in the early morning, things are calm, the dust from the previous day has had the conditions of the nighttime air to settle, and the early morning air is still. The chance that some errant microbe is floating around ready to land in your culture is less likely. If you are working in a kitchen, doing this before people have begun to move about the house in a busy way can be helpful.

If you are interested in learning more about growing spores, we recommend *Miso Production* by William Shurtleff and Akiko Aoyagi, in which they thoroughly cover the process, starting with growing the koji on rice mixed with a trace amount of ash to increase spore yield and survival rate while discouraging contaminants.

Once koji grabs your soul, there's no turning back.

JEREMY UMANSKY

Making Koji: An Overview

Koji is the backbone on which you can build so many ferments. A weak koji (poor growth, not enough enzymes) will show itself down the line in unsatisfactory ferments. This is most obvious when making amazake and sake. We want to help you both grow great koji and enjoy the magic.

When we decided to try to make koji, Kirsten bought cultures and, honestly, they sat in the freezer for a year before we were ready to try it. A big reason why we hesitated was that we were intimidated by the idea of an incubation chamber, wooden trays, and managing the temperature for a number of days. Everything we read left us feeling daunted. We finally made a batch with a heating pad and blankets, using our cheese thermometer to monitor the koji. We won't lie: managing that first batch was frustrating, but in the end, we had koji. It wasn't beautiful. It didn't look at all like the soft, pillowy koji we saw on Instagram, but it smelled good — floral, like a spring day, yet mushroomy and earthy. It was sweet. Our next batch was a barley koji and we think we did everything wrong, but nevertheless, in the end, we had barley koji. Again, the fragrance was bloomy and the kernels were sweet, so we made miso. After aging for a year, it was tasty.

Now, after visiting Jeremy and making countless batches, we can assure you that while koji is a bit finicky, it wants to grow. This stuff grows wild — it is a living organism that is programmed to grow, and it will. You don't need to get bogged down in equipment or technique. You can grow koji in in your own kitchen with simply a casserole dish, plastic wrap, and a warm place. If you stay within its comfort zone (between 80°F/27°C and 95°F/35°C), chances are the koji will turn out great, whether you fuss over it and stir it regularly or not, and you will be able to use it to make great miso, amazake, and other fun things. Koji grown for sake requires the most persistence in getting the conditions just so; koji grown for everything else is more forgiving. Trust yourself and don't compare your koji to what you think it should be. If it's tasty to you and does its job, then it's all good. You've got this! Start small and learn as you go — it's a wonderful, rewarding journey. Some batches will be more or less successful, and that is part of the journey. We have talked to various chefs and home chefs doing hardcore "fermentation R&D" and they all agreed that the results can be variable — which we can tell you we found comforting, as we had amazing batches followed by meh batches or complete fails. It happens to all of us. But remember: the worst that could happen is that you have to send some to the compost (your pile won't complain, as you're just feeding it more wonderful microbes). Bad koji will let you know it's bad by looking and smelling foul, or off, and you will know it's off when the aroma is not beguiling.

We will walk you through the process. In this first part, we give you a detailed road map and offer choices so that you can select which production method works best for you. We've streamlined the step-by-step instructions that follow to a simple process that works quite well for home production and ultimately (once you have your system in place) requires relatively little babysitting. The step-by-step instructions are for growing a basic rice koji. It is the process we use, but feel free to find the best method for your situation. We want you to have the knowledge and tools to explore any substrate and join the koji renaissance.

The first few times you make koji, be prepared to babysit it for a few days. Like anything that's alive, it doesn't follow a mechanical clock. While it is generally pretty consistent, you'll likely encounter some batches that follow a different schedule. Be sure to observe, smell, and touch as needed to make sure you are on track. Once you get into the groove of making it and know the particulars of your incubation equipment and technique, the process becomes much easier.

Getting Ready: Sanitary Considerations

When working with koji, you want to keep your space, your implements, and your hands clean. Remember that there are many other microbes out there (like bread mold and bacillus) that would welcome the opportunity to join the party. This is not to say that you need to drive yourself crazy with sterilization; you just want to make sure your work area is clean. Think of that extra bit of surface cleaning you do when

someone has a cold. For extra measure, we sanitize all the utensils we will be using by one of three methods: misting a bit of 190-proof (at least) alcohol or brewing sanitizer across the surface and air-drying, or pouring boiling water over the surface and keeping it there for at least 30 seconds and then air-drying, or using the sanitation cycle of our dishwasher. You can also sanitize glass casserole dishes or jars (for miso) by washing them in hot soapy water and putting them in a microwave for 45 seconds, or until dry.

Choosing the Right Grain

White rice is a good grain choice for beginning koji makers because it's most likely to be successful. Jeremy Umansky (page 184) recommends a long grain and Betty Stechmeyer from GEM Cultures (page 16) recommends a polished sushi. We agree that both are great starter rices, and we will add haiga (a short-grain rice with its germ still intact) to the list. Pearled barley is also a good choice for anyone just getting the hang of growing koji.

Koji Builds Community

We have met people all over the country who are connected, not unlike threads of mycelium, by their love of koji and what it can do for food and flavor. It is truly a community of people learning and freely sharing their discoveries and recipes. People are using koji both traditionally and unconventionally — for charcuterie, for example, and to ferment their plant-based "cheeses." It is an exciting time. As we reached

out to folks, we were taken into the fold. If you fall in love with this alluring fungus, you can start exploring the social media world with the hashtag #kojibuildscommunity (that hashtag, incidently, was started by Rich Shih, whom you will meet in chapter 9). There are also quite a few Facebook groups that have rich conversations and support around growing koji and the various ferments that it initiates.

After you've familiarized yourself with the process of growing koji on these easy grains, move on to brown rice, other whole grains, and soybeans if you wish. With a little experience under your belt, you will have a better idea of how to troubleshoot any problems that might arise. Brown rice, or any whole grain for that matter, can be a challenge to grow koji on. One reason is that the hull of brown rice is impenetrable to the koji mold, so the grain exterior needs to be roughed up a bit. Soaking to expand the kernel is often enough to give the koji a chance to reach the starch. Another reason is that the little grooves in the hull of the brown rice can retain soil bacteria that also want to consume the starch. As Betty points out, "Polishing the rice removes this variable, so there is less competition for aspergillus to use the rice yummies."

Soybeans are a little more susceptible to bacillus growth with temperature fluctuations. Since part of the learning curve is also learning how to control the temperature, you might want to avoid a soybean koji when you are just starting out.

We cover the relationships between koji and different types of rice and grains in more depth throughout this chapter.

Cooking the Grains

There are two schools of thought on rinsing the substrate grain before cooking it. Traditionally, the grain is thoroughly rinsed to remove the dusty starches left on the surface from the polishing (hulling) process. The thought is that the starches can cause the individual grains, once they're steamed, to stick together, which can impede oxygen, making it difficult for the fungus to convert the grain as completely and

efficiently. The other school of thought is that it is better not to rinse your grain at all because that surface starch is food for the fungus and will help it get going. We did many side-by-side experiments and found that we had equal success with robust mycelium growth and good sugar conversion when using unrinsed grain. But you can experiment, too, and find out for yourself what you prefer.

The grains are traditionally steamed, which renders them al dente, with the individual kernels easily separated. They will feel slightly rubbery. When steamed, the grains are more likely to give you a koji that resembles what you might purchase, with separated grains that have a light dusting of mycelium. We've found that cooking the grains in a rice cooker or boiling them in a pot also works, so long as the cooked grains are not wet but al dente. This cooking method creates more of a soft, matted koji where the individual grains are hard to distinguish, which we happen to prefer, but that is a personal preference. Some recipes actually require a wetter rice (see the mirin hack on page 222), so it also depends on your goal. See page 67 for full instructions on cooking grains.

If you are growing koji on beans, you will want to cook or steam them to be al dente (see page 56). The cooked beans should maintain their shape and be soft enough to easily crush between your thumb and ring finger.

Inoculation

Place the cooked substrate into a casserole dish or tray or onto butter muslin. Break up the lumps and spread it out so steam can escape. Allow the substrate to cool to around 110°F/43°C.

While the substrate cools, you can sterilize the flour. We use the generic term "flour" in the recipes, which is meant to cover a number of starchy flours, such as all-purpose flour, rice flour, tapioca flour, and even cornstarch — choose your preference. Sterilizing the flour ensures that the koji mold will not have to compete with other microbes in the early stages of its growth. To sterilize the flour, place it in a small skillet and toast over medium heat for about 5 minutes, or until lightly browned. Remove from the heat and let it cool.

Add the spores to the toasted flour. You place koji in flour for two reasons: The first is to give the koji easy food (think of it as an energy drink) so that the enzymes can get a head start on the medium you are trying to break down. The second is to help scatter the spores. The spores were designed to be airborne and can be a challenge to mix evenly through the entire substrate; by adding some flour, you can more easily disperse them into a heavier material.

Sprinkle half of the koji-flour starter into the substrate and mix. The substrate will take on a light olive green coating. Add the rest of the starter and mix until you are confident that your substrate is evenly coated.

Aspergillus likes a bit of humidity, so at this point you may need to cover your fermentation vessel. Your incubation method will determine what type of covering will work best. It may be a damp, clean cotton cloth in a proofing box, oven, or other system that produces dry heat. In a dehydrator, the airflow is so dry that you will want to use plastic wrap poked with a couple of diagonally opposite holes at the ends of the tray to allow air to flow. If you are using a water bath incubator, you won't need to cover it at all. See chapter 2 for details on all of these incubation setups.

Incubation

This is the part of the fermentation that can feel the most intimidating, especially if you have read about muros, the traditional rooms or wooden cabinets used for incubation. *Muro* means "room" in Japanese, and these rooms had to be well insulated against the cold of winter, when sake is traditionally made in Japan. Modern muro cabinets come with microcomputers that allow for precise control of heat and humidity and are typically made of cedar to absorb moisture without rotting. Modern methods of making koji are to spread it in a thin layer on a covered tray in a precision-controlled environment.

It's true that koji does need specific conditions to grow, but if they are the right conditions, your koji won't know if it is on a traditional wooden tray in a cedar cabinet or a heating pad. Koji thrives in a similar environment to what we ourselves prefer, if maybe a little on the warmer side. To grow and metabolize, it likes to be between 80°F/27°C and 95°F/35°C.

For small batches of koji, we have found an immersion circulator or ferment terrarium (see page 28) to be ideal. We set the water temperature at 88°F/31°C and don't have to monitor the internal temperature of the koji. A bread proofing box is our second favorite, as the temperature setting, in our experience, is very responsive and on target. And while you still have to monitor the internal temps of the koji, you aren't dealing with the drying air of a dehydrator. (A caution: If you try to fit too many trays in a proofing box, you may lose some airflow and your koji may not do well.)

For larger batches, we use a professional-style cabinet dehydrator. You need a dehydrator

Be Spore Aware

Have you ever stepped on a puffball mushroom in the woods? If so, you know that spores are designed by nature to scatter and propagate. They do this by being ultralight, so when working with them be aware of any puffs of flyaway spores. It is important that you avoid breathing the spores. The spores can clump in your lungs, where they cannot be expelled, making you sick.

One strategy to avoid airborne spores is to carefully measure the spores into a small bowl and then gently add the flour on top; the flour will weigh down the spores when you stir them. If you will be doing a lot with the spores, like grinding up a homemade starter in a blender, wear a dust mask and allow the dust to settle before opening the lid. If you have a compromised immune system, wear a dust mask at all times when working with spores.

SCIENCE SAYS . . .
A Closer Look at Temperature

By controlling koji's temperature, you also control its enzymatic action. Koji grown at the lower end of its comfort range (80°F/27°C to 85°F/29°C) will have more protease enzymes, which will result in more savory flavors. This koji is better for things like making miso and shoyu or for curing and marinating meats.

Koji grown in the higher temperature range (85°F/29°C to 95°F/35°C) will have more amylase enzymes. These will turn those starches into sugar, giving this koji the sweeter flavor that is desired for amazake and sake. However, where there is available sugar, there are lactobacillus and yeast waiting for the feast, and they are quick to move in.

If you are working in this higher temperature range, keep a thermometer in the body of the substrate to monitor the internal temperature. As the fungus grows, it generates its own heat. If it goes above 95°F/35°C, like a compost pile that gets too hot, it will self-terminate. If this happens, your koji will not be viable and other microbes may move in.

that can be set to 85°F/29°C or lower. We use a cabinet-style dehydrator, as taught to Kirsten by Jeremy, in the step-by-step instructions on page 196. For the first few hours, monitor the temperature of your dehydrator as well as the interior temperature of the koji to make sure it is on point. For example, our dehydrator always runs cool, so for an 85°F/29°C environment we must set our dehydrator to 90°F/32°C.

If you are making koji in a traditional manner, mound the inoculated substrate in muslin-lined thick towels or a pot of some kind, or, in large production, in cloth-lined wooden cribs. Mounding the substrate will help control the temperature and humidity and keep the young mycelium warm, which makes a lot of sense in large batches. This process doesn't require an outside heat source and is perfect for batches that have the mass to keep themselves warm.

For the first 12 to 18 hours, the goal is to keep the koji at around 90°F/32°C. It is a good idea to check the temperature a few times over the first few hours, so that you can make sure your system is adjusted properly. The internal temperature of the inoculated substrate should stay between 80°F/27°C and 95°F/35°C to give you optimal enzyme development. Once the koji has started to metabolize, it keeps itself warm; in fact, it generates a lot of heat, which must be dissipated. If the temperature gets above 95°F/35°C, it will not ruin the enzymes, but extended periods of overheating will kill the mold and open the door to other microbes. Overheating is, perhaps, the most common reason for failure.

After the first 18 to 24 hours, you will notice a fuzzy white coating on your substrate. Grains will start to have a pleasant, yeasty, floral scent,

To Stir or Not to Stir

Among the modern worldwide koji community, there is discussion as to whether koji should be stirred regularly to keep the grains separated. Stirring disperses the heat, gives the koji some oxygen, and helps the koji penetrate the substrate.

Traditionally, koji is stirred regularly because it is believed to help the koji penetrate the substrate more effectively, thereby creating greater enzyme activity. However, many modern makers get the results they are looking for by not stirring and letting the koji form a mat. In our own experiments with very small batches of koji, we have found that since the amount is so small, stirring can result in one of two things: the grains

lose moisture and the substrate becomes too dry, or if the grains are on the moist side, the grains become a bit clumpy and wetter. In both these scenarios, we prefer to let it make a beautiful fuzzy mat that we break apart to use. We tend to do what feels right for each individual batch. The exception is barley koji; it forms so differently and doesn't create a thick fuzzy mat, so we always have better results stirring it.

From all the conversations we have had, we think that it comes down to your setup, what you are making, and your personal preference. Don't let too much information out there confuse you. Experiment, and you will find the way that works for you and your fermentation.

and beans will have a more mushroomlike yeasty smell; both smell distinctly sweet.

At this point, whether you have a large batch or a small one, you must dissipate enough heat to prevent the koji from overheating and self-destructing. Heat can be dispersed in several ways: the koji can be divided among trays and spread out, the temperature in the incubation system can be lowered, and the koji can be stirred (or not; see To Stir or Not to Stir, page 194). In a traditional large, mounded batch, the koji is taken out of the crib or cloth-wrapped ball at this point, spread in trays, and stirred a few times to keep it from overheating. It is traditionally stirred once more over the second 24 hours. When we stir our koji, we have found our hands to be the perfect tool to gently break up the clumps of grains or beans without damaging the individual kernels. However, you can use a clean wooden spoon or rubber spatula if you prefer. After each stirring you can make furrows in the bed, which offer more surface area for dispersing heat. If you've spread your koji in a thin layer on trays, you don't need to stir it.

After 40 to 50 hours of incubation, the koji will be done. Koji is finished when it is cottony and fuzzy and almost looks a little blurry, as if your eyes aren't quite focusing. If the koji has begun to produce spores, it will also have fuzzy patches of yellow-green. For sweet ferments, a little bit of sporulation is not ideal but is still okay. For amino ferments, you have a little more leeway. If your koji has gone too far, you may want to adjust the plan for what you are making with it. Sporulation won't hurt you but it can create musty flavors, and as Jeremy said, "dank wet-dog notes." In some ferments, like Black Soybean Spicy Douchi (page 207), it is desirable and delicious — no wet dogs, we promise.

When the koji is ready, harvest it for:

▸ Immediate use
▸ Immediate refrigeration for use within the next 10 days
▸ Immediate freezing for use within a month or two
▸ Immediate drying for longer-term storage of up to a year

A final note: the fresher the koji, the better it is for sugars. If you are making amazake or sake, you may want to use your koji as soon as possible. As it ages, it becomes better for producing amino ferments. When we have a batch of old koji, we "save" it by making shio koji (page 223), which we find we can never have too much of.

Storage

To store your koji in the refrigerator, place in a sealable food storage container. While some argue that koji should never be frozen, many others do so successfully. Freezing fresh koji is probably the simplest way to extend its shelf life. The enzymes stay intact in the freezer, and frozen koji is just as potent as fresh koji. Thaw it in the fridge and then bring to room temperature before using.

If you wish to dry your koji for longer-term storage, spread it out on dehydrator trays. Set in a dehydrator at 85°F/30°C for about 7 hours, or until thoroughly dried. If kept cool and dry, dried koji will keep at room temperature for about 6 months in the refrigerator or freezer for up to a year. The longer you keep it, the less potent it will become.

Alternative Incubation

If you don't have an incubator, dehydrator, or bread proofing box, you can still incubate koji. GEM Cultures provides a comprehensive set of instructions with their koji spores; this is a modified version of that process.

Lay out three or four clean, thick towels. Lay a square yard or so of a clean, tight-weave cloth on top of the towels. Place the steamed grain in the center of the cloth. Break up the lumps, spread it out, and allow the steam to escape. Allow the substrate to cool to body temperature. Sprinkle in half of the koji-flour starter and mix well. The substrate will be coated in a light olive green layer of spores. Mix in the rest of the starter mix. When you are confident that the starter is distributed evenly, mound the inoculated substrate, place a thermometer in the middle of the mound, and wrap the cloth around it, allowing the thermometer to show through. Continue to wrap the rest of the towels around your bundle. Remember, your goal is to keep the koji at 90°F/32°C. Use more towels or blankets as needed to maintain this warmth. You can put this bundle in the oven with the pilot light on, though some ovens will be too warm. You can also use hot water bottles or heating pads to keep it warm. Use the thermometer to guide you and make modifications as needed.

After 24 hours, open the bundle and transfer the grains to a stainless-steel or wooden tray or a glass casserole dish. Wrap the tray or dish with plastic wrap or your tight-weave cloth, and again wrap the tray in the bundle of towels and monitor the temperature until the koji is done.

STEP-BY-STEP

Let's Grow Koji!

What You Need

- ▸ Rice (or other substrate)
- ▸ Sanitizing equipment (we often use a spray bottle filled with 190-proof alcohol)
- ▸ Rice cooker or other means to cook or steam rice
- ▸ Small skillet
- ▸ Small bowl
- ▸ Koji culture
- ▸ Metal spoon
- ▸ Glass or stainless-steel casserole dish for fermenting the rice
- ▸ BPA-free plastic wrap or cheesecloth
- ▸ Incubation chamber (see chapter 2)
- ▸ Instant-read food thermometer

1. Sanitize your tools and work surface before getting started.

2. Cook or steam the rice until the grains are al dente, with a fluffy consistency, so that the individual grains maintain their shape.

3. While the rice is cooking, toast all-purpose wheat flour or rice flour in a small skillet over medium heat to sterilize it.

4. When the flour is cool, stir in the koji spores or create a small crater in the flour into which you place the spores. Take care. They are very light and will want to float away.

Instructions continue on next page

5. When the rice has finished cooling, spread it out in a glass casserole dish and allow it to cool to 110°F/43°C.

6. Sprinkle half of the koji-flour mixture onto the rice.

7. Fold in the koji gently to evenly disperse the spores, gently breaking up clumps as you go. Then sprinkle on the rest of the koji-flour mixture and gently stir in again. It will take on a soft olive green hue from the spores, which helps you see that it is even.

8. Cover the dish with plastic wrap or moistened layered cheesecloth. If you're using plastic wrap, make a few small holes on diagonally opposite corners for ventilation.

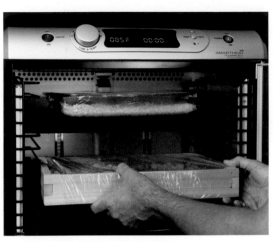

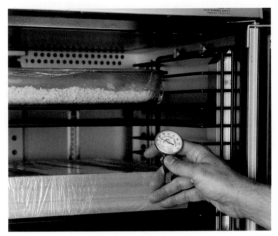

9. Set your incubation chamber to 85°F/29°C. If you're using a fermentation terrarium or other water bath setup, set the water to 88°F/31°C. Place your dish in the incubator.

10. Check the internal temperature of the koji every few hours during the day. It should be between 80°F/27°C and 95°F/35°C. Monitor moisture level and add a dish of water or re-moisten cloth covering as needed.

11. After 24 hours, it will smell faintly aromatic in a sweet, floral, mushroomy way, with the beginning blush of a soft white covering and partly bound together. Check the temperature to ensure the correct range. If your batch is deeper than than 1½ inches and the temperature has risen, you may need to stir it to break up the lumps to dissipate some of the heat. After stirring, make furrows in the substrate, cover the dish, and return it to incubation. Continue monitoring, adjusting, and stirring as needed.

12. After around 40 hours of incubation, the koji will be finished; it will have a cottony appearance and will be knit together in a mycelial mat. If you see areas of yellow-green coloration, it means that the koji is producing spores. Pull the koji from the incubator.

BLACK SOYBEAN
SPICY DOUCHI,
page 207

RICE KOJI

PURCHASED DRY
RICE KOJI

SOYBEAN KOJI,
page 206

BARLEY KOJI,
page 203

Rice Koji

YIELD: ABOUT 3 CUPS

We prefer to make small batches of koji so that we have fresh koji to work with more often. This recipe makes just a little over 1 pound of koji (maybe a little more if the koji is wet). If you want to make more koji at a time, try our Barley Koji recipe (page 203). These recipes are guidelines; use them as a foundation for experimenting with other grains to make whatever kind of koji you want.

PROCESS **koji (page 196)**	INCUBATION **40–50 hours**

3 cups (525 g) rice of your choice (see page 202)

2 tablespoons (20 g) all-purpose or rice flour

1 teaspoon (1 g) koji starter

Note: Confirm the quantity of koji starter against what is recommended by the manufacturer of your starter.

1. Sanitize your tools and your workspace. If you are using brown or black rice, soak it overnight in water to cover by 2 inches, then strain and rinse.

2. Cook or steam the rice to a fluffy al dente consistency, so that the individual grains maintain their shape, following the instructions on page 67.

3. While the rice is cooking, toast the flour: Place the flour in a small skillet and toast over medium heat until lightly browned, about 5 minutes. Remove from the heat and let cool.

4. Place the koji spores in a small bowl. When the flour is cool, add it to koji spores. Stir carefully — the spores are very light and will want to float away.

5. When the rice has finished cooking, immediately spread it out in a glass casserole dish and allow it to cool to 110°F/43°C.

6. Sprinkle half of the koji-flour mixture onto the rice. Gently fold in the koji to evenly disperse the spores, breaking up clumps as you go. Sprinkle on the rest of the koji-flour mix and gently stir again.

7. Cover the dish with plastic wrap or moistened layered cheesecloth. If you are using plastic wrap, make a few small holes on diagonally opposite corners for ventilation.

8. Set your incubation chamber to 85°F/30°C. If you are using a fermentation terrarium or another water bath setup, set the water to 88°F/31°C. Place your dish in the incubator. Check the internal temperature of the koji every few hours during the day. It should be between 80°F/27°C and 95°F/35°C. Incubate for a total of about 40 hours.

9. After 24 hours, the koji will smell faintly aromatic in a sweet, floral, mushroomy way. The surface will appear to have the beginning blush of a soft white covering. The rice will be partly bound together. Check the temperature again to make sure it is in the correct range. If your batch is relatively thick (deeper than than 1½ inches) and the temperature has risen, you may need to stir it to break up the lumps in order to dissipate some of the heat. After stirring, make furrows in the substrate before covering the dish and returning it to the incubator. Continue to monitor the temperature and moisture, adjusting as needed and stirring the koji as needed.

Recipe continues on next page

Rice Koji, *continued*

10. Pull the koji from the incubator when it is finished; it should have a cottony appearance and will be knit together in a mycelial mat. If you see areas of yellow-green coloration, pull the koji from the incubator immediately (and see the troubleshooting tip on page 377).

Variation: Oat Koji

Make this exactly as you would the rice koji on page 201, but use 3 cups oat groats in place of the rice. Soak the groats for 6 to 8 hours, then steam as you would rice. We experimented with steel-cut oats, thinking the koji would have an easy go at the starches within, but it was too difficult to cook or steam the steel-cut oats without them becoming too sticky for good growth. So be sure to get oat groats.

When this oat koji is ready, it will smell like koji meets an oatmeal cookie. We have quite a few recipes to help you get to know this koji, such as a delicious scone (page 331), oat amazake (page 218), and granola (page 330), yet strangely we haven't made this koji into cookies yet. Let us know how yours turn out.

Rice Options

Here is a small sampling of the different kinds of rice you can use as a substrate for growing koji. We've ordered them from lighter to darker.

Sushi. This is a polished rice and a good starter variety.

Arborio. This Italian-style polished rice is easy to find and makes a wonderful base for koji. When it is boiled or steamed, it is not quite as sticky as sushi rice, and we find that makes it a bit easier to work with.

Basmati. Like other long-grain rice, it is easy to work with and gives a nice flavor.

Long grain. We have found this rice to make a very pleasant all-purpose koji.

Parboiled. This rice has some of the nutritional content of brown, but there is no bran to slow down the koji. It is a good starter rice.

Haiga. The bran of this rice has been removed, but it retains its germ. We find that it makes a lovely koji. It is a good substitute for brown rice if you find that difficult to work with.

Short-grain brown. This rice is a standard for koji and for good reason: it retains the vitamins in the bran and germ. Most koji available for purchase is brown rice koji. Just be sure you have soaked the rice long enough or roughed it up a bit to break into the bran.

Wild. This is technically not rice but an aquatic seed native to North America. Wild rice has to be cooked until the grain opens up in order for its starches to be available to the koji. Steaming is not a good option. Instead, boil wild rice until you see the white interior. The nutty flavor is interesting and fun in miso.

Forbidden (black). This is one of our favorite rices to kojify. Koji growing on the black grains is a striking sight. We simply soak and steam the rice in our rice cooker with good results. It has a nice sweet nuttiness.

Barley Koji

YIELD: ABOUT 7 CUPS

Barley koji is wonderful and is the traditional base for certain types of misos. We've found that it finishes a bit differently than rice koji: it doesn't mat together in the same way, and the fuzzy exterior is shorter. Be sure to watch it carefully during the incubation, as barley koji is more prone to overheating once it starts metabolizing. Finished barley koji should have that same sweet mushroomy smell, but it will be enhanced by a nutty chestnut aroma.

This batch will give you enough to make a batch of Barley Miso (page 265) and some Hishio and Nattoh Miso (page 260) for fresh eating while you wait for the Barley Miso.

PROCESS **koji (page 196)** | INCUBATION **40–50 hours**

7 cups (1,225 g) pearled barley

¼ cup (35 g) all-purpose or rice flour

2 teaspoons (2 g) koji starter

Note: Confirm the quantity of koji starter against what is recommended by the manufacturer of your starter.

1. Soak the barley for 6 to 8 hours. Drain and rinse.

2. Steam the barley as you would rice (see page 67), but let it steam longer (about 1½ hours on the stovetop or 30 minutes in a pressure cooker). Or, boil the soaked barley in plenty of water over high heat in a 6-quart saucepan for about 1 hour, stirring and adding more water as it is absorbed and evaporates. The barley is done when the kernels are soft all the way through and popped open but not mushy. The mass will be quite rubbery. (You will break up the lumps when you inoculate it.)

3. While the barley is cooking, toast the flour: Place the flour in a small skillet and toast over medium heat until lightly browned, about 5 minutes. Remove from the heat and let cool.

4. Place the koji spores in a small bowl. When the flour is cool, add it to the koji spores. Stir carefully — the spores are very light and will want to float away.

5. Pour the cooked barley into a strainer and let it drain for a few minutes, then spread it out on a tray to cool to 110°F/43°C.

6. Sprinkle half of the koji-flour mixture onto the barley. Gently fold in the koji to evenly disperse the spores, breaking up clumps as you go. Sprinkle on the rest of the koji-flour mix and gently stir again.

7. Cover the dish with plastic wrap or layered cheesecloth. If you are using plastic wrap, make a few small holes on diagonally opposite corners for ventilation.

8. Set your incubation chamber to 85°F/30°C. If you are using a fermentation terrarium or another water bath setup, set the water to 88°F/31°C. Place your dish in the incubator. Check the internal temperature of the koji every few hours during the day. It should be between 80°F/27°C and 95°F/35°C. Incubate for a total of about 40 hours.

Recipe continues on next page

203

KOJI

Barley Koji, *continued*

9. After 24 to 36 hours, the surface will appear to have the beginning blush of a soft white covering, and the aroma will be sweet, mushroomy, and beguiling. The koji will begin to produce its own heat. Monitor it closely at this point, since barley is more susceptible to rising temperatures. Check the temperature again to make sure it is in the correct range. Stir it to break up the lumps in order to dissipate some of the heat. After stirring, make furrows in the substrate before covering the dish and returning it to incubation. Continue to monitor the temperature and moisture, adjusting as needed and stirring the koji as needed.

10. Pull the koji from the incubator when it is finished. The grains will look as if they've been dusted with confectioners' sugar, so it will appear much less fuzzy than rice koji.

Variation: Millet Koji

Millet-based misos are wonderful, with sweetness and a fun texture. Millet is natural choice, given its long history with koji. It makes a cheery amazake as well. We also really like using millet koji with heirloom common beans (see our recipe using Good Mother Stallard beans on page 273).

For most alternative grains, we tend to default to a light rice koji starter; however, with millet and sorghum, we have also gotten good results using a barley koji starter.

To make millet koji, soak the millet for 6 to 8 hours, then steam until al dente, 30 to 35 minutes (don't boil, as it will quickly become mush). Alternatively, you can give the millet a quick toast before soaking to bring out some nutty flavors. Follow the same process as for barley koji (page 203).

Experimenting with Substrates

When you are experimenting with making koji using new substrates or procedures, we recommend working with small batches of around 1 pound (450 to 500 grams) dry weight of grain. For this amount of grain, use 1 teaspoon (1 gram) of koji starter, or the amount recommended by the manufacturer of the spores you will use, dispersed in about 2 tablespoons (20 grams) of flour.

While you need a certain amount of mass for successful growth, if you are fermenting in casserole dishes, your amounts can be small. Each substrate reacts differently to koji, so working with small amounts as you are learning will help minimize the amount you might need to send to the compost pile. And each tane koji starter will treat the substrate differently. It is fun to see how each tane koji reacts to the same substrate under the same conditions.

Many times we have fermented a series of microbatches (about 1 cup/175 grams dry weight) of each variety, all at the same time. We line them up on a tray so that they are distinguishable and separate but can keep each other warm. At first they look like a rainbow of diverse grains, then the koji grows and they all look quite similar on the outside. When we do this, we make a larger batch of dispersed spores (1 to 2 teaspoons/1 to 2 grams of spores for ¼ to ½ cup/35 to 70 grams of flour). We use a teaspoon or two of this koji-flour mixture for each test batch.

We used light rice koji starter for these grains, unless otherwise indicated. Our experiment notes:

Buckwheat. We steamed the grains, which takes only 5 to 10 minutes. The flavor is sweet, reminiscent of toasted buckwheat with maple syrup drizzled on top.

Emmer. Like wheat or brown rice, the bran of emmer must be roughed up to make sure the koji can get in and do its job. Usually soaking and steaming are enough to break it up. A few pulses in a food processor after soaking and before steaming will make a wetter but significantly sweeter mat. Not surprisingly, emmer is similar to wheat; we made amino sauces with it.

Popcorn. This is a fun one to play with. Instead of cooking the corn, you pop it. The starch in the popped corn is right there — it's the turnkey for the koji to move in and do its magic. We recently discovered popped sorghum and suspect it, too, would respond well.

Quinoa. This makes a fluffy koji, and the flavor is decidedly quinoa but very sweet. The quinoa breaks down more quickly than rice. Keep tasting it after about 24 hours and pull it when you like it. As it incubates, the grains begin to liquefy inside. We've found that quinoa koji makes fun tasty pastes.

Rye berries. We soaked the berries for 24 hours, steamed them, and then inoculated them with barley koji. It worked quite well and has a unique flavor.

Sorghum. We made a number of batches of sorghum koji and found that steamed al dente sorghum was still too tough, and it dried out before the koji was able to create enough enzymes. Instead, soak the sorghum then boil in a pot with three times as much water as sorghum to break it apart and make it moist enough for the koji to access the sugars.

SORGHUM KOJI

Soybean/All-Bean Koji

YIELD: ABOUT 3 POUNDS

Be sure to purchase soybean koji starter (the Japanese name is *mame koji*, "bean koji"). It has the most protease production, which is needed for the soybeans. GEM Cultures has an approachable-size starter kit with enough for two 3-pound batches. You can use your soybean koji to make hatcho-style miso (page 270) or Wheat-Free Tamari (page 297). This is also the base recipe for any legume koji.

PROCESS **koji (page 196)**	INCUBATION **40–50 hours**

4 cups (700 g) soybeans or other legumes

2 tablespoons (20 g) all-purpose or rice flour

2 teaspoons (2 g) soybean koji starter

Note: Confirm the quantity of koji starter against what is recommended by the manufacturer of your starter.

1. Soak the soybeans overnight, then drain and rinse.

2. Steam the soybeans, following the directions on page 54. They are done when you can easily crush them between your thumb and ring finger.

3. While the soybeans are cooking, toast the flour: Place the flour in a small skillet and toast over medium heat until lightly browned, about 5 minutes. Remove from the heat and let cool.

4. Place the koji spores in a small bowl. When the flour is cool, add it to the koji spores. Stir carefully — the spores are very light and will want to float away.

5. When the soybeans are cooked, drain them and let them cool to 110°F/43°C.

6. Place the beans in a large bowl. Sprinkle about one-third of the koji-flour mixture over the beans and stir gently. Repeat until all of the mixture has been added and it covers the soybeans evenly.

7. Transfer the mixture to a casserole dish, taking care to make sure it is spread evenly and not more than 1½ inches thick. If it is thick, be sure to run a furrow or two through the surface to increase the surface area. Cover with plastic wrap and poke a few very small holes in diagonally opposite corners for ventilation. It is important to keep the moisture of the beans in.

8. Set your incubation chamber to 85°F/30°C. If you are using a fermentation terrarium or another water bath setup, set the water to 88°F/31°C. Place your dish in the incubator. Check the internal temperature of the koji every few hours during the day. It should be between 80°F/27°C and 90°F/32°C. (Soybeans are a little more susceptible to bacillus, so the temperature can't go quite as high as it does with rice koji.)

9. After 24 hours, you should smell a slight mushroomy aroma (the beans don't have the same sweet smell as grains). The surface will have a white fuzz. Stir to break it up and allow oxygenation, then replace the cover and continue to incubate for a total of 40 to 50 hours. The koji is finished when it has a full coverage of cottonlike mold. It is okay if there is a bit of sporulation, indicated by some yellow-green colored patches. Significant amount of sporulation will change the flavor, which you will see in the recipe for douchi.

Black Soybean Spicy Douchi

YIELD: ABOUT 3 CUPS

If you have ever been to a traditional Chinese restaurant, you've likely experienced the salty powerhouse of umami, pungency, and even a little sweetness that is douchi, also known as fermented black beans or Chinese salted beans. (Any vegetable or meat "in a bean sauce" is likely made with douchi.)

There are many different types of douchi, and many different methods for preparing each type, but most start with growing koji on soybeans that are yellow and sometimes black. According to Sandor Katz, it is the process that turns the beans black. The interesting thing about this ferment is that the beans are fermented to full sporulation (see Be Spore Aware on page 193 for precautions you may want to take). You will let them turn entirely green, then dry them out completely before washing off the koji. Use this pungent bean ferment to flavor sauces or vegetables; it lets you add a bit of flavor without having to resort to salt.

PROCESS **koji (page 196)**	FERMENTATION **50–60 hours, aging 5 months**
2½ cups (450 g) black soybeans	**1.** Soak the soybeans overnight, then drain and rinse.
¼ cup (35 g) all-purpose or rice flour	
1 teaspoon (1 g) soybean koji starter	**2.** Steam the soybeans, following the directions on page 54. They are done when you can easily crush them between your thumb and ring finger.
½ cup (135 g) salt, plus extra for dusting the jar and topping the ferment	
AFTER FERMENTATION	**3.** While the soybeans are cooking, toast the flour: Place the flour in a small skillet and toast over medium heat until lightly browned, about 5 minutes. Remove from the heat and let cool.
4 Fresno or other red chiles, seeds removed if desired	
2 garlic cloves	**4.** Place the koji spores in a small bowl. When the flour is cool, add it to the koji spores. Stir carefully — the spores are very light and will want to float away.
2 teaspoons (12 g) salt	
¼ cup (50 g) sugar	
½ cup (118 mL) rice wine	**5.** When the soybeans are cooked, drain them and let them cool to 110°F/43°C.

Recipe continues on next page

Black Soybean Spicy Douchi, *continued*

6. Place the beans in a large bowl. Sprinkle about one-third of the koji-flour mixture over the beans and stir gently. Repeat until all of the mixture has been added and it covers the soybeans evenly.

7. Transfer the mixture to a casserole dish, taking care to make sure it is spread evenly and not more than 1½ inches thick. If it is thick, be sure to run a furrow or two through the surface to increase the surface area. Cover with plastic wrap and poke a few very small holes in diagonally opposite corners for ventilation. It is important to keep the moisture of the beans in.

8. Set your incubation chamber to 85°F/30°C. If you are using a fermentation terrarium or another water bath setup, set the water to 88°F/31°C. Place your dish in the incubator. Check the internal temperature of the koji every few hours during the day. It should be between 80°F/27°C and 90°F/32°C. (Soybeans are a little more susceptible to bacillus, so the temperature can't go quite as high as it does with rice koji.)

9. After 24 hours, you should smell a slight mushroomy aroma (the beans don't have that same sweet smell as grains). The surface will have a white fuzz. Stir to break it up and allow oxygenation, then replace the cover and continue to incubate for a total of 50 to 60 hours. The koji is finished when you see yellow-green coverage over all the beans.

10. Now it's time to dry the beans. Set your dehydrator for 125°F/57°C. Dry in the dehydrator for 10 to 12 hours. Or, if you have the hot summer sun available to you, set the beans out in the sun for several hours. Stir as needed for even drying. Dry the beans until they resemble the dry beans they started as.

11. Wash the beans until the mold is pretty much gone. (It takes more effort than you would imagine.) Take care not to break the individual beans.

12. Bring 2 to 3 cups of water to a boil, then let the water cool to 80°F/26°C. Pour enough of this warm water over the beans to cover them and let sit for 30 minutes, then drain the water.

13. Transfer the rehydrated beans to a bowl and add ½ cup salt. Mix thoroughly.

14. Prepare a widemouthed quart jar by rinsing it with boiling water and then dusting the inside of the jar with salt. Press the salted bean mixture into the jar and top with a ¼-inch layer of salt.

15. Add a weight (see page 247), then screw the lid on tightly. Set aside to ferment in a room-temperature spot (around 70°F/21°C to 77°F/25°C) for 4 months. Check the ferment after a few days; if the lid is bulging and air pockets are building between the beans, you may need to burp the jar and press out any CO_2. To do so, push on the weight until the air gaps are gone. You will generally only need to do this once, but it's a good idea to check it again in another 2 weeks or so.

16. After the 4-month fermentation period, open the jar and scoop out the salt layer on top. Transfer the beans to a large bowl. Combine the chiles, garlic, salt, and sugar in a food processor and grind into a paste. Add the fresh paste to the bowl and mix with the beans. Place the bean mixture in a clean quart jar and top with the rice wine. Add the weight and put the lid on to seal. Let ferment in a room-temperature spot for 1 month, burping the jar as needed.

BLACK SOYBEAN
SPICY DOUCHI

The Little Ferment That Started It All . . .

Thanks to some douchi ginger that was found in a grave dating back to 200 BCE in Changsha, the capital of Hunan province in central China, we know that douchi has been around for thousands of years. Chinese scholar Li Shizhen included a salt-free version called *dan douchi* in his sixteenth-century materia medica as treatment for ailments like diarrhea, irritability, restlessness, and insomnia. Douchi is likely the first fermented soybean product to move (slowly, in concert with Chinese Buddhism) to other parts of Asia, where it evolved into products like doenjang, miso, and soy sauce.

There are many different types of douchi that are specific to different regions of China. They are defined by the type of soybean (yellow or black), salt content, water content (dried to watery), and the main microbe used to ferment. There are four variations: those fermented with mucor (a fungus), those fermented with rhizo-pus (a fungus), those allowed to spontaneously ferment with *B. subtilis* (a bacteria), and those fermented with aspergillus (a fungus), which is a koji ferment, the oldest known variety, and the one we will be using as a jumping-off point for the douchi-like ferments in this book.

CHAPTER 8

AMAZAKE

AND OTHER TASTY KOJI FERMENTS

Koji is a building block for many other foods, offering enzymes that, over time, change the composition of whatever substrate we ask it to work on. In this chapter, you will have a chance to explore koji's superpowers as it is used in things like amazake, shio koji, and country sake; in the next chapter, we'll talk about using koji to make miso and other tasty pastes. We also will invite you to explore new ideas and possibilities with a few recipes that grow koji on unconventional substrates like bread and pork.

Most of the recipes in this chapter can be made with purchased precultured rice or barley koji, which means that you can dive right into this chapter and experience the enchanting power of transformation that takes place when we apply enzymes to food. Once you are hooked, you probably will want to make your own koji. Let's get started.

Meet Amazake

Amazake is a sweet, aromatically rich mash or — more commonly — beverage. Its name translates as "sweet sake." Just as sweet, fresh-pressed apple cider is the nonalcoholic first step to making hard cider, amazake is sake's virgin sister. To make amazake, you combine a cooked grain with koji and provide an environment for the amalyse and protease enzymes to do their work. Traditionally it is rice-based, which is delicious, and it is enjoyable to use the amazake technique to bring out the nutrients and interesting flavors of other grains. One of our favorite alternatives is oat amazake (see page 218). Amazake can be enjoyed warm and thick or thinned out like a rice milk. It is an excellent dairy-free addition to smoothies.

Amazake gains its sweetness from the natural process of the koji breaking down starches into their simple sugars. Let's say you make the brown rice amazake (page 217). The carbohydrates in the rice are now half simple sugars and half complex carbs. The simple sugars (mostly maltose) will take your body about 1½ hours to assimilate, while the complex carbs will take around 3 hours to digest; this means extended fuel for you. So, drinking this amazake in the midafternoon will supply the energy you need to get you through that long slump at the end of the day.

Because its complex nutrients have been predigested, or broken down into simpler forms, amazake is an especially good food for anyone with compromised digestion. If you drink unpasteurized amazake, you will also get the benefits of the live digestive enzymes that help your body break down the proteins, fats, and carbohydrates you eat.

Amazake can be used as a sweetener in cooking and baking. We think of it as being similar to brown rice syrup. The difference, besides the fact that it is not a clear syrup, is that amazake has a more complex flavor. It is deliciously well suited for use in scones, muffins, brownies, blondies, some cookies, and rich spice cakes. Treats baked with amazake will have good body,

Using Precultured Koji in the Recipes

The recipes in this chapter assume homemade, fresh, moist koji, though, as we noted, they can be made with purchased koji. Commercial precultured koji will mostly likely be dry (unless you are lucky to have a local maker who shares or sells their koji). Most of the folks we talked to suggested adding a little more liquid to the recipe when using dry koji to make up for the missing moisture, though they all agreed it wasn't a big deal. We follow Jeremy Umansky's advice and keep the koji amounts the same for wet and dry koji because we'd rather have a bit of extra koji than not enough.

If you wish to hydrate your dry koji, a good rule of thumb is to add about 10 percent of the weight of the dry koji in boiled and cooled water. Stir in the water, cover, and let it sit for a least 30 minutes to allow the moisture to penetrate the koji. When we use dried precultured koji in our amazake, we simply add an extra ½ cup of water when we cook the rice to make it wetter.

More Rice Ferments

It is no wonder that where there is rice, there are alcoholic ferments. All over Asia you will find variations on this theme. In the tempeh chapter you met *tape* (page 129). If you have an Instant Pot, you may see a setting for *jiu niang*, a Chinese sweet, glutinous rice ferment that becomes alcoholic. *Makgeolli* is a rice liquor from Korea.

While these are similar in many ways to amazake and sake, they are not started with koji. Makgeolli uses yeast and *nuruk* (amylase enzymes that parallel the koji). *Jiuqu* for jiu niang are sold in little balls in Asian markets and are combinations of aspergillus, yeast, and bacteria.

richness, and moisture, rather like they were made with buttermilk.

When cooking with amazake, a good rule of thumb is to use ¼ cup of thick amazake for each tablespoon of sweetener in the recipe. If you are using it to replace a dry sweetener like sugar, you'll also need to take out 3 tablespoons of liquid from the recipe to account for the extra liquid you're putting in with the amazake.

Amazake is a great starting place, a gateway if you will, to the wonderful world of koji ferments. We will show you how to make amazake with your own homemade koji as well as with purchased dry koji, if you want to try this ferment without the commitment and learning curve of making koji.

Making Amazake: An Overview

As you've learned, amazake starts with koji. You can use koji grown on any grain or you can use rice koji, purchased or homemade, to start any grain, as you will see in Koji Polenta (page 221).

When working with enzymes, think of Goldilocks and the Three Bears. You are looking for the temperature to be just right: too

low and the enzymes aren't active enough and opportunistic spoilers can move in; too high and the enzymes deactivate. The sweet spot for the enzymes to do their work is 138°F/59°C. When we make amazake, we set our dehydrator to 135°F/57°C. This gives us some wiggle room for fluctuations, because as soon as the temperature hits 140°F/60°F, it deactivates the enzymes.

If you have set up an incubator for other ferments and can reliably set your temperature to that range between 135°F/57°C and 138°F/59°C, use that. If you are looking to make amazake but have not yet committed to an incubation setup, here are a few simple alternatives.

Double-boiler curing method. This is the most labor-intensive method for making amazake, but it is the simplest in terms of setup and equipment. The first time you try it, do it in the morning so that you will have time to supervise the temperature. Once you have established a good process, you should be able to leave it overnight and awaken to fresh, warm amazake mash for breakfast.

Prepare a water bath in a large pot by filling it halfway with water and warming it to a temperature between 135°F/57°C and 138°F/59°C. Set a smaller lidded stainless-steel pot inside the larger pot, so that it sits in the water. Using a thermometer, monitor the water temperature, adjusting the heat or adding heat diffusers under the large pot as needed to keep it in the specified range. Once you have stabilized the heat, you can put the grains and koji into the smaller pot.

Now move the thermometer into the amazake mixture itself to monitor the temperature closely (this would be the labor-intensive part). Remember that it's critical that the temperature not exceed 140°F/60°C, or else the enzymes will deactivate. Once you feel confident that your setup is keeping the amazake's temperature stable, your workload lightens considerably.

The oven. Some ovens have a setting that allows for a long, slow cook. If your oven does, try it. Just be sure to keep a thermometer in the amazake mixture and, as with the double-boiler method, check the temperature frequently to make sure it is staying in the right range. Once

you have dialed in the settings and can trust your setup, you are free to forget about it for a few hours.

Electric cooker. Some of the new electric slow cookers can be programmed to specific temperatures. If you have a slow cooker that you can set to that magic temperature zone, then this is a very convenient method. We have made amazake in an electric pressure pot with the normal yogurt setting, which seems to range around 105°F/41°C to 108°F/42°C (with a jar in a water bath it seems to max out at 110°F/43°C). This isn't the ideal range, but it was sufficient to get an okay amazake after 10 hours. (The fermented rice [*jiu niang*] setting on an Instant Pot is not well suited to amazake, as it too low.)

Sous vide method. If you happen to have a sous vide system, place the koji and cooked rice in a sealed plastic bag and immerse it, with the immersion circulator set to 138°F/59°F.

SCIENCE SAYS . . .

A Deeper Look at Temperature

You might have noticed that amazake is incubated at 138°F/59°C, but koji is incubated between 80°F/27°C and 95°F/ 35°C. This begs the question: how does the koji fungus survive the higher temperature during amazake preparation?

It doesn't. There are two processes at work, with different functions. The first is the growth of the koji mycelium on the cooked rice; this growth produces the enzymes. After you grow the koji, you are basically harvesting the enzymes to work at breaking down the next substrate, whether it is beans in miso, more grains in amazake or

sake, or meats to tenderize. The fungus (spores and mycelium) can't survive above around 113°F/45°C, but the enzymes can persist up to 140°F/60°C.

If amazake is fermented at lower temperatures, say 90°F/32°C to 100°F/38°C, the enzymes will work much more slowly. In this situation, lactic acid bacteria will often move in and begin the souring process before the amazake has had a chance to get sweet enough. If you know you are going to ferment at a lower temperature, use a ratio of 1:1 or 2:1 of koji to cooked grain.

STEP-BY-STEP

Let's Make Amazake!

What You Need

- Rice cooker or other means to cook rice or other grain
- Bowl and colander for soaking rice or other grain
- Rice or other grain koji

- Metal spoon
- Mason jar or other large jar with a lid
- Incubation chamber (see options in this chapter, page 213)

1. If you are making amazake with a grain that requires soaking, like brown rice, soak it overnight.

2. Drain the rice in a colander.

3. Cook the rice as you normally would on the stovetop or in a rice cooker.

4. Place the rice in a 2-quart mason jar and let cool to 135°F/57°C.

Instructions continue on next page

5. Add the rice koji and stir to incorporate.

6. Screw on the lid.

7. Place the jar in your incubator (we use a dehydrator set to 135°F/57°C). Let incubate for 6 to 8 hours. If convenient, stir after the first hour and once or twice again during the incubation. (Don't get out of bed for this.)

8. When finished, the amazake will be liquidy and have a sweet aroma and taste. Eat as is or blend with water for a sweet beverage.

Rice Amazake

This recipe is for a very basic amazake that has a fairly thick porridge consistency. It can be eaten as is, or you can blend it with water to make about 4 quarts of amazake beverage (see page 219). When it comes out of the incubator, it is warm and sweet. When topped with a little butter and some dried fruit, you have a fermented rice porridge. This recipe calls for 1½ cups koji rice. If you have more koji, you can play around with the ratios. We have found that a higher ratio of koji to fresh rice will cause the amazake to liquefy more and become sweeter. It will also ferment more quickly; be sure to catch it before it starts to sour. Of course, if that happens not all is lost, as you'll see when you read about sour amazake (page 218).

You can use any rice you like or have on hand, even leftovers if you reheat them. You may also enjoy the oat variation that follows.

Note: Don't use distilled water, as the fermentation needs the trace amounts of calcium. For those of you with hard water, look at that — it's good for something after all.

PROCESS **koji (page 196)**	FERMENTATION **6–12 hours**

1½ cups (285 g) brown, sweet brown, or white rice (to make 4 cups cooked rice)

1½ cups (260 g) fresh light rice koji or dried koji rice grains

1. If you are using brown rice, soak it 8 hours or overnight. If using white rice, soaking is optional.

2. Drain the rice, if you soaked it. Cook the rice in a rice cooker or on the stovetop, following the instructions on page 67, then let it cool to 135°F/57°C.

3. Add the koji rice to the cooked rice and mix well. Pour the mixture into a 2-quart mason jar and tighten the lid.

4. Set the jar in your incubation chamber at a temperature between 135°F/57°C and 138°F/59°C. We set our dehydrator to 135°F/57°C. Incubate the mixture for 6 to 10 hours, or until it has a floral aroma and a mild, sweet taste. (The cooler the incubation, the longer it will take.

Also, fresh koji will take less time than dry koji.) Stir the mixture after the first hour and again once or twice during incubation. If you want it sweeter and a little more liquidy, let it ferment a little longer. It will continue to sweeten, but only up to a point. When it hits its limit, the flavor will start to turn slightly sour and have some bitter or alcoholic notes.

5. When the amazake is finished, enjoy some warm and put the rest in the refrigerator, where the enzymes will get cold enough to halt the process. If you prefer a smooth texture, blend the amazake in a blender or food processor. It will keep in the refrigerator for 3 to 4 months, or for 6 months in the freezer.

Note: Many recipes recommend boiling the amazake, essentially pasteurizing it, when it is finished. But when you do this, you are deactivating all of the enzymes (its superpower). We prefer not to boil it.

Variation: Oat Amazake

Waking up to a thick, warm, freshly fermented oat amazake porridge is a darn good way to start the day. Fresh oat amazake smells like an oatmeal cookie just out of the oven — it's pretty intoxicating. And, if you skipped reading about the benefits of oats on page 62, you might take a look at all the reasons why this is a powerful breakfast.

If you're not an oatmeal kind of person, don't look away yet. This amazake can be used in scones (page 331) or in granola (page 330) for an incredible power breakfast or hiking provisions.

You'll need about 2½ cups rolled oats or 3 cups oat groats (enough to make 4½ cups cooked oats) and 1½ cups fresh oat koji (page 202) or dried koji rice grains. If you're using whole oat groats, soak them overnight. Cook the oats in a rice cooker or on the stovetop as you would rice, for 20 to 30 minutes, then follow the instructions for making rice amazake (page 217).

Sour Amazake

Amazake will naturally become sour if it has fermented too long or been left out of the fridge. Bacteria and yeasts are all waiting to eat that free sugar and provide their own conversion. Sour amazake is a nice flavor enhancer and works well for pickling or marinating vegetables. If you want to sour your amazake on purpose, you can place a bit of your sweet amazake in a jar, add 1 part water to 2 parts koji, put the lid on but don't tighten it, and leave at room temperature for a week.

Warm Amazake Beverage

Amazake has been consumed as a drink for centuries. Like hot milky tea, a frothy latte, rich broth, or hot drinking chocolate, this sipping beverage is about comfort and inner warmth — a sweet sip to get you through the afternoon or to enjoy with a friend. At its simplest, mix one part fresh amazake with equal parts piping hot water, then finely grate a small amount of fresh ginger on top. We have found so many delicious ways to enjoy this drink, especially for breakfast; many of our recipes feel like nourishing warm smoothies. Head over to page 325 to check out some of the possibilities.

BROWN RICE
SWEET AMAZAKE

OAT AMAZAKE, page 218

KOJI POLENTA, opposite

BROWN RICE SHIO KOJI, page 223

BREAD AMINOS, filtered, page 227

Koji Polenta

This recipe was inspired by a recipe from Betty Stechmeyer (see page 16) for koji polenta, a dish she makes with leftover polenta. Betty described it as having a sweet beehive flavor, which was spot-on. The corn is crazy sweet (guess that is why corn syrup is so sweet; there are tons of simple sugars in that starch) with a hint of honeycomb. We took her recipe one step further and used masa harina (corn that has been nixtamalized) and came out with an enticing creamy breakfast porridge.

As a side note, Betty also shared with us that she adds koji to leftover oatmeal or mixed-grain cereals to make a sweet drink that can be added to a smoothie. (Good idea, right? Don't forget, koji is a fun way to use plain grain leftovers.) Her basic measurement is to add ¾ cup koji to every 2 cups leftover porridge. Now there's a way to use up that lumpy leftover oatmeal that sits around in the fridge!

FERMENTATION **3–4 hours**

---

3 cups (705 mL) water

1 cup (120 g) masa harina

¾ cup (118 g) rice koji, or other comparable grain koji

---

1. Bring the water to a boil in a small saucepan over high heat. Turn the heat down to medium-low and let the water simmer.

2. Slowly add the masa harina, a little at a time, whisking constantly. Keep whisking until all lumps have been worked out. Carefully bring the pot back to a simmer; the thick texture can cause the bubbles to "erupt." Stir frequently until you have a thick porridge, about 5 minutes. Remove from the heat.

3. Allow the porridge to cool to about 130°F/ 54°C, then add the koji, stirring until well mixed. Add a little water if necessary; the porridge may thicken quite a bit while cooling.

4. Transfer the mixture to a clean quart jar and seal with the lid. Set the jar in your incubation chamber at 135°F/57°C and let incubate for 3 to 5 hours, depending on your sweetness prefer-ence. It will be quite sweet after 3 hours but will become sweeter as it continues to incubate.

5. Eat immediately, or refrigerate and warm when ready to eat. This will keep for 2 days in the refrigerator.

AMINO SAUCES

Amino sauces build umami by breaking down proteins to amino acids like glutamate. The protein can come from anything, from mashed beans to day-old bread to egg whites. There are two basic ways to achieve an amino sauce. One is to grow the koji on rice, barley, or other grain and add the resulting koji to the protein source. The other method is to grow the koji directly on the protein source.

Salt is an essential ingredient in amino sauces, but the amount varies depending on your goal. Chef Jeremy Umansky advocates a lower salt range (3 to 5 percent) because you get more flavor in the sauce thanks to the extra microbes — bacteria and yeast — that can join the party. These low-salt amino sauces have powerful flavors with decidedly more funk, but are a little more unpredictable and don't keep for as long. We've found that a shio koji that is 3 percent salt needs to be refrigerated early on to maintain flavor, whereas at 5 to 8 percent salt it will keep well at room temperature. At Emmer & Rye in Austin, Texas, they like the clean, crisp finish of 10 percent salt solution in their amino sauces.

The higher the salt ratio, the more shelf stable the ferment is. At 10 percent salt and higher, you are killing all but the hardiest bacteria. Some people pasteurize their amino sauces for more stability.

What Is Mirin? And a Mirin Hack . . .

Authentic mirin is a sweet rice wine that begins with rice koji. The first mirin was made by combining sake with sweet glutinous rice. This drink had almost no shelf life, as it was high in sugar and susceptible to spoilage by all the yeasts in the combination. At some point, brewers started distilling this beverage, which resulted in *shochu*, an 80-proof alcohol. Brewers continued to experiment through the centuries, adding sweet glutinous rice and enzymes back into the shochu and letting it age for long periods. Mirin became an exclusive beverage. Later still, it was adopted as an ingredient in cooking, as it has superior flavoring qualities that add subtle sweetness and tang. Today it is used primarily as a cooking wine.

True mirin has a number of health benefits, partly due to all the transformation that happens in a multifaceted double-fermentation process that takes place over the course of a full year. It is hard to find outside of Japan. The traditional mirin industry took a hit during the rice shortages of World War II, leaving very few makers. Most of the mirin you will find in the supermarket is a blend of syrups, glucose, ethyl alcohol, amino acids, and salt — a far cry from rice koji, glutinous sweet rice, and water.

True mirin requires a lot of time and many processes, including distillation. But Jeremy Umansky has a hack. To make a fermented mirinlike sweet, syrupy ingredient, grow koji on rice that is slightly wetter than al dente. As the enzymes break down the rice, the insides of the rice will liquefy and ooze a yellow liquid (sometimes you have to pinch the rice to see it). Once the rice koji has this quality, wring it out in a piece of muslin to release as much of the "mirin" as you can.

Shio Koji

YIELD: 1 PINT

If you make only one amino sauce, this is the one. The more we tried using it, the more it became part of our routine. Now there is always a jar of shio koji next to the stove, right next to the salt jar. It is easy to make, lasts for months, and is simple to incorporate into your cooking.

Shio means "salt," so shio koji is humble salted koji, but the flavor it imparts to dishes is far from humble. It is outrageously delicious and tenderizing when used as a marinade for both vegetables and meats. It brings out the natural sweetness and umami in foods and delivers that special something-something that you cannot name when used as a flavor-enhancing sauce — it's literally the secret sauce.

The salt halts the aspergillus, but the enzymes remain, just like in miso or shoyu (see chapter 9). This is an open ferment, and if it is not overly salty, the environment is also hospitable to lactic acid bacteria as well as some salt-tolerant yeasts. We generally make our shio koji with 3 to 6 percent salt by weight. This recipe is for the 6 percent shio koji, which lasts longer than the 3 percent and has a sharper, cleaner flavor.

Feel free to make this sauce with any type of koji, such as barley, oat, or millet. Play with the flavors the different grains impart.

FERMENTATION **7–14 days**

1½ cups (250 g) rice koji

3 tablespoons (50 g) sea salt

2–3 cups (470–705 mL) water, boiled and cooled, to fill the quart jar

---

1. Combine the koji, salt, and water in a bowl. Break up the koji as needed. Mix well.

2. Place the mixture in a quart jar. Cover with cheesecloth and secure the cloth with a rubber band. Let sit at room temperature for 7 to 14 days. (The speed of fermentation will vary based on the temperature.) Stir every day.

3. Use the shio koji as is, or if you prefer a smooth consistency, you can press it through a sieve with a wooden spoon or blend in a blender. Store at room temperature; it will keep for 6 months to indefinitely.

Variation: Shoyu Koji

Same idea, new flavor. Think soy sauce with a kick. Replace the salt and water in the shio koji recipe with soy sauce, and use 3 parts koji to 2 parts soy sauce. Follow the same fermentation process. The consistency will be thicker than that of shio koji; shoyu koji is more of a paste.

Using Shio Koji

- Use shio koji anywhere you would use salt. A general conversion is to use double the amount of what you might use with salt — for example, if you are salting a pot of cooked beans and would normally add 1 teaspoon salt, add 2 teaspoons shio koji. (For folks concerned about salt intake, shio koji can actually reduce your sodium levels because along with its saltiness, it brings enzymatic action and flavor; with that punch of flavor and umami, you'll end up using less salt overall.)

- Infuse flavor into tempeh with a 20-minute soak in shio koji, using it like you would a marinade (see pages 335 and 336 to see how it is used to flavor tempeh bacon and sausage).

- Mix a little bit of shio koji with butter and brush it on bread or rolls, either before baking them or when they are fresh from the oven.

- Make quick pickled vegetable salads and snacks. See Koji Pickles on page 316.

- Replace the salt in dressings and dips with shio koji.

- Try a little shio koji in salsa or guacamole. It's delicious, and as a helpful side effect, the shio koji will help keep the avocado from browning while it's out on the snack tray.

- Make creamy, umami-rich tofu by marinating it in shio koji: Press the tofu block under weight as you would for Tofu Misozuke (see page 319), then put it in a ziplock bag with 1/3 to 1/2 cup shio koji and marinate it in the refrigerator for 1 week. Make sure the entire surface of the tofu block is in contact with the shio koji.

- Replace the salt in soups and stews with shio koji. Not only will it improve the flavor, but you will get a slightly richer texture.

- Coat meats such as salmon, chicken, or pork in shio koji and marinate overnight in the refrigerator. Go protease! Yum!

SHIO KOJI,
strained, page 223

BREAD AMINOS,
whole

BREAD AMINOS,
blended

BREAD AMINOS,
strained

BROWN RICE
SHIO KOJI, page 223

SHOYU,
page 298

Bread Aminos

YIELD: 1 QUART

This recipe was inspired by Emmer & Rye restaurant in Austin, Texas, where they make amazing amino sauces. It yields an amino sauce that is 5 to 8 percent salt. We encourage you to play around with the technique.

This recipe calls for 24 ounces of bread. We freeze all the ends and pieces that get too stale for toast until we have enough for this recipe.

KIRSTEN WRITES: A few years ago, I visited Emmer & Rye in Austin, Texas, at about the time that Jason White (now part of the research team at the Noma Fermentation Lab in the acclaimed Danish restaurant Noma) was setting up the restaurant's foraging and fermentation larder. Chef Kevin Fink brought out amazing amino sauces that they had been working on. These sauces were born of necessity as a way to use the restaurant's leftover egg whites (a by-product of fresh pasta making) and day-old bread (a by-product of fresh in-house baked bread). Jason started with the process for making soy sauce, but instead of growing koji on wheat and soy and fermenting in a brine solution, he grew the koji on wheat, then placed the egg whites in a salt brine solution and added them to the koji. The koji broke down the proteins in the egg whites into amino acids to make that rich umami. The resulting sauce had a sweet, nutty aroma.

A few years later, I had an opportunity to taste these sauces again and hear chef Page Pressley describe the process of making them as I scrawled notes on the only paper I had in my purse — an envelope. He added rice koji to lightly charred bread pieces, along with a 5 to 20 percent salt brine solution.

In our own experiments, we found that when we grew the koji directly on stale bread, it had much more umph and flavor than when we added inoculated koji to the bread and brine mixture.

PROCESS koji (page 196)	FERMENTATION 40–54 hours, plus at least 6 months for aging

24 ounces (680 g) bread

½ cup (70 g) all-purpose or rice flour

1 teaspoon (1 g) koji starter

2 quarts (1.9 L) water

Salt: ⅓ cup plus 2 teaspoons (100 g) for a 5% solution, or ½ cup plus 1 tablespoon (165 g) for an 8% solution

1. Break the bread into roughly ½-inch cubes and place in a bowl.

2. Place the flour in a small skillet and toast over medium heat until lightly browned, about 5 minutes. Remove from the heat and let cool, then combine with the koji spores in a small bowl.

3. Mix the koji-flour mixture into the bread cubes, making sure that all the bread has a dusting of the mixture. Stir carefully — the spores are very light and will want to float away.

Recipe continues on next page

Bread Aminos, *continued*

4. Transfer the inoculated bread to a casserole dish that has a lid. Leave the lid off, but cover the dish with plastic wrap. Make a few small holes on diagonally opposite corners for ventilation.

5. Set your incubation chamber to 85°F/30°C. If you are using a fermentation terrarium or another water bath setup, set the water to 88°F/31°C. Place your dish in the incubator. Check the internal temperature of the koji every few hours during the day. It should be between 80°F/27°C and 95°F/35°C.

6. After 20 to 24 hours, you will begin to notice the telltale sweet smell and the beginning growth. Stir the bread pieces so that they all receive some fresh oxygen, then replace the plastic wrap and continue to incubate until you see nice koji growth, another 20 to 30 hours.

7. When the koji is done, place in a 2-quart jar. Combine the water and salt in a large saucepan and bring to a boil over high heat. Remove from the heat and allow to cool to 130°F/54°C. Pour the brine into the bread koji and mix thoroughly.

8. Seal the jar with its lid. Let sit at the warm end of room temperature (80°F/27°C to 85°F/29°C) for 24 hours. Stir a few times as convenient during this time.

9. After the first 24 hours, let the mixture sit at room temperature for a week, but now regular room temperature is fine. Open the jar to release gases and stir thoroughly at least once a day. After a week, the ferment should have settled down. If not, continue burping and stirring until it does. When it is no longer extremely active, seal the container and let it age at room temperature for at least 6 months.

10. When you are ready to use the bread aminos, strain out the bread solids, or blend in for a thicker sauce. Use the liquid for seasoning. Because it has a lower salt ratio, store this amino sauce in the refrigerator, where it will keep indefinitely.

SIMPLE UNREFINED SAKE

Whatever your view on alcohol consumption, you will have to admit that it has been a part of the human story for a very long time. Sake is an alcoholic beverage made from rice. While its origins are unclear, the earliest reference to it can be found in the Book of Wei, a third-century Chinese text that mentions the Japanese and their "drinking and dancing."

Making sake is a two-part process. In the simplest form, koji first converts steamed rice into sugar (making amazake), and then wild or introduced yeast converts the sugar into carbon dioxide and alcohol. It's likely that sake was first made accidentally when a pile of boiled rice was inoculated with wild koji spores and then left to sit long enough to produce the liquid we know as sake. Someone was brave enough to taste this unforeseen liquid and was rewarded with a kick that they must have liked, because the process was quickly adopted and documented. It's likely that another unforeseen adaptation happened when koji from this sake process landed in a pile of freshly boiled soybeans. The koji did what it does best, and the result was a much more pleasant version of soybeans — miso. (For more on that topic, see chapter 9.)

CHRISTOPHER WRITES: *The first time I tasted sake, I was in the private dining room of a restaurant on the fiftieth floor overlooking Tokyo at night. I was a very young engineer with Hewlett-Packard on my first trip to Japan, and my hosts and colleagues were at least twice my age. I had gotten the sense that they weren't sure why someone so young had been sent to do the training I was charged with, but when they invited me to dinner on my last night and I told them I was vegetarian, I think that sealed their opinion of me.*

From the miles we walked through the downtown looking for an appropriate vegetarian restaurant, I got the feeling that mine wasn't a popular choice in Tokyo in the early 1990s. The restaurant we did find, however, was excellent. Over the next 2 hours we were served the most amazingly beautiful and flavorful food, and it changed my world forever. I am not sure my colleagues were that impressed, however, so they took to one-upping each other when it came to ordering the next bottle of sake. Bored men with a company credit card is always a bad thing, and I became the amusement for the evening as I was plied with endless small cups of what I suspect was some pretty great sake. That night, I was told, sometime in the early hours, a strong earthquake hit the area and set off alarms. I slept through it all, thanks no doubt to my new friend, sake.

Meet the Maker
Chef Ken Fornataro

While many people in this country are just now discovering koji, Ken has been using it for over 40 years to preserve and transform food and flavor. Early in his career as a chef, he befriended whole-foods advocates, immigrants from eastern Europe and Japan, and macrobiotic practitioners including Aveline and Michio Kushi, William Shurtleff and Akiko Aoyagi, and many others who shared traditional preservation, fermentation, and cooking techniques with him. This set the course for his journey of acquiring profound knowledge in both the cooking and science of these foods. He is now the founder and executive chef of culturesgroup (culturesgroup.net). The organization focuses on how cultures and societies are related to food and how they function in relation to their resources and environment, and the ever-increasing scientific knowledge base that makes it all possible. Here is a conversation we had with Ken.

Q: People are defining koji in a number of ways these days. To some it is a technique, to others it is an ingredient. How do you define koji?

A: Koji is a hardworking, versatile organism that, when exposed to the right temperature and humidity, becomes an effective tool for accomplishing hundreds of different things — from degrading the bonds in complex carbohydrates such as whole grains and beans to creating sweet nutrient-enhanced foods to using acidic secretions during its second metabolism to accomplish substrate-specific feats like soy sauce, vinegar, animal feed, and even fertilizer.

Koji can modify itself based on its environment, enabling it to grow using the energy it takes from the substrate to create different enzymes, which break down proteins and things into absorbable, enhanced nutrients and tastes. Koji is a facilitator of transformation. It's a catalyst, enzyme, and substrate.

Koji can be used like salt or sugar. It can break down proteins, fibers, and cellulose. Or it can transform other things in its environment — like bacteria and yeasts — to create tastes and textures.

Koji is the tour guide down a continuous pathway, a journey of interactions between many elements, coexisting, collaborating, competing, and leaving powerful and often very tasty metabolites and enzymes in its wake. It's a living thing that performs brilliant feats in exchange for consideration and accommodation, like any other living thing.

Q: What is your advice for the novice maker?

A: Create the environment koji needs to survive and thrive. Do a trial run and monitor the temperature and humidity over a 3-day period. Remember the basic principle of heat retention and creation. Like a good farmer or parent, you will learn through experience, and through the skills and observations of those who have come before you.

Be vigilant in maintaining hygiene, temperature, oxygen, and density by piling it up or spreading it out. The more things you make with koji, the more skills you gain. Ultimately your ability to regulate temperature, microbial diversity, and koji-friendly communities will determine your success. After 40 years, I make everything from miso to soy sauce to grilled meats and fish differently based on what I learned from using koji the day before. It's hard to imagine making great

foods or beverages without it, and the health benefits are substantial.

Q: **Christopher has been an amateur cider maker for many years, and so he thought that adding doburoku (see page 232) to his repertoire would be fairly easy. He was mistaken. Would you share your thoughts and advice for someone who has a wine, cider, or beer brewing background and would like to attempt doburoku?**

A: It's a starch breakdown first, not a typical fruit ferment. The goal when making beverages is to create enzymes, primarily amylases, that allow the yeast to get at the sugars from a wide variety of starches. Doburoku is much closer to beer making. Then yeast is added to ferment, which is a little closer to how cider is made. A higher temperature is crucial, however, and it's a good idea to add yeast and typically lactic acid after the enzymatic activity has a good head start and before dramatically lowering the temperature. Otherwise, the old-school approach is to first make bacteria create lactic acid, then use the yeast and the favorable environment you've created to make alcohol.

Q: **It seems like sake is booming in many countries, yet unlike, say, wine exports from France, Japan exports only 4 percent of its production. What are your thoughts on feeding the growing demand for sake in the United States in terms of domestic production? Do you think we are poised for a craft sake industry?**

A: Every year Japan sends more sake to the United States than before. There is no shortage of Japanese-made sake, but perhaps a deficit of interest and understanding. What there is, as with every food item or beverage, is fierce and sometimes very ugly competition.

Artisanal sake is definitely not a really huge growth area in Japan, unless you consider the use of new machines for every step as artisan sake making. It should be.

More sake makers will pop up in the United States. Every couple of years we'll see a new kura — unless inns and small farms realize how easy and profitable sake making can be.

Q: **We would like to share your fermentation journey in our book. In a couple of paragraphs, would you tell us more about the role that food, flavor, and exploration of new tastes has played in your life?**

A: Food is survival. Health is defined by metabolic capacity. Fermentation helps preserve food for times when survival may be harder because access to seasonal local food can be more difficult. By using the techniques of preservation, including agents such as koji, whole countries can survive during lean times because koji also helps turn otherwise inedible or hard-to-digest things into food or drink.

Often, preserved and fermented foods don't depend on electricity or refrigeration. Most of the techniques of fermentation have been driven by hunger and necessity over thousands of years. Chefs are mandated to create new ways to expand edible food supplies, utilize sustainable domesticated and wild resources, and create new foods from what was once considered inedible. Knowing how to use yeasts, bacteria, and fungi to create or preserve food and to keep people healthy is a survival skill and a cooking skill — a requisite one for any chef — but also a way to season, enhance, or manipulate food's texture and taste for pleasure.

231

Doburoku
(Country Sake)

By chef Ken Fornataro, culturesgroup.net

YIELD: 1 GALLON

KEN WRITES: Brown rice has been used for centuries to make sake. People have come up with all different ways of milling the fats and bran off the rice to get to the pure starch center (*shinpaku*) that you need to make sake. These days, the mechanization or semiautomation of all aspects of sake making has created a nostalgia for how sake used to be made.

Enter doburoku, sake's country cousin. The techniques used to make it are typically cheaper and faster than brewing sake and don't sacrifice the integrity or quality of the result. Doburoku is done when it's the sweetness you like. This recipe is written so you can cadge some amazake up front before going all doburoku.

There are basically five ingredients in doburoku and sake: water, rice, yeast, *Aspergillus oryzae* (koji), and some type of acid. The quality and processing of these ingredients are what really matters. When and how each item is added, and at what temperature, makes the difference between a spectacular *junmai daiginjo nihonshu* — really premium sake — and a cloudy, less refined but tasty doburoku.

Typically, doburoku is sieved or mashed to a creamy white beverage. When chilled, the sake will separate out and rise to the top. Use the sediment, or lees (*kasu*), for other purposes, from pickling to marinating food. As with all brewing, be sure to sanitize your equipment and area before starting. See my tips on page 234.

PROCESS **koji (page 196)**	FERMENTATION **48 to to 96 hours**
3¾ cups (550 g) rice koji 6⅔ cups (1,600 mL) water 4 cups (500–750 g) uncooked sweet rice or sake rice ½ teaspoon (1 g) sake yeast (see Ken's tips, page 234) ¼ teaspoon (1 g) lactic acid for brewing, or other acid	**1.** If you are using chilled or frozen koji, let it come to room temperature (68°F/20°C to 72°F/22°C). Heat the water to 140°F/60°C in a pot and keep it warm by covering it and perhaps laying a towel over the pot. **2.** Rinse the rice very well. Soak nonglutinous short-grain rice (sweet rice) for about 2 hours, then drain and steam it (page 67). If you are using a rice cooker or pot, cook the rice in a ratio of 1:1.25 rice to water. If you don't soak the rice, increase the ratio to 1:1.5. Let the rice cool to 120°F/49°C to 140°F/60°C.

3. While the rice cooks and cools, place 3⅓ cups (800 ml) of the 140°F/60°C water in a crock or gallon jar. Add the koji and mix well. Let sit, covered loosely with a clean cloth, for 2 hours. By the second hour it should be pretty warm. Stir the koji mixture, then add another 3⅓ cups (800 ml) of the 140°F/60°C water.

4. Check the temperature of the cooked rice; it should fall in that 120°F/49°C to 140°F/60°C range, and it can be reheated if necessary. Once you're sure it's in the right temperature range, stir the rice into the koji mixture.

5. Cover your vessel semitightly, allowing for some gas to escape, or put an airlock on it. Set the vessel into an incubation chamber set to 120°F/49°C. (You can keep it at temperature by placing it in an oven with the light on, a bread proofing box, or a dehydrator, or by carefully warming it with a heating pad. The fermentation will work at temperatures as low as 72°F/22°C, but it will take a lot longer.) Stir the mixture once or twice a day.

6. On day 2, you can either leave the ferment as it is or, if you'd like some sweet amazake to enjoy (while you wait), you can extract as much amazake as you like. Replace any amazake you take with a similar amount of warm water (140°F/60°C) or more cooked rice. Stir.

7. Add the yeast and lactic acid to the amazake (or the new amazake, rice, and water mixture). Let ferment for an additional 48 hours or more, depending on how tangy you want the doburoku to be. Keep it between 50°F/10°C and 65°F/18°C, or even refrigerate it during this time.

8. After 48 hours of fermentation with yeast and lactic acid, what rises to the top of the rice and water mixture is fairly clear sake. The rice should all be pretty mushy and there shouldn't be a whole lot of yeast activity — there shouldn't be any crackling or heavy bubbling. It can be drunk as is or further clarified by straining out the lees, which will give you kasu or sake lees to use for pickles.

9. Serve warm, reheating it gently if needed. It will become less sweet the longer it ferments. Store doburoku in the refrigerator to slow the fermentation. It will keep in the fridge for up to a week, but be careful of carbonation buildup.

Chef Ken's Tips for Making Doburoku

▸ Cleanliness and temperature control throughout the process are key, as is the quality of the water. You really want potassium, phosphate, magnesium, and calcium. However, if the water levels are high in iron, use soft water.

▸ Cook the grains well.

▸ While sake yeast is the best choice, you can substitute wine or champagne yeast, or bread yeast as a last resort, if sake yeast is not available. If you want it wild-style, you can put ¼ cup of raisins in warm water and let it sit until bubbly and use that instead of packaged yeast. Once you have made doburoku, you can use unpasteurized dregs from a previous batch in place of commercial yeast.

▸ Don't suffocate your brew! Sake is typically stirred pretty frequently in open vats until the yeast has gorged itself. The alcohol it produces slowly kills it off, or at least slows it way down, especially when chilled. Depending on your goals, though, you may barely stir your doburoku once or twice over 5 days. Most people are trying to create a slightly tangy, fizzy, sweet, but also probiotic brew.

▸ Although most people don't ferment their doburoku to make a high-alcohol drink, you could. When it is left unpasteurized, the alcohol, and thus carbon dioxide, production continues, so be careful if you are capping your container tightly.

▸ Check after 4 days if you like a sweeter beverage; after 5 days it will be a little more alcoholic and less sweet. If you add water after 2 days, it will get sourer faster.

Brown Rice Vinegar

Brown rice vinegar's acidity and amino acids make it not only tasty but also incredibly healthy. Rice vinegar starts with koji — as you may be noticing, this is how many traditional ferments begin. Koji is mixed with cooked brown rice and water and starts to convert carbohydrates into sugar. Sugar feeds the yeasts, which make a highly alcoholic sake. In roughly 2 months, this sake is mixed with more water and good vinegar from a previous batch and, traditionally, partially buried in ceramic pots. The microbes found in the inoculating vinegar, which are mostly acetic acid bacteria, transform the sake into vinegar over the next few months. The vinegar is then diluted a final time with water, put into tanks, and allowed to age in order to mellow and gain complexity of flavor.

We made cider vinegar and other fruit vinegars for years, but our first batch of rice vinegar failed miserably. At first we were perplexed, but then we realized we simply hadn't considered that since the sake's alcohol content was so much higher than that of cider, we would have to add water to dilute it first. It takes decent alcohol (much to the chagrin of many alcohol makers) to make noble vinegar. Vinegar making isn't a way to "save" an unsatisfactory sake. If you want to make vinegar and don't want to make doburoku, simply procure some well-made sake and follow the directions on page 235 for making vinegar.

Doburoku Vinegar

By chef Ken Fornataro, culturesgroup.net

YIELD: ABOUT 3 PINTS

KEN WRITES: You can make vinegar with just about anything that has ethanol (alcohol) in it. Doburoku is a great start, and it is helpful if you measure the alcohol content (alcohol by volume, or ABV) beforehand, which can be done with a hydrometer. If you don't want to go to the trouble of making doburoku first, you can substitute a well-made sake. Mix with water as needed to reduce the ABV to between 6 and 8 percent.

This is a typical acetic fermentation that should produce acetic acid and other interesting things. You'll need either a vinegar mother or unpasteurized vinegar, which has a concentration of acetobacter bacteria, unless you have a lot of time to allow for wild yeasts and bacteria to colonize your brew.

You can dramatically accelerate fermentation by introducing oxygen because yeasts love it and produce more alcohol and CO_2, and ultimately acetic acid. In any case, it's an aerobic process that craves oxygen, so don't use a tight cover!

Surface area also contributes to the speed of the ferment — the larger the surface area, the faster the ferment. Just make sure it's covered to prevent insects from getting into it.

FERMENTATION **6 to 9 months**

---

5 cups (1,183 mL) doburoku or sake

5 cups (1,183 mL) water

½ cup (75 g) coconut sugar (or brown, date, or raw sugar)

2 cups (473 mL) unpasteurized apple cider vinegar or, alternatively, a vinegar mother

---

1. Combine the doburoku, water, and sugar in a stainless-steel pot and stir really well until the sugar has dissolved. Gently heat to 85°F/29°C, if necessary, to dissolve the sugar, then remove from the heat.

2. Transfer to a clean gallon jar. Add the vinegar or vinegar mother. Cover with muslin or any cloth that will allow air circulation. Secure the cloth around the rim with a string or rubber band. Let ferment at room temperature for 6 to 9 months, or until you feel the vinegar tastes right. Try not to harass your mother (i.e., don't disturb it too often).

KOJI PORK TWO WAYS

JEREMY WRITES: *I love good pork and I love koji, and the two of them work so well together. The fatty pork easily picks up the amazing, nuanced flavor and aroma of the koji.*

In the first recipe, you'll use a koji-based product, in this case amazake. In the second, you'll actually grow koji on the surface of the meat. Both recipes take about the same amount of time to culture, so you can start them at the same time and compare the results side by side.

You'll notice that neither of these recipes call for any spices or herbs. That's intentional. The first time you try koji-treated meat, you really want to experience the taste, flavor, and aroma that koji brings to the table. After that, feel free to add any seasonings or flavorings that you want. I'm willing to bet that you won't even want to, due to the amazing experience koji alone provides your palate.

These recipes will feed four people if they are paired with salads and side dishes. Grilled corn and a slaw dressed with oil and vinegar are great choices.

Amazake Pork Ribs

By Jeremy Umansky, Larder Delicatessen & Bakery, Cleveland, Ohio

YIELD: 4 SERVINGS

If you are curious about the effect koji has on meat, here is an easy first foray into this culinary technique. Simply marinate the ribs in amazake in the refrigerator, just as you would with any barbecue sauce. The amazake's sweetness also serves to enhance the umami flavor and aroma that the enzymes bring to the table as they break down the proteins.

JEREMY WRITES: I like using amazake instead of shio koji (page 223) because I feel that I have more control over how salty the final product is. Salt levels can vary in commercial and homemade shio koji, so I simply salt the ribs before cooking them. If you make the Koji Cultured Pork (page 238) at the same time, I recommend grilling the ribs over indirect heat so that they stand in contrast to the pan-seared chops.

INCUBATION **35 hours**

1 slab (3–4 pounds) pork ribs

2 cups (480 mL) Amazake (page 217)

Salt

1. Place the ribs in a 2-gallon ziplock bag and add the amazake. Seal the bag, then massage it so that the amazake evenly covers the ribs. Place the ribs in the refrigerator and let marinate for 32 hours.

2. Fill a pot with warm water and submerge the bag of ribs in the water. Use an immersion circulator to keep the water at 107°F/42°C (this is the temperature at which the protease enzymes are most effective), or add warm water as needed to maintain the temperature. Keep the ribs in the water for 3 hours.

3. Remove the ribs from the bag and wipe off as much of the amazake as you can. Don't rinse them (if you do, you will wash off the residual sugars that will give the ribs a nice char when they'e grilled).

4. Season the ribs with salt to taste and grill them over indirect heat until they reach an internal temperature of 130°F/54°C.

Koji Cultured Pork

By Jeremy Umansky, Larder Delicatessen & Bakery, Cleveland, Ohio

YIELD: 2 SERVINGS

JEREMY WRITES: This is by far my favorite way to pair koji with meat. By now you should already be convinced of the seemingly magical properties of koji. That being so, it should be of no surprise when I tell you that not only is this process beyond fascinating, but it will yield the most amazing pork you have ever eaten. In a way I feel like the harbinger of bad news; this will ruin pork for you in the most magnificent way. After you eat koji cultured pork, you will forever look upon pork without koji as subpar.

INCUBATION **36 hours**

2 boneless pork chops, ½ to 1 inch thick

Salt

Sugar

1 teaspoon (1 g) light rice koji spores

½ cup (70 g) rice flour

1. Combine the pork chops with equal parts salt and sugar to taste in a large bowl. Let sit at room temperature for 20 minutes, or until the chops start to weep.

2. Mix the spores and rice flour together in a small bowl, then sprinkle over the pork. Toss to fully and evenly coat the pork. Use more rice flour as needed to ensure that the pork is completely covered.

3. Place a footed cooling rack in a stainless-steel pan or casserole dish. This will allow air to circulate around the meat. Place the coated chops on top of the rack.

4. Culture the pork at 90°F/32°C and 90 percent relative humidity for 33 to 36 hours. This will work best with an incubation terrarium setup (see page 28). At this point, the pork should be fully cultured. The pork will be fully covered in the mycelium, which may look like a cottony layer across the top (see facing page).

5. Transfer the cured chops to the refrigerator to chill, covered, for at least an hour, or until you are ready to use them. They will keep in the fridge for 2 to 3 days. After that they will start to dry out and will start to resemble charcuterie. Alternatively, wrap the chops tightly and store in the freezer. They will hold for at least 6 months this way.

6. To cook, heat a neutral oil in a skillet over medium-low heat. Pan-fry the chops to medium doneness, or until the internal temperature is 140°F/60°C.

What to Expect When Growing Koji on Meat

Jeremy shared a few tips:

▶ As the meat cures it won't have any off odors. After 30 or so hours, you may (or may not) get a fleeting whiff of an ammonia odor. If you get that whiff, the meat is finished. Put it in the refrigerator until you are ready to cook it. If the meat is going off, the smells of spoilage will be overwhelming. The difference will be clear.

▶ In some cases, the meat will look almost brainlike with valleys and ridges as the mycelium pulls away from the meat. This uneven texture is fine.

▶ There may be a few patches of pink where there is excess moisture. This is okay.

▶ Spore-producing patches (yellow-green tinge) are okay. Sometimes charcuterie makers will allow about 40 percent of the microbes to produce spores in order to bring out dank, musty, cave-age flavors, but you aren't doing that here.

CHAPTER 9
MISO
AND OTHER FERMENTED BEAN PASTES, PLUS TASTY SAUCES

Miso is more than just soup stock. It is an umami-rich, salty fermented paste, traditionally made with soybeans and a grain koji, generally rice koji or barley koji (with the exception of an all-soybean miso). It is a quintessential Japanese seasoning that is also a nutritional powerhouse. It plays a vital role in the kitchen and in the health of the people who use it. It is both very simple and very complex. As you read through this chapter, we think you will see what we mean. If you are new to miso or have only had it as a broth served alongside sushi, we encourage you to find some good-quality unpasteurized miso and start getting to know its many flavors and nuances. That will give you something to enjoy while you wait for your own batches to be ready. And don't let the time or the making intimidate you; the hands-on portion is quite simple, and the waiting time is actually over before you know it.

We have provided traditional fermented bean paste recipes from different traditions, but we focus on miso to acquaint you with the process. This will give you a foundation on which to build your own repertoire of flavors and will leave you with ideas and tools to make your own unique tasty pastes. Throughout the world, folks are using the ancient knowledge, technique, and art of miso to create savory, umami-rich, complex, and healthful pastes. Some pastes in this chapter are made with traditional ingredients and ratios while other recipes use just the method with nontraditional starches and proteins and move into the realm of tasty pastes.

Fermented bean pastes are a wonderfully complex cooperation of fungi, yeasts, and bacteria that create a fascinating, rich, complex food that is both flavorful and wildly nutritious. As such, fermented bean pastes are an essential source of flavor and nutrition throughout East Asia. More than just condiments, these pastes are a concentrated, inexpensive, shelf-stable source of protein and play a significant role in people's diets. In short, they are a staple part of traditional cuisines from this region.

The beauty of miso is in its forgiveness. You can almost do anything to it and it will still ferment.

NANCY SINGLETON HACHISU, *PRESERVING THE JAPANESE WAY*

History of Miso and Fermented Bean Pastes

Bean pastes are ancient foods made using a wide range of techniques that grew out of the different regions from which they hail. It is generally agreed that the predecessor to all of these pastes, hailing from China, was *chiang* (*jiang*), which first came into existence at least 2,500 years ago, give or take. In China today, *chiang* denotes hundreds of fermented products. They include douchi, which are fermented, salted, and dried black soybeans (see our recipe on page 207) and hoisin. The hoisin you'll find on the supermarket shelf is usually an unfermented soy-based sauce with garlic and spices, but its history goes way back. Traditionally it is a fermented product, and it was actually one of the first foods produced using *shih* (semidry fermented things, primarily soy).

Recipes migrate with people, and so did chiang. It's said that a Chinese Buddhist priest who came to Japan in the seventh century to promote Buddhism brought chiang along with him. There it came to be called *shi* (see Hishio, page 260, which is said to be directly linked to this ancient ferment). It is believed that at relatively the same time, chiang spread to Korea as well, where it became known as *jang*.

The creation story of miso is not entirely clear. Since Neolithic times, peoples of East Asia have made fermented grain and fish known now as *jōmon miso*. But this didn't necessarily become the forerunner of the miso we know today. It is likely that prior to the eighth century in Japan, miso was made only by Buddhist priests in their temples and consumed by the elite of the country. At the beginning of the eighth century, the emperor of Japan established a bureau to regulate the production, trade, and taxation of miso. Over a century later, miso had made it out of the temples and capital city and into the rest of the country, where over the next centuries it would become very popular with the common people.

As Koreans, Chinese, and Japanese immigrated to other countries, they not only brought their fermented bean pastes with them, they also brought the knowledge of how to prepare them. That knowledge eventually took seed in future generations and spread far outside the family tree. For example, in Indonesia their fermented soybean, called *tauco*, varies from a chunky sauce to a smoked dried version. It is mainly consumed in West Java, where the most popular variety is a sweet, soft version that is made with a short fermentation time and 25 percent palm sugar.

Meet the Microbes

Miso is a multicultural cooperation of many microbes — so many, in fact, that there are likely hundreds of microbes scientists haven't isolated yet. We won't make you wade through the guest list, but we will highlight the who's who at the party. The truth is, you don't need to know any of these guys to make miso, but it is fascinating.

You've already met many of the key players in previous chapters, such as *Aspergillus oryzae* and friends, as well as an unwelcome guest: *Bacillus subtilis* (of natto fame). This bad boy is not on the invitation list. However, the reality is that *B. subtilis* is ever present everywhere, so it is usually present in miso at a constant concentration throughout the process without causing a problem. If *B. subtilis* takes over during the making of the miso koji, the koji is called *nebari-koji* or *natto-koji*, and it can cause the miso to be discolored and off flavored. It can be more of a problem than it sounds because the bacteria will continue to live in the miso and contaminate foods that are integrated with the miso later. (Clearly *B. subtilis* is not invited to the miso party, and if it shows up, we keep it in the corner so it doesn't interact with the other guests — and no drinks!)

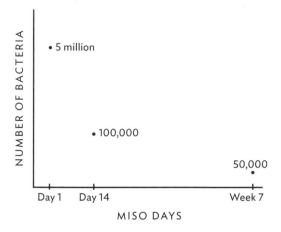

Bacteria are either aerobic or anaerobic, meaning either they require oxygen to survive or they thrive in the absence of oxygen. On day 1 of miso making, you'll find in the mix nearly 5 million aerobic bacteria comprising about 800 different bacterial strains. Their numbers drop dramatically over the first 2 weeks to about 100,000, and then they continue to drop but much more slowly, stabilizing at about 50,000 around week 7 (about halfway through the fermentation).[25] So the miso party starts with the population of the Greater Boston area or Guadalajara, Mexico, and by the end it's only the population of Coeur d'Alene, Idaho. Now let's look at who shows up and who stays for the full fermentation.

In the beginning, the dominant aerobic bacteria are *Kocuria kristinae, Staphylococcus kloosii,* and *S. gallinarum.* Combined, they make up 65 percent of all the bacteria in the fermenting miso. *B. subtilis* and closely related species, including *B. amyloliquefaciens,* make up the other 35 percent. The dominant groups must party pretty hard because most of them are gone by the fourth week and there is no trace of them by week 12, when only *B. subtilis* and its relatives remain. This is pretty surprising, right? The ones we didn't invite outlast all the other bacteria, and yet this isn't a problem? The reason is due to the presence of the anaerobic bacteria famous for yogurt and vegetable fermentation: lactobacilli. Lactobacilli are anaerobic bacteria that are able to assert their influence and control over the aerobic partiers. They and other lactic acid bacteria produce bacteriocins, or antibiotics, which can be thought of as biopreservatives. They promote microbial stability in the fermented foods, and in particular with fermented bean pastes, they keep the *Bacillus* gang in check.

The most important lactic acid bacteria involved in miso fermentation are salt-tolerant (halophilic), with *Tetragenococcus halophilus* dominating. It produces acid to decrease the pH and create flavor in what is called the *shio-nare* effect. Many lactic acid bacteria species, like *L. plantarum,* start out strong but fade out by week 4, as do many of the other aerobic bacteria. Week 4 seems to be a critical point when the aerobic bacteria go through a shake-up and die off. Miso starts out with a lot of oxygen and aerobic bacteria, but as the miso develops liquid (which will become tamari), the tables tilt toward the anaerobic team and most of the aerobic bacteria die off. Bacillus are amazingly tough and survive the harsh conditions, but in much smaller numbers, and they are kept in check by the natural antibacterial properties of the bacteriocins produced by the lactic acid bacteria.[26]

Health Benefits

Like other foods made from fermented soybeans, miso inherits a number of health benefits from the microorganisms predigesting the legumes. Unique to miso is the fact that there is a whole team of enzymes and microbes — the koji aspergilli, the lactic acid bacteria, the yeasts, and the bacilli — working together, or at least with tolerance for each other, to produce amazing health benefits.

Specifically, miso is high in proteins that have been broken down into amino acids and further into peptides, which our bodies can more easily turn into energy. We should quickly note that despite miso's high sodium content, rodents in the studies showed stable blood pressure.

Lower cholesterol. We've noted that raw soybeans contain saponins, which produce a soaplike froth when you wash them. Well, once we consume them in miso, they continue to act like a washing agent but this time inside our blood vessels, which can become blocked with cholesterol. The saponins are joined by lecithin and linoleic acid, which help dissolve the cholesterol clinging to our blood vessels.

Reduced cancer risk. A daily miso soup reduces your risk of gastric cancer. Also, miso made from fermented soy contains high levels of isoflavones that are known to help in the prevention of breast cancer in women. There have even been a few well-documented studies noting that the consumption of miso mitigates the effects of radiation.

After World War II, Japanese doctor Shinichiro Akizuki spent years caring for atomic bomb victims a few miles from ground zero; however, neither he nor his staff suffered the typical effects of radiation. He postulated that they were protected from the radiation because they drank miso soup every day. Akizuki's theory was confirmed in 1972 when researchers determined that miso has dipicolinic acid, an alkaloid that combines with heavy metals and expels them from the body.[27] Two decades later, a report published in Japan by Dr. Akihiro Ito described a study of mice that had been exposed to radiation and then fed a diet containing 10 percent miso or soy sauce for 13 months. They found that the female mice on the miso diet experienced a significant decrease in the frequency and number of liver tumors.[28]

More B vitamins. The vitamin B_2 content is increased by fermentation, likely from the koji. B_{12} is also increased, likely from the lactic acid bacteria. B_{12} is the one vitamin that does not occur in plants, so anyone following a plant-based diet needs to find a source for B_{12} to avoid deficiency, which can lead to irreversible nerve damage.

Strong bones. Miso contains lysine, which helps our bodies better absorb calcium, and it also contains high levels of magnesium, which helps build strong bone structure.

Traditional medicine views aged miso as having medicinal qualities. It is clear when you smell or taste these misos that they are in an entirely different class from their younger miso cousins and are potent medicine.

Making Miso: An Overview

As you will see, making miso is pretty simple — it's simple to put together and simple to manage. It is actually the easiest ferment in this book. The most important thing is to keep it weighted and to occasionally stir smaller batches. You can neglect your miso and it will still ferment. And if there is mold on top, just scrape it off.

Start with Koji

Making miso is a two-part process that begins with koji (*Aspergillus oryzae*). Traditionally (and this is your quick review), steamed grain (generally rice or barley) or soybeans are inoculated with the koji spores and spread in shallow wooden trays. The trays are kept in a humid, warm environment (we are growing mold, after all) for 40 to 48 hours. During this time the fungus blooms and binds the substrate together. Remember, in this part of the fermentation, the bloom produces an abundance of enzymes that break the more complex proteins, fats, and starches into the more digestible and simpler amino acids, fatty acids, and sugars. The beauty of making miso is that you can make your own koji or you can purchase koji and get started. Purchased koji is usually dry; to get the desired texture, you can premoisten it with a bit of the bean liquid or wait until after you've mixed it with beans or grains and some of their cooking liquid to determine whether more liquid is needed (more about this on page 246).

If you make your own koji, you can save yourself time by planning out what you want to make with the fresh koji — whether it is one or a number of ferments. If you plan in advance, you can soak and cook your other ingredients (beans, for example) partway during koji's fermentation, so they are ready to mix with the salt and fresh koji the moment that the koji has reached the desired growth.

GOOD MOTHER STALLARD
MILLET TASTY PASTE
(page 273)

Using Precultured Koji in Miso

The recipes in this chapter assume that you are using homemade, wet koji, but they can also be made with precultured dry koji. Simply premoisten the koji with some of the bean pot liquor (cooled to below 135°F/57°C). You will use about 10 percent of the weight of the dry koji in the added liquid. Add about a tablespoon at a time, letting the water absorb for a few minutes before adding another tablespoon. Allow at least 30 minutes for the moisture to penetrate the koji grains.

Cook the Beans

As with all the ferments in this book, give the beans a nice long soak. Unless the recipe indicates otherwise, you do not need to cook the beans al dente. It's best to steam them, just be sure to save the water, as you will need to add some later to get the right consistency.

Combine Koji with Other Ingredients

Break up the koji if it is bound together in a mat. If you are using precultured dry koji, you can premoisten it by adding a little bean water (see box above). The koji should be just hydrated, not wet. Then mix the koji with the cooked soybeans, some of the bean cooking liquid, salt, and some unpasteurized finished miso, often called seed miso. The seed miso isn't entirely necessary, but it will introduce the unique bacteria and yeasts that not only jump-start the process, which shortens the wait time, but also ensure that the flavor is repeated from batch to batch. In Japan, where miso shops have been making miso in the same place for centuries, the barrels, the walls, and that small bit of seed miso pass on the heritage of each shop's unique flavor. This is no different than the Roquefort cheese caves of France, where the unique environment and microbial culture conspire to make something distinctive and repeatable.

The personality (or variety) of a traditional miso is controlled in the beginning by the proportions of koji, beans, and salt. The proportions determine not just the flavor but how long you will need to wait to eat the miso. The more koji, the sweeter your miso will be. If you bump up the koji and lower the salt, it will ferment faster and be lighter — think sweet white miso. The lower-salt misos also have a relatively short shelf life. Conversely, less koji (fewer enzymes adding all that simple sugar), more beans, and more salt (which slows down the fermentation) will produce a long-aging, dark, savory miso. Besides providing a control for the aging time in the miso, the salt also halts the growth and sporulation of the koji fungus. The salt doesn't stop the digestive enzymes in the koji, however — they remain active.

Don't worry, you don't have to remember all this right now. We've included a section called Choose Your Own Modern Miso Adventure (see page 268), which gives you simple formulas so that you can easily choose what you want to make and get all the proportions right the first time.

Mash the Ingredients

The koji, cooked beans, bean cooking liquid, salt, and seed miso are mixed and mashed together and pressed tightly into wooden vats, crocks, or jars. There are a few options for grinding or mashing your miso, and which is best for you will depend on the texture of your beans. For larger batches of harder beans, particularly soybeans and chickpeas, we like to run the mixture through a meat grinder. You can also use a potato masher or, in some cases, a hand blender. For smaller batches, especially with smaller, softer beans, we often mash the mixture with our hands because we like the chunkier texture.

Many homemade misos often do not have the same texture as commercial varieties. As you experiment with different recipes, you will find that some misos are wetter and softer while others are pastier, almost like toothpaste. That is normal. That said, too wet is an issue — soupy miso will have problems, including acetone smells and flavors.

━━━ PRO TIP ━━━

Watch That Bean Water

Most miso recipes call for you to add some of the bean cooking water to hydrate the miso mixture as you are making it. It turns out that soybean cooking water, more than that of other beans, is vulnerable to souring and turns quickly. So it must be used immediately, especially in warm weather. Christian Elwell of South River Miso (see page 252) shared that in the summer, they don't use the bean liquid for their sweet white miso because it is too risky in this lower-salt recipe. Instead, they use boiled (and cooled) water.

Place the Miso in Vessels

Prepare your jar or crock by putting boiled water (you boil it to sanitize it) or some of the bean cooking liquid in the vessel and swishing it around to moisten the sides. Pour out the liquid, then use your fingers to pinch and cast the salt to dust the sides and bottom of the vessel. Then put the mash in the vessel. If the mash is soft, put it in a little at a time and take care to avoid creating air pockets. Use a butter knife to work out any air pockets that develop and press with your hands or a masher, like one used for sauerkraut.

If the mash is a thick doughlike paste, it can be difficult to get it into the vessel without creating air pockets. A traditional strategy is to first roll portions of the miso mixture into small balls (these can be golf balls to tennis balls in size, depending on the size of your batch and vessel) and then throw these balls into the jar with a bit of energy. The balls will stick together, and with careful pressing, air pockets will be minimal.

Add Weight

The miso is covered with a layer of cotton cloth, such as muslin, cut to fit the circumference of the vessel, and salted around the edges. The miso is then weighted quite heavily to preserve the anaerobic environment, which is crucial to a good ferment. A good rule of thumb is to use a weight that is equal to the weight of the miso. So, 5 pounds of miso would get 5 pounds of weight. (Some makers feel that the weight should be 125 perent of the miso's weight, meaning that for that same 5 pounds we would need 6¼ pounds of weight.) For crocks and jars, you can stack ceramic weights, clean (boiled) rocks, or the glass disk weights that have become popular. Another option is a plastic ziplock bag filled with marbles, pebbles, salt, or even water, which will provide both weight and cover. If you weight your vessel with the filled plastic bag, you will directly salt the edges but you won't need the cloth cover.

Monitor the Early Fermentation

Over the first few weeks, the salt will draw more liquid and move it through the weighted miso, forming a liquid brine layer — the tamari. It is a good idea to check on the ferment in the first few weeks to make sure that you are getting at least a small pool of liquid, even if it is minimal. If not, add more weight. Conversely, if during those first weeks you see a fair amount of tamari pooling, your mixture may be too soft, in which case you will want to add more weight. In this instance you can also remove a bit of this tasty liquid to cook with — just don't remove it all. This briny liquid acts like a shield to ward off any undesirable microbes that want to move in.

Age

The enzymes are now free to digest and break down the beans. Aging is the second step of the fermentation. The natural rhythms and temperature fluctuations of the seasons are part of what helps build the nuance of flavor.

Traditionally, miso production begins in either the fall or spring months, when the air is fresh and the weather is just warm enough to grow koji but not too cold. The fall is also when the soybeans and rice are just harvested and at their freshest, and the spring occurs right before the heat of the summer months, when the fermentation is most active.[29] In the case of longer-aged misos, this cycle of less activity in the cooler months and higher activity in the summer months repeats.

The good news for the home fermenter is that you don't have to worry much about maintaining any specific temperature as the miso ferments. Just let it sit at room temperature. Christian Elwell of South River Miso (see page 252) told us that he finds 78°F/26°C to be ideal for miso fermentation. The main thing is to keep it warm enough for the microbes and the enzymes they produce to remain active; at 50°F/10°C or below you are not going to get a good ferment because it is simply too cold for the enzymes to carry out

A Miso Year

In Japan, a miso that is a "year" old has not necessary existed for 12 months. In other words, a 2-year miso isn't necessarily 24 months old. This can be confusing to people who are not familiar with the lunar year. In Japan, everything (people, dogs, cats, miso) becomes a year older at the lunar year (a big birthday party which is usually celebrated in February). So miso that was created in the fall, which has been fermenting for just 3 or 4 months, officially becomes a 1-year miso at the lunar new year. A miso that is created in the spring, after the lunar new year, does not turn 1 until the following year's lunar new year. In this book, we calculate miso age based on months, as is standard in the West (so a 2-year miso is in fact 24 months old).

their work, but most people's homes do not get that cold, even in winter. That said, be aware that modern heating systems may dry your miso out. Find a location for your fermentation vessel that is protected from the direct onslaught of moving air. Small batches are especially susceptible to drying, as they don't have a lot of mass, so any micro amounts of evaporation show up more dramatically. If you are fermenting in an opaque container, like a crock, light is not an issue. If you are making a multiyear miso in a glass jar, we suggest making sure the spot you choose is dark as well.

If you are fermenting in a clay crock or wooden vat, microscopic pores in the material will allow natural oxygenation of the miso. If you are fermenting in a glass jar, however, your miso will have little contact with fresh air. In this case, it is important to stir the miso every once in a while, or to put it into a new jar; you want to release gases and inject a little oxygen, which you will press back out as you tuck the miso into place and weight it again. (If you see any signs of yeast or mold on your miso, remove them before stirring or transferring your miso to a new jar.) A wet miso may need to be stirred in the first month, but a drier miso may not need to be stirred for 5 or 6 months. The best policy is to check on your misos periodically. We generally set aside one Sunday a month when we check all of our long-aged ferments. This gives us a chance to catch anything that is going awry, but it is also a great learning opportunity. Watching, smelling, and tasting your ferments throughout the process can provide you with a great deal of valuable information and experience.

PRO TIP
Climate Control

If you want to play around with mimicking the seasons, *The Book of Miso* by William Shurtleff and Akiko Aoyagi (see page 41) lays out a plan: Start by setting your miso fermentation vessel in a location that is between 70°F/21°C and 75°F/24°C for the first 2 months, then find a warmer area (think summer) that is around 85°F/29°C (or up to 90°F/32°C for a quicker ferment) for the next 2 months. Then return your ferment to the original, slightly cooler temperature.

As the ferment sits silently on your counter, in a closet, on a water heater (if your home runs cool and you want to give it a little more warmth), or under a bed, the yeasts begin to turn the simpler sugars into alcohol. The bacteria are also at work, converting sugars into acids. At some point while you are waiting (maybe more or less patiently), these acids and alcohols begin to react with one another to produce esters, which are key to the complex flavors. At the same time, the colors are changing. The salt is mellowing out in conjunction with all that is happening, and the layers of aroma are building. Cool, right?

You may have read that miso isn't good unless it has aged for a full year and that it only gets better from there. That is true for some varieties, but for others 1 to 2 months of aging — or even less — is perfect. You will see that Cheryl Paswater's recipe for Macadamia Tasty Paste (page 281) is aged for only a few days. As with most fermentations, the point at which the ferment is done depends on a lot of things, such as temperature, the quality of the koji, and the proportions of salt, koji, and beans. If you want to experiment with multiyear aging, it is perfectly acceptable to divide a batch of miso and place one jar in the larder and keep the other jar aging.

Any type of miso is a good candidate for aging. Aging miso is a wonderful phenomenon. Even if you don't make miso yourself, you can purchase a good-quality miso and stash it in a cool place to see what happens. (Kitchen science for the kids — just sayin'.) The thing to know is that it *will* change but it *can't* ferment too long; it's just a matter of flavor. If you really want a sweet miso (Sweet Miso Formula, page 269, or Sweet White Miso, page 262), then you should harvest it early, as indicated in the recipe, but if you don't, it will still be good — it just may not be sweet.

We found a study[30] that explains how the flavor of miso changes as it ages. At 10 days, the flavor is still beany (not a surprise). By 3 months, salt is still the dominant flavor and the miso aroma begins to develop. Sugar increases over the first 5 months, as does the nitrogen and protein hydrolysis. At this point, the miso is close to the desired flavor yet plain and flat. By 11 months (a typical "miso year"), the flavor is on target, with slight complexity that fills the mouth. As the miso continues to age, it develops a more intense taste.

Miso begins to change color around the 5-month mark. The deeper the color, the older the miso. A sweet white miso is just that: nearly white and can be as young as 3 to 4 weeks. A red miso has usually aged a year or more, and pure soybean hatcho miso that is sometimes aged for 9 years is nearly black. This is due to the Maillard reaction, a chemical reaction between amino acids and sugars that creates hundreds of exceptional flavor molecules and browning. (We usually see this reaction when we are

Can a Miso Be Too Old?

The longer a miso ages, the more its flavor profile changes, but it won't turn rancid or spoil. Christian Elwell told me about what he affectionately calls South River Miso's "miso museum." In it, he has a 10- to 12-year-old sweet white miso that is 4 percent salt with a 1:2 bean-to-koji ratio (in other words, the type of miso that supposedly has the shortest shelf life). He described the miso as "black, with a rich deep earth fragrance. Not the same but akin to the smell of healthy humus in the soil, like when you pull back the leaves on the forest floor, revealing that deep dark fertility."

The texture is smooth, with the individual ingredients no longer distinguishable. He said that the flavor was no longer sweet but bitter — in a pleasant, bright way — with no signs of degeneration. He also shared that in his experience, the sweet white miso was the least likely to develop mold over the long term. And if you are wondering, the oldest miso in the museum is from the Ohio Miso Company (circa 1979–80). Christian said it is quite stable, like a bouillon cube.

caramelizing foods or browning meats over high heat.) As the ferment sits around, the flavor compounds continue to break down and form entirely new flavor compounds that are rich in umami and deeper in color.

Harvest

When the miso is ready to harvest, the layers are carefully removed. You may or may not see yeast or mold growth on top. If you do, start slowly and work your way down. First wipe away any mold or yeast growth off the edges of the crock, then scrape off the layer that is a happy microbial jungle; it may smell, but don't let that deter you. Once the surface is clean of any yeast or mold, stir the miso deeply from top to bottom to allow the air back into it. It is like giving the miso a breath of life — it went in as a sum of its parts, and now it emerges as something entirely new. You can experience this by tasting the miso a couple of times. The flavor will change in those first few minutes from green and unripened to a rich, rounded flavor. If at this point the flavor is still flat, or too salty, or a bit sour, or the color is still light, cover it back up and let it keep on fermenting.

If the miso is deemed done, yay, it's time to enjoy! If it fermented in a jar, press a piece of parchment paper on top of the miso (to minimize oxidation and dehydration) and place that jar in the fridge to keep the flavor more stable. If it fermented in a vat or crock, you have a couple of choices. You can collect a pint or so for your daily use and tuck the rest back in for more aging, or you can go ahead and put all of it in jars, top with parchment paper, and place them in the fridge. If you keep your miso in airtight containers that have little air space, with a piece of parchment paper on top, and you use

clean utensils to scoop it out, chances are it will keep indefinitely. If you suspect it has gone bad (maybe it smells foul or the flavor is off), let your senses be the judge and trust your gut.

Store

Miso is quite shelf stable. Though most people do refrigerate it out of habit, some people just find a cool spot. You be the judge. A refrigerator simply slows down the fermentation, which is helpful if you want your sweet miso to stay sweeter longer. If you want your miso to continue to change and age, you can leave it out. If you are having issues with a batch, put it in the fridge to better control the conditions. Fall can be a challenging time for storing miso at room temperature because the organisms that promote decay (like fruit flies) are so much more active then. Wherever you choose to store your miso, it is important to press a piece of parchment paper on top to slow oxidation.

Meet the Maker
Gaella and Christian Elwell
South River Miso

It was the mid-1970s and the back-to-the-land movement was in full swing. Young people were searching for something that felt missing from their modern, suburban upbringing — from spirituality to connection with land. Gaella and Christian Elwell were no exception. Having both lost their fathers to cancer at a young age, Gaella and Christian pursued paths to deepen their own well-being. They met while studying macrobiotics in Becket, Massachusetts, with Michio and Aveline Kushi. They also met their future — miso. The Kushis served miso daily for breakfast.

252

Soon Gaella and Christian were on their way west in a red Volkswagen van decked out with curtains that Gaella had made. They studied briefly with Alan Chadwick and then went to Asunaro, a school of Oriental medicine and fermented foods run by Naboru Muramoto, author of *Healing Ourselves*. They spent the winter and spring with Sensei Muramoto, not just eating miso but making it.

They returned to Massachusetts, got married, bought 60 acres of bare farmland outside Conway, and were expecting their first child. In this whirlwind, they got a call from Thom Leonard, whom they'd met at Asunaro. Thom and two partners had started the Ohio Miso Company 18 months earlier, but there was a falling out and Thom was looking for a change.

After deliberating for a few days, the Elwells decided to become miso makers. A few months later, Christian went to Ohio to pick up a few vats, some fundamental equipment, and 13 tons of miso, which landed in an unheated barn on the far side of the South River that divided their land. It was 1980. That first winter, they packed orders in bulk containers — 45, 18, or 9 pounds (miso sold for 80 cents a pound then) — and hauled

the boxes across the shallow river to the road with a horse-drawn wagon.

In 1981, they broke ground on the timber-frame building that is still their production facility today. The heart of the building is the wood-fired masonry stove with a stainless-steel kettle that is built into the bricks. It is in this vessel that the grains are steamed and the beans slowly cooked for 20 hours. The Elwells believe that it is this slow wood heat that brings out the best in the beans.

South River Miso is made in the old farmhouse style of Japan. It is an entirely handmade process that is now run by 12 employees. First the grains are steamed, cooled to 112°F/44°C, and inoculated, and then they are mounded in a crib to rest and hold the dissipating heat. Overnight, the temperature climbs from 85°/29°C to 100°F/38°C, as the koji comes to life. The koji is stirred to disperse the heat and gases and to stir the outside grains to the middle. It is then scooped into American cypress trays that are tucked into the koji room, where they are stirred twice and monitored for temperature over the next 24 hours. As the mold metabolizes, it continues to create heat, and on the second day any augmentative heat is removed from the room so that the koji stabilizes at around 100°F/38°C. On the third morning, it is ready.

The koji is salted, and then is ready to meet the slow-cooked beans. It is a marriage, Christian says, as these two ingredients will never be the same again. We think the ceremony may well be the best part.

The wedding venue is the large stainless-steel tub. The beans are shoveled into the tub. Seed miso is mixed with some of the cooking liquid, and the slurry is poured over the beans. Meanwhile, one of the employees puts on tall, clean organic cotton socks, wraps them in plastic, and tapes this footwear into place right below the knee. When the tub is full, the employee steps in and begins to stomp. It is a methodical process of tight sideways steps back and forth across the mixture four times until the beans are crushed. Then half of the salted koji is added, tread upon until thoroughly mixed, leveled, and the rest of the koji is added and again mixed by foot. During the entire process, liquid is added to bring the mixture to the right moisture content. Are you feeling the Japanese farmhouse now? The employee treads on about 500 pounds of future miso in tub-sized batches; South River produces over 120,000 pounds of miso a year.

After visiting the plant, Gaella took Kirsten to the daily community lunch, which is cooked fresh for the employees by a hired cook. This allows the Elwells to connect with their staff every day, as they are no longer active in the hands-on production. Gaella says that it has always been important to them that everyone working there has a good day, as their business decisions have been based on quality of life. It shows, as some of their employees have been at South River for over 20 years.

The tour ended at the giant American cypress vats. The larger barrels hold 7,200 pounds of miso. It takes 3 weeks of 5-day-a-week production to fill each vat. Once a vat is filled, the miso is covered with muslin cloth, topped with a wooden pressing lid, and weighted down with 2×4s and concrete blocks. Finally, the vat is covered and begins its long, slow ferment. South River makes 10 varieties of miso. The light misos are aged for less than a year. The darker misos are aged for 3 years.

Miso making in Japan has deep roots. Some Japanese miso businesses have been making miso continuously for centuries — such as Maruya Hatcho Miso, which opened its doors in 1337. Everything at South River Miso is, by comparison, shiny and new, yet the sense of stability, intention, patience, and a good life permeates the farm in the same way that miso and its microbes permeate the cypress vats that have been at work for just a few decades, so far.

253

Let's Make Miso!

Once you have made or purchased the koji, the rest of the process is quite simple and quick (the active part at least). When you read through these step-by-step instructions, we think you will agree. If you are daunted by the thought of making your own koji, don't let that stop you from discovering the joy of making your own miso. We encourage you to dive into the vat and make miso with purchased koji. This process is the same no matter what ingredients you choose. Use the chart on page 269 as a guide for making your own recipes.

We all have different styles in the kitchen. As the conductors of this symphony of microbes and enzymes, we orchestrate the ingredients and they make the music. Some of us want to have very precise control over the ingredients, measuring them to the gram. Some of us are very comfortable with just putting it all together and letting the sound track play as it will. The beauty of miso is that the microbes will do their thing either way. This is not to say that measuring isn't important, but rather that you have latitude to work in your comfort zone. Because measuring by volume varies greatly (based on how tightly you pack the koji, what type of salt you use, and so on), we've included gram and volume measurements to suit your style.

What You Need

- Beans
- Koji, homemade or commercial
- Bowls, various sizes, for soaking the beans, catching the bean cooking liquid, moistening the koji, mixing the beans and koji, and so on
- Large pot or electric pressure cooker for cooking the beans
- Colander or strainer
- Metal spoons
- Scale for weighing the koji and beans
- Salt
- Meat grinder or potato masher (optional)

- Seed miso
- Crock, jar, or other fermentation vessel
- Butter knife or chopstick
- Muslin
- Weights
- Cloth or paper cover
- String or rubber band
- Painter's tape
- Permanent marker
- Notebook and pen

1. Soak your beans for 8 to 24 hours before you will be ready to make miso. If you are making your own koji, refer to the step-by-step instructions on page 196.

2. After the beans have soaked, boil or steam them until they are soft. They should give way to pressure when squeezed between your thumb and ring finger.

3. Strain the beans, reserving the cooking liquid. Let the beans and their cooking liquid cool to body temperature (below 100°F/38°C), gently stirring the beans from time to time to release steam.

4. Place the koji in a bowl. If the koji is dry, add a small amount of the bean cooking liquid, a little bit at a time, to moisten. You want it to be well hydrated (each kernel moist all the way through) but not wet. Once it is the right consistency, weigh the koji.

Instructions continue on next page

255

MISO AND OTHER FERMENTED BEAN PASTES, PLUS TASTY SAUCES

5. Combine the koji, beans, and salt in a bowl. Mash the mixture into a chunky paste with uniform small pieces. You can use a meat grinder, potato masher, or your hands. If it gets too thick, add a little of the bean cooking liquid.

  Note: A doughlike consistency will give you a paste-style miso, whereas a chunky texture will give you a "country" or "cottage" miso. Which style you choose is up to you.

6. Add the seed miso and mix again until everything is evenly blended.

7. Prepare the fermentation vessel by rinsing the sides of the jar with a little of the bean liquid or boiled water. Pour out the excess and then sprinkle salt evenly over the inside of the vessel.

8. Pack the miso into the vessel, a spoonful at a time, tamping as you go to remove any air pockets. Or form the miso into small balls and toss them firmly into the vessel to minimize air pockets.

9. If needed, run a butter knife or chopstick along the sides of the vessel to remove any air pockets. Smooth the surface of the paste.

10. Cut a piece of muslin to fit perfectly across the top of the miso. Place it on top. Sprinkle about ½ tablespoon of salt along the edges of this cover to seal any gaps.

11. Set a weight on top of your miso to press out gases. Ideally, the weight on top will equal the weight of the miso.

12. If your vessel has a lid, place it on top. Otherwise, set a cloth or paper cover over the fermentation vessel and secure it with string or a rubber band. Label it with painter's tape and a permanent marker (it's easy to forget what you made after a few months). Let sit at room temperature on your countertop for the first few weeks. This will give you a chance to make sure it's settling properly before you move it to its more permanent home.

Instructions continue on next page

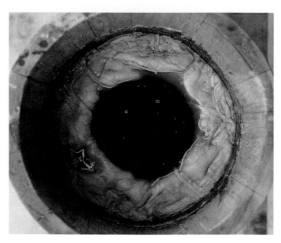

13. After a few weeks to a month, you should see liquid forming on top. If you do not see any liquid, add more weight. Once you see the tamari, even just a little, you are in good shape and can forget about the miso for months at a time.

14. After the miso has aged for a while, you will see more liquid on top. This is the tamari, and it is very flavorful. You can strain off some for eating, but the rest should be mixed right back into the paste.

15. A long-aged miso can look a little like something that you've been warned against. It's okay. Keep going; your miso will beautiful underneath. When your miso is ready, you'll remove the layers on top that have been exposed to oxygen. You will know when you reach the good miso, because it will no longer be discolored and will look and smell as it should.

16. Finished miso can be eaten as is (at right), or you can run it through a food grinder, mill, or processor if you want a smoother paste (at left). Commercial miso is often heat-treated to halt the fermentation, which renders the magic of the enzymes nonexistent. There is no need to pasteurize your miso. Simply store it in the refrigerator, where it will be stable for years.

Taste-Testing Your Tasty Paste

It's fine to taste your miso (or any bean paste) partway through the ferment as long as you are able to tuck everything back in place afterward. Carefully lift off the weights, clean them, and set them on a clean plate. If the surface of the ferment is scummy, carefully remove that top layer, wiping the edges of the fermentation vessel clean as you go. To taste, put a butter knife in through the center and pull out a bit that is deeper. It will give you a better sense of the flavor because the surface is saltier. After you have tasted, smooth the surface, sprinkle additional salt, and tuck it all in as before.

Blending Your Misos and Tasty Pastes

Different fermented pastes have different flavor elements, even if they are made from similar ingredients, depending on proportions, timing, and other factors. The sweet quick-fermented styles and the deep, earthy long-term ferments have vastly different offerings for your culinary use. But what if you want sweet *and* earthy? You can mix and match your misos and bean pastes to get the flavors you are after — and that's a good reason to have multiple types in your fridge. Whether you are buying or making miso, having multiple varieties gives you incredible possibility, whether you are blending for soup broth, preparing a marinade, or making a sauce. Winemakers do it with wine; cider makers do it with cider. Pit masters do it with barbecue sauce. Why not with tasty pastes?

One-year-old Soybean Miso (page 270)

Hishio and Nattoh Miso

YIELD: A LITTLE MORE THAN A PINT

Hishio is an ancient ferment that comes from Japan. Many believe that it is the forerunner of shoyu. Nattoh miso, a much newer ferment, has more whole soybeans that are suspended in syrupy liquid, giving it the look of natto, but that is where its similarity to natto ends. This is not a bacillus ferment, and there is no natto in this condiment. Both of these ferments are part miso, part shoyu, and part pickle. GEM Cultures aptly describes them as "chutneys." They serve that same purpose on a noodle, rice, bean, or grain bowl — they lend that syrupy, thick, bold note. Though less sweet and pungent than what we think of as chutney, these condiments are salty and umami-rich with a sweetness in the background from the barley koji base.

Hishio and nattoh miso have a much shorter fermentation time than traditional miso — only 2 to 4 weeks. They are also a great place to experiment with seasonal vegetables — the possibilities are endless. In fact, the recipe here, which comes from GEM Cultures, is more just guidelines. It calls for barley koji, but we have played with other grain kojis, like sorghum, with some good results.

These recipes don't call for a sweetener (such as rice, barley, date, or maple syrup), but it can be tasty. Add to taste after fermentation.

FERMENTATION **2–4 weeks**

Hishio

Ratios: Approximately 75% barley koji, 15% soybeans, 10% vegetables

⅓ cup (60 g) dry soybeans

1½ cups (262 g) Barley Koji (page 203)

½ cup (approx. 88 g) diced vegetables (see note below)

1 cup (235 mL) naturally fermented soy sauce or shoyu

Note: Have fun here! Traditional vegetables include sea vegetables, eggplant, uri melon, celery, and root vegetables like carrot, daikon, and burdock, but don't stop there. Why not try celeriac, parsnip, turnip, or parsley root? We have also used peppers, green tomato, green apple . . . you get the idea.

Nattoh Miso

Ratios: Approximately 60% barley koji, 35% soybeans, 5% vegetables

⅔ cup (115 g) dry soybeans

1⅓ cups (230 g) Barley Koji (page 203)

¼ cup (approx. 43 g) sliced kombu, ginger, or chile peppers

1 cup (235 mL) naturally fermented soy sauce or shoyu

1. Soak the soybeans for 8 to 24 hours, then cook following the instructions on page 54.

2. Combine the soybeans, koji, vegetables, and soy sauce in a quart jar and ferment at room temperature for 2 to 4 weeks. You will know the ferment is ready when it develops a misolike aroma. It will keep for many months at room temperature or indefinitely in the refrigerator.

Sweet White Miso

YIELD: ABOUT 1 QUART

White misos are the gateway misos. Sweet and gentle, they are easy to love and easy to eat; we found that we wanted to put white miso in all of our recipes. This basic recipe is similar to a type of sweet miso that is traditionally made in Kyoto, Japan. It is like a soft cheese — delicious, delicate, and not as shelf stable as some of its cousins. This miso has a high concentration of koji rice, the least amount of soy, and a low-salt brine, and during fermentation it becomes almost custardlike.

Use this recipe as written or use it as a template for your own sweet miso. We have made a delicious white miso with white tepary beans and Carolina Gold rice for an heirloom, born-in-the-Americas version.

PROCESS **miso (page 254)**	FERMENTATION/AGING **3–5 weeks**

1 cup (175 g) dry soybeans

3½ cups (600 g) light Rice Koji (page 201)

2½ tablespoons (42 g) salt, plus extra to prep the vessel and top the miso

1 tablespoon (16 g) unpasteurized miso (see note below)

Note: Any unpasteurized miso will work, but we prefer to use white if it is available.

1. Soak the beans for 8 to 24 hours. If you're making your own koji, time the soaking with the koji cycle; you will want your beans to be cooked when your koji is ready to come out of the tray.

2. Boil the soybeans in plenty of water until they are soft, about 1 hour, or steam them in an electric pressure cooker for 45 minutes. Drain, reserving the bean cooking water, and then spread them out on a tray to cool slightly and steam off moisture.

3. When the beans have cooled to below 100°F/38°C, combine them in a bowl with the koji and 2½ tablespoons of the salt. Mash together with a potato masher, or put through a meat grinder if you have one. Add the miso and thoroughly mix. Add enough of the bean cooking water to achieve the desired consistency — it should be chunky and dryish, like mashed potatoes.

4. Using a bit of the bean cooking water, rinse the inside of your fermentation jar or crock, making sure to coat all of the surface. Then sprinkle 1 tablespoon of the salt into the jar, making sure to coat all the sides and the bottom of the vessel.

5. Spoon the bean mixture into your jar or crock, doing your best to remove as many air bubbles as possible.

6. Set a small piece of unbleached cotton cloth or parchment paper cut to fit the diameter of your vessel on top. Sprinkle about ½ tablespoon of the salt along the edges of this cover to seal any gaps. Weight the miso as best you can; see page 247.

7. Cover the entire vessel with cloth or paper, securing it in place, and follow the instructions for aging on page 248. We have found that the flavors become perfect in 3 to 5 weeks.

8. When you are ready to harvest your miso, open it up and follow the harvesting instructions on page 251. You may need to scrape off the top surface of the miso until you get to something that looks nice and rich in color. You can either strain off the tamari (the liquid pooling on the top of the miso) or you can mix it back into the miso and eat it as is. Your miso may be chunky; if you prefer a smoother paste, process it in a grinder or food processor. Store in an airtight container. The miso will keep indefinitely in the refrigerator.

Variation: Light Miso

In this variation for a white rice type miso, you see the ratio of soy to rice koji is brought into balance and the salt is increased. This changes the nature of the miso a bit; it ages longer and is more stable when finished. It is a good all-purpose mellow miso. It is also smooth and perfect for baking. We have found that the added saltiness gives foods a salted caramel appeal.

FERMENTATION/AGING **6–12 months**

---

2 cups (350 g) dry soybeans

2½ cups (440 g) light Rice Koji (page 201)

½ cup (135 g) salt, plus extra to top the miso

1 tablespoon (16 g) unpasteurized miso

CHICKPEA MISO, page 266

SWEET WHITE MISO, page 262

RED MISO, page 267

QUINOA RED BEAN (ADZUKI) MISO

Barley Miso

This is a basic, longer-aged barley miso. It is rich, brown, and earthy with a classic flavor. Of course, you have to wait a while before you can enjoy it.

If you have a nice big batch of barley koji, you can make a mellow barley miso at the same time so that you have one to eat soon and one to wait for. For a mellow barley miso, simply increase the amount of koji and decrease the amount of beans, salt, and aging time. If you follow the ratios for the Sweet White Miso (page 262), you will end up with a delicious mellow barley miso.

PROCESS miso (page 254)	FERMENATION/AGING 6 months–2 years

3 cups (525 g) dry soybeans

4½ cups (830 g) Barley Koji (page 203)

½ cup plus 1 tablespoon (150 g) salt, plus extra to prep the vessel and top the miso

1 tablespoon (16 g) unpasteurized miso

1. Soak the beans for at least 8 to 24 hours. If you're making your own koji, time the soaking with the koji cycle; you will want your beans to be cooked when your koji is ready to come out of the tray.

2. Boil the soybeans in plenty of water until they are soft, about 1 hour, or steam them in an electric pressure cooker for 45 minutes. Drain, reserving the bean cooking water, and then spread them out on a tray to cool slightly and steam off moisture.

3. When the beans have cooled to below 100°F/38°C, combine them in a bowl with the koji and ½ cup of the salt. Mash together with a potato masher, or put through a meat grinder if you have one. Add the miso and thoroughly mix. Add enough of the bean cooking water to achieve the desired consistency — it should be chunky and dryish, like mashed potatoes.

4. Using a bit of the bean cooking water, rinse the inside of your fermentation jar or crock, making sure to coat all of the surface. Then sprinkle

1 tablespoon salt into the jar, making sure to coat all the sides and the bottom of the vessel.

5. Make golf ball–size balls with the entirety of the bean paste. Layer your vessel as you go, tossing the balls in with a bit of energy to help minimize air pockets.

6. Set a small piece of unbleached cotton cloth or parchment paper cut to fit the diameter of your vessel on top. Sprinkle about ½ tablespoon of salt along the edges of this cover to seal any gaps. Weight the miso with about 8 pounds, or as best you can; see page 247.

7. Cover the entire vessel with cloth or paper, securing it in place, and follow the instructions for aging on page 248. This miso is tasty after 6 to 8 months, but it does get much better with time (a year or more).

8. When you are ready to harvest your miso, open it up and follow the harvesting instructions on page 251. You may need to scrape off the top surface of the miso until you get to something that looks nice and rich in color. You can either strain off the tamari (the liquid pooling on the top of the miso) or you can mix it back into the miso and eat it as is. Your miso may be chunky; if you prefer a smoother paste, process it in a grinder or food processor. Store in an airtight container. The miso will keep indefinitely in the refrigerator.

Chickpea Miso

YIELD: ABOUT 6 CUPS

Chickpea miso has a mellow flavor and a pleasing light yellow color. It is a fantastic all-purpose miso and is ready when fairly young. The ratio of chickpea to koji is nearly equal. This miso can also be made as a chickpea-barley miso — simply replace the rice koji with barley koji.

PROCESS **miso (page 254)**	FERMENTATION/AGING **2 months**

2 cups (350 g) dry chickpeas

2½ cups (440 g) light Rice Koji (page 201)

½ cup plus 1 tablespoon (150 g) salt, plus extra to prep the vessel and top the miso

1 tablespoon (16 g) unpasteurized miso

1. Soak the chickpeas for 8 to 24 hours. If you're making your own koji, time the soaking with the koji cycle; you will want your chickpeas to be cooked when your koji is ready to come out of the tray.

2. Boil the chickpeas in plenty of water until they are soft, about 1 hour, or steam them in an electric pressure cooker for 35 minutes. Drain, reserving the cooking water, and then spread them out on a tray to cool slightly.

3. When the chickpeas have cooled to below 100°F/38°C, combine them in a bowl with the koji and ½ cup of the salt. Mash together with a potato masher, or put through a meat grinder if you have one. Add the miso and thoroughly mix. Add enough of the bean cooking water to achieve the desired consistency — it should be chunky and dryish, like mashed potatoes.

4. Using a bit of the bean cooking water, rinse the inside of your fermentation jar or crock, making sure to coat all of the surface. Then sprinkle 1 tablespoon salt into the jar, making sure to coat all the sides and the bottom of the vessel.

5. Spoon the chickpea mixture into your jar or crock, doing your best to remove as many air bubbles as possible.

6. Set a small piece of unbleached cotton cloth or parchment paper cut to fit the diameter of your vessel on top. Sprinkle about ½ tablespoon of salt along the edges of this cover to seal any gaps. Weight the miso with about 6 pounds, or as best you can; see page 247.

7. Cover the entire vessel with cloth or paper, securing it in place, and follow the instructions for aging on page 248. We find that the flavors become perfect in about 2 months.

8. When you are ready to harvest your miso, open it up and follow the harvesting instructions on page 251. You may need to scrape off the top surface of the miso until you get to something that looks nice and rich in color. You can either strain off the tamari (the liquid pooling on the top of the miso) or you can mix it back into the miso and eat it as is. Your miso may be chunky; if you prefer a smoother paste, process it in a grinder or food processor. Store in an airtight container. The miso will keep indefinitely in the refrigerator.

Red Miso

Red miso seems to be the most ubiquitous type of miso on market shelves in the United States. This miso is made with red rice koji, but despite its name, the koji itself isn't red and it's not what gives this miso its color. The deep red hue comes from the way the beans are cooked. The soybeans are simmered for 6 to 8 hours, then cooled overnight before being brought back to a boil the next day.

PROCESS **miso (page 254)**	FERMENTATION/AGING **6–12 months**

2 cups (350 g) dry soybeans

2½ cups (440 g) red rice koji or Rice Koji (page 201)

½ cup plus 1 tablespoon (150 g) salt, plus extra to prep the vessel and top the miso

1 tablespoon (16 g) unpasteurized miso

1. Soak the soybeans for 8 to 24 hours. If you're making your own koji, time the soaking with the koji cycle; you will want your soybeans to be cooked when your koji is ready to come out of the tray.

2. Early in the day, place the soybeans in a pot covered with plenty of water and bring to a boil. Skim off any foam and lower the heat to maintain a simmer. Simmer the beans in a covered pot for at least 6 hours and up to 8 hours. Be sure to check your water level regularly to maintain submerged beans. Turn off the heat. Keep the pot covered and let them sit overnight. The next day, bring the beans to a boil and boil for 10 minutes. Drain, reserving the cooking water.

3. When the beans have cooled to below 100°F/38°C, combine them in a bowl with the koji and ½ cup of the salt. Mash together with a potato masher, or put through a meat grinder if you have one. Add the miso and thoroughly mix. Add enough of the bean cooking water to achieve the desired pastelike consistency.

4. Using a bit of the bean cooking water, rinse the inside of your fermentation jar or crock, making sure to coat all of the surface. Then sprinkle 1 tablespoon salt into the jar, making sure to coat all the sides and the bottom of the vessel.

5. Spoon the mixture into your jar or crock, doing your best to remove as many air bubbles as possible.

6. Set a small piece of unbleached cotton cloth or parchment paper cut to fit the diameter of your vessel on top. Sprinkle about ½ tablespoon of salt along the edges of this cover to seal any gaps. Weight the miso with about 6 pounds, or as best you can; see page 247.

7. Cover the entire vessel with cloth or paper, securing it in place, and follow the instructions for aging on page 248.

8. When you are ready to harvest your miso, open it up and follow the harvesting instructions on page 251. You may need to scrape off the top surface of the miso until you get to something that looks nice and rich in color. You can either strain off the tamari (the liquid pooling on the top of the miso) or you can mix it back into the miso and eat it as is. Your miso may be chunky; if you prefer a smoother paste, process it in a grinder or food processor. Store in an airtight container. The miso will keep indefinitely in the refrigerator.

CHOOSE YOUR OWN MODERN
MISO ADVENTURE

A modern miso or tasty paste is basically the combination of four ingredients: cooked legumes, koji-fied grains, salt, and time. The last two ingredients are intertwined, as the amount of salt needed is proportional to the amount of time the miso will age. The longer the aging, the more salt you need to keep it fermenting properly. We created a guide to help you navigate your first experiments. As you get the hang of it, you can take the ratios up or down. The less salt, the more "interesting" and "funky" the flavors, which can be fun or heading toward spoilage. Just remember: if it is full-on off (see troubleshooting on page 378), it is food for the compost pile, not for people.

We've tried to give a spectrum of options from which to launch your tasty paste pantry. As you may remember, the traditional recipes call for higher salt ratios than the modern misos and tasty pastes, partly because we don't have the same control challenges (seasonal temperature and humidity fluctuations) and partly because our tastes have changed — though as a society we tend to ingest more salt because it is hidden in so many processed foods. Miso adds more flavor with less sodium. In modern misos and tasty pastes, we are looking to capture both flavor and health.

You will need a gram scale to create your own recipe. We did all the math for you in the recipes that follow, but you will need to know the weights of your ingredients to determine the correct proportions when customizing your own misos.

To create your own recipe, start by asking how much miso you want to make, and let that answer guide your quantity of legumes. Next, pick which type of miso you want to make; this will determine the amount of koji you need. The salt quantity is determined by the weight of the legumes and koji, so add up their weights and multiply by the ratio specified for the type of miso you want. Don't forget to write the date on your container and when it should be ready.

Another way to customize your miso is by adding botanicals. This practice is not without

precedence. The traditional *namemiso* "finger lickin' misos" ("finger lickin'" comes from the Japanese *nameru*, which means "to lick") are distinguished by these inputs. You can add your botanicals when you are packing the paste into the fermentation vessel or after some aging has taken place. We advise you to keep the percentage of herbs and veggies to 10 percent or less of the total weight. One of our favorite botanical misos is South River Miso's Dandelion Leek. For this miso, they add ramps, dried nettles, and sea vegetables to a combination of 2-year brown rice miso and 1-year brown rice miso and then age the whole batch for another year.

For the most adventurous, try kojified grains and a substrate of your imagination — chocolate donuts, pizza, leftover zucchini bread, the whole enchilada, the list goes on. We are only half joking; many foods can be kojified and turned into tasty amino-rich pastes, as you will see when you read about Miso Method Cheese on page 286. For unconventional ingredients, you can also use the following guidelines. For example, weigh a dozen donuts, add two-thirds of that weight in rice koji to the donuts, grind them together with 4.5 percent of their weight in salt, pack it all in a fermentation vessel, and weight it down.

SWEETEST AND SWEET MISO FORMULAS

DRY LEGUMES 1 pound per about 3 quarts of finished miso

KOJI 3 times the amount of legumes

SALT 4.5 percent of the total weight of legumes + koji (for sweetest); 6 percent of the total weight of legumes + koji (for sweet)

TIME Age for 2–3 weeks for sweetest, or 2 months for sweet

Example: For about 3 Quarts of Sweet Miso
1 pound (450 grams) dry legumes
3 pounds (1,350 grams) koji
⅓ cup plus 1 tablespoon (108 grams) salt

Let's do the math:
Sweetest: 450 (grams of legumes) + 1,350 (grams of koji) = 1,800 grams

1,800 × 0.045 = 81 grams of salt

Sweet: 450 (grams of legumes) + 1,350 (grams of koji) = 1,800 grams

1,800 × 0.06 = 108 grams of salt

MELLOW MISO FORMULA

DRY LEGUMES 1 pound per a little over 2 quarts of finished miso

KOJI 2 times the amount of legumes

SALT 10 percent of the total weight of legumes + koji

TIME Age for 6 months

Example: For about 1¼ Gallons of Mellow Miso
2 pounds (900 grams) dry legumes
4 pounds (1,800 grams) koji
1 cup (270 grams) salt

Let's do the math:
900 (grams of legumes) + 1,800 (grams of koji) = 2,700 grams

2,700 × 0.10 = 270 grams of salt

MEDIUM MISO FORMULA

DRY LEGUMES 1 pound per about 2 quarts of finished miso

KOJI Same amount as legumes

SALT 12 percent of the total weight of legumes + koji

TIME Age for 12 months

Example: For about 2 Gallons of Medium Miso
4 pounds (1,800 grams) dry legumes
4 pounds (1,800 grams) koji
1½ cups plus 1 tablespoon (432 grams) salt

Let's do the math:
1,800 (grams of legumes) + 1,800 (grams of koji) = 3,600 grams

3,600 × 0.12 = 432 grams of salt

SALTY MISO FORMULA

DRY LEGUMES 1 pound per about 1¾ quarts of finished miso

KOJI Half the amount of legumes

SALT 16 percent of the total weight of legumes + koji

TIME Age for 24 months

Example: For about 2½ Quarts of Salty Miso
1½ pounds (680 grams) legumes
¾ pound (340 grams) koji
½ cup plus 2 scant tablespoons (122 grams) salt

Let's do the math:
680 (grams of legumes) + 340 (grams of koji) = 1,020 grams

1,020 × 0.16 = 163 grams of salt

Soybean/All-Bean Miso

YIELD: ABOUT 1½ QUARTS

This type of miso is made entirely without grains. Instead, the koji spores are grown straight on the beans. (Purchase *Aspergillus hatcho,* which is intended for growing on soybeans; see the source guide on page 388.) Use this hatcho koji as a base for a pure soybean hatcho-style miso, or for any all-bean ferment using other legumes. The hatcho koji spores will also grow directly on sunflower seeds. We have adapted this recipe from GEM Cultures.

Deep and earthy, hatcho miso is a top-shelf example of a pure soybean miso and has been made in the Okazaki area of Japan for over 600 years. Chef Josh Fratoni described this paste as "Vegemite on steroids." It is made with very little liquid, so it is also quite firm.

PROCESS **miso (page 254)**	FERMENTATION/AGING **24–30 months**

4 cups (700 g) koji grown on soybeans or other legume (page 206)

½ cup plus 1 tablespoon (135 g) salt, plus extra to prep the vessel and top the miso

1 cup boiled water, plus extra for rinsing out the fermentation vessel

1 tablespoon (16 g) unpasteurized miso

1. Mash the koji with a potato masher or run it through a meat grinder if you have one.

2. Stir ½ cup of the salt, 1 cup of the boiled water, and the miso into the koji. The mixture will be quite thick and pasty.

3. Rinse the inside of your fermentation jar or crock with just-boiled water, making sure to coat all of the surface. Then sprinkle the remaining 1 tablespoon salt into the jar, making sure to coat all the sides and the bottom of the vessel.

4. Make golf ball–size balls with the entirety of the paste. Layer your vessel as you go, tossing the balls in with a bit of energy to help minimize air pockets.

5. Set a small piece of unbleached cotton cloth or parchment paper cut to fit the diameter of your vessel on top. Sprinkle about ½ tablespoon of salt along the edges of this cover to seal any gaps. Weight the miso with about 3 pounds, or as best you can; see page 247.

6. Cover the entire vessel with cloth or paper, securing it in place, and follow the instructions for aging on page 248. This miso is typically ready in 24 to 30 months, but if you are like us, you can't possibly wait that long. We start tasting ours after about 3 to 4 months so that we can enjoy the evolving flavor.

7. When you are ready to harvest your miso, open it up and follow the harvesting instructions on page 251. You may need to scrape off the top surface of the miso until you get to something that looks nice and rich in color. You can either strain off the tamari (the liquid pooling on the top of the miso) or you can mix it back into the miso and eat it as is. Your miso may be chunky; if you prefer a smoother paste, process it in a grinder or food processor. Store in an airtight container. The miso will keep indefinitely in the refrigerator.

Shiitake Tasty Paste

YIELD: ABOUT 1 QUART

We live in the Pacific Northwest, where mushrooms abound, so it is only natural that we thought of making mushroom miso. Despite all our attempts to connect with the elusive fungi ("to find a mushroom you must *be* a mushroom"), it turns out that we are not among those people who find pounds and pounds of morels, boletes, or chanterelles. Our hauls can be counted by the number of specimens — "Woo-hoo, we found six! We can make an omelet!" We remain ever optimistic; this year is always *THE* year. Meanwhile, we make our mushroom tasty paste with dried shiitakes that we buy at the store.

We have found that the flavor of dried mushrooms is superior in this application, so if you use your found mushrooms, we suggest drying them first. With dried mushrooms, the flavor is brought forward — it is pure mushroom umami, squared.

Use this tasty paste like mushroom broth; we use it in anything that needs a little something-something. It will be your secret when people rave about your cooking. This paste also makes a wonderful mushroom tamari. In early trials of this recipe, we used much higher salt ratios, which dominated the mushrooms but was incredible for salting dishes. If you want to try a higher salt concentration, use 80 grams of salt instead of 48.

We suggest you measure the ingredients by weight first and by volume second. We find that mushy mushrooms are hard to measure consistently in cups. The weight is a better way to go.

PROCESS **miso (page 254)**	FERMENTATION/AGING **1–2 months**

2 cups (40 g) dried shiitake mushrooms

2 cups (350 g) Rice Koji (page 201) or Barley Koji (page 203)

2 tablespoons plus 2 teaspoons (48 g) salt, plus extra to prep the vessel and top the miso

1 tablespoon (16 g) unpasteurized white miso

1. Bring a kettle of water to a boil. Put the mushrooms in a bowl and pour just enough boiling water over them to cover them. Soak until fully hydrated, about 5 minutes. Then wring out the mushrooms, as you don't want them too wet. Save the soaking liquid.

2. Place the mushrooms in a food processor. Pulse a few times until the mushrooms are finely chopped but not mushy.

3. Break up koji, if necessary, and add it to the food processor. Pulse one more time to mix it in.

4. Add 2 tablespoons plus 2 teaspoons of the salt and the miso and mix until evenly mixed. It will be soft and a little chunky, but it shouldn't be overly wet. If you used dried koji, you may need to add a little of the mushroom soaking liquid.

5. Rinse the inside of your fermentation jar or crock with a bit of the mushroom soaking water, making sure to coat all of the surface. Then sprinkle 1 tablespoon of the salt into the jar, making sure to coat all the sides and the bottom of the vessel.

6. Spoon the mushroom mixture into your fermentation vessel, doing your best to remove as many air bubbles as possible.

Recipe continues on next page

Shiitake Tasty Paste, *continued*

7. Set a small piece of unbleached cotton cloth or parchment paper cut to fit the diameter of your vessel on top. Sprinkle about ½ tablespoon of the salt along the edges of this cover to seal any gaps. Weight the tasty paste as best you can; see page 247.

8. Cover the entire vessel with cloth or paper, securing it in place, and follow the instructions for aging on page 248. This tasty paste is ready enough after the first month, but we find that the flavor becomes perfect after about 2 months.

9. When you are ready to harvest your miso, open it up and follow the harvesting instructions on page 251. You may need to scrape off the top surface of the miso until you get to something that looks nice and rich in color. You can either strain off the tamari (the liquid pooling on the top of the miso) or you can mix it back into the miso and eat it as is. Your miso may be chunky; if you prefer a smoother paste, process it in a grinder or food processor. Store in an airtight container. The miso will keep indefinitely in the refrigerator.

Good Mother Stallard Millet Tasty Paste

YIELD: ABOUT 1 QUART

There are so many amazing beans that it is hard to know where to start. We were lucky enough to have a conversation with a connoisseur of heirloom beans, Steve Sando of Rancho Gordo (see page 50). Steve introduced us to Good Mother Stallard. It's a dense, meaty (and delicious) bean that holds its own and doesn't need many inputs (like salt) to make it shine.

This sweet-style paste is one of our favorites — and that is saying a lot after the dozens we made while writing this book. If you can't find Good Mother Stallard beans, good substitutes are orca, cranberry, Anasazi, or Appaloosa beans.

PROCESS **miso (page 254)**	FERMENTATION/AGING **1–2 months**

1 cup (175 g) dry Good Mother Stallard beans

1½ cups (225 g) Millet Koji (page 204)

2 tablespoons (34 g) salt, plus extra to prep the vessel and top the miso

1. Soak the beans for 8 to 24 hours. If you're making your own koji, time the soaking with the koji cycle; you will want your beans to be cooked when your koji is ready to come out of the tray.

2. Boil the beans in plenty of water until they are soft, about 1 hour, or steam them in an electric pressure cooker for 35 minutes. Drain, reserving the bean cooking water.

3. When the beans have cooled to below 100°F/38°C, combine them in a bowl with the koji and 2 tablespoons plus 1 teaspoon of the salt. Mash together with a potato masher. Add a bit of the bean cooking water to achieve the desired consistency — it should be chunky and dryish, like mashed potatoes.

4. Using a bit of the bean cooking water, rinse the inside of your fermentation jar or crock, making sure to coat the entire surface. Then sprinkle 1 tablespoon salt into the jar, making sure to coat all the sides and the bottom of the vessel.

5. Spoon the mixture into your jar, doing your best to pack it firmly and remove as many air bubbles as possible.

6. Set a small piece of unbleached cotton cloth or parchment paper cut to fit the diameter of your vessel on top. Sprinkle about ½ tablespoon of salt along the edges of this cover to seal any gaps. Weight the paste as best you can; see page 247.

7. Cover the entire vessel with cloth or paper, securing it in place, and follow the instructions for aging on page 248. This tasty paste is ready enough after 1 month, but we find that the flavors become perfect after about 2 months.

8. When you are ready to harvest your paste, open it up and follow the harvesting instructions on page 251. You may need to scrape off the top surface of the paste until you get to something that looks nice and rich in color. You can either strain off the tamari (the liquid pooling on the top of the paste) or you can mix it back into the paste and eat it as is. Your paste may be chunky; if you prefer a smoother paste, process it in a grinder or food processor. Store in an airtight container. The paste will keep indefinitely in the refrigerator.

Meet the Maker
White Rose Miso

Sarah Conezio and Isaiah Billington are former pastry chefs who are passionate about sourcing ingredients locally (in their case, in the mid-Atlantic region) and using microbes to transform these everday ingredients into powerful flavor. In an effort to maximize local flavors, they got into making vinegar and miso. They knew that each of these foods could accept a wide variety of inputs, making it easy for them to work closely with a few small growers. Sarah and Isaiah make vinegar under the name Keepwell Vinegar and miso under the brand White Rose Miso.

When referring to the fermenting work they do, Sarah and Isaiah use the term "chasing sugar." Given that many microbes, like many of us, are all about sugar, we were intrigued. We asked them to tell us about "chasing sugar."

Q: You have amazing-looking ferments that are definitely not made with traditional ingredients. We find that exciting. We would love to hear about how you come up with ideas for these misos.

A: The flavors are 100 percent our attempts at re-creating what has worked before, but within the constraint that we have to buy every input directly from a responsible mid-Atlantic farmer. That means that rice is often just too expensive (if it's available at all) for us, so miso as we are used to defining it is too difficult for us to make. We look for all of the members of the koji family tree that we can re-create using local inputs. Hoisin sauce became a tempting project when I found a tidbit of research saying that it was a miso-type fermentation using charred sweet potatoes, soybeans, and wheat — all of which we have in abundance. Sunchoke miso is just another iteration of the same thought. In this case, we first wanted to try it because we knew a few guys with some extra to sell, and the flavor of the hard-roasted sunchoke is so good already.

If it's delicious, we keep doing it. We have to be really honest with ourselves and decide when our commitment to sourcing the food locally has led us into a space where we're making decisions that disrespect the tradition and technique surrounding these foods. There's a lot of bad miso out there, now that it's fashionable.

Q: How (and for how long) do you age your miso?

A: We age in 12-gallon barrels weighted with river rocks. The rocks need to be as heavy as the miso in each barrel. Miso is all over the place. There are plenty of floating variables — the ratio of koji to substrate, the salt and hydration levels — that affect the time necessary for the proper product. We do a benne seed miso, for instance, that has a short fermentation. The seed has oil that will eventually spoil, so that miso ferments for about 6 weeks. Our longest ferments right now are at about 12 months.

Q: Is this a good time to be an artisan maker of vinegar and miso in the United States? If so, why?

A: It's never been better than right now, certainly. There is an outsize focus on the food we eat, how it's prepared, what spice is used, what new technique is absorbed. We're dying to invest meaning in our feeding rituals, and we overdo it with this food media, like it's empty calories. It takes a high level of effort to have a real connection to how your food is grown, prepared, cleaned, and delivered to you, to see your own values reflected in and corroborated by the food systems that you support, to even be able to infuse any kind of real meaning into the biological sustenance imperative. And none of us have the time. And it's hard. And our lives are hard in other ways, and who wants to invite a challenge to what is supposed to be a basic ability to feed and nourish ourselves with a similarly basic level of competence? . . .

To be a specialty food producer in 2017 is to have a chance maybe to engage that conversation, put it in a bottle, and sell it for enough money to be able to afford to do it again tomorrow. It means to stand opposite the table from the lady at the farmers' market who wants a meaningful gift for her newlywed friends who have just gotten really into cocktails. To be at the farmers' market is to be like a cleric at a church who hears people's confessions and identity constructions and just moves the needle toward grounding some of this self-expression and values corroboration. There really is a need for hard work to be done in good land stewardship, and it won't be cheap to get behind it, but we can probably find a way to afford it.

Q: We remember inviting people to taste locally produced fermented vegetables at farmers' markets way before that was hip. How do you introduce your miso to people who have never tasted miso?

A: This is a bit of a tough question, mostly because all of the people who buy our miso are hip to using it in a professional chef manner, rotating on the David Chang–Sean Brock axis. These are people who are used to breaking recipes, techniques, and traditions down into their component parts of taste, texture, temperature, et cetera, and rebuilding them within the context of their culinary milieu. Our miso is a huge win for them for the following reasons: 1. It's hip. 2. It delivers umami easily. Even pros don't often have a grasp on miso, at least in my (admittedly 100 percent Western hemispherical) experience. 3. It's made using the inputs they already want, from good farmers. 4. And this one is sneaky — it's sweet. Successful restaurant cuisine, more than anybody on our side of the aisle likes to admit, is based on feeding people food that is sweet but doesn't seem to be.

Sea Island Red Pea Miso

By Isaiah Billington and Sarah Conezio, White Rose Miso

YIELD: ABOUT 4 POUNDS

The Sea Island red pea is a cowpea, brought from Africa, that was grown in the Carolina Sea Islands before the American Civil War. They were part of the crop rotation with rice in the Carolina Lowcountry, supporting the health of the soil as well as the nutrition of the people. Once on the brink of extinction, these seeds survived in gardens and feral patches and are slowly being brought back into the fields and on tables.

ISAIAH AND SARAH (SEE PAGE 274) WRITE: "We work hard to balance our respect for traditional methods against the exigencies of our local ingredients. Heinz Thomet and Gabriela LaJoie, of Next Step Produce in Newburg, Maryland, supply us with all of our barley. They are passionate about delivering nutrition through the food that they grow, and they will not polish the bran off the grain that they sell. We first saw this as a problem to solve, but now we embrace whole grains in our miso. Most instructions for cooking grains for koji stress that they must be steamed, not boiled, but you'll find that boiling works quite well for unpolished barley. The little kernel pops open to absorb just the right amount of water and to allow the mold to spread during incubation."

PROCESS **miso (page 254)**	FERMENTATION/AGING **72 hours for koji; 6–9 months for miso**

2½ cups (500 g) whole barley

½ teaspoon (1 g) koji starter

2½ cups (500 g) Sea Island red peas

¾ cup plus 1 tablespoon (108 g) sea salt, plus extra to prep the vessel and top the miso

1. Try to time the making of the koji with readying the peas. Soak the barley in at least 2 quarts of water for 6 hours to 12 hours. Drain and rinse after soaking.

2. Combine the soaked barley with plenty of water in a 6-quart saucepan. Boil over high heat for about 1 hour, or until cooked through and popped wide open. Stir and add more water as it is absorbed and evaporates, and skim off any scum or flotsam that comes to the surface. If the water starts to thicken, the grain has begun to overcook and release too much starch; remove it from the heat immediately.

3. Pour the cooked barley into a strainer and let it drain for at least a few minutes. Spread it out on a tray to cool slightly and steam off more moisture.

4. When the barley has cooled to below 100°F/ 38°C, transfer it to a bowl. Using a fine-mesh strainer, shake half of the koji spores evenly over all of the barley, then mix and repeat until you've mixed in all of the spores. Gently stir the barley for another full minute to distribute evenly.

5. Line a half sheet tray with a large kitchen towel. Transfer the inoculated grain onto the towel and spread it out in an even layer about 1 inch thick. Fold the towel over the barley to cover.

6. Place the tray in an incubation chamber set to 85°F/29°C and high humidity (see page 192 for options). After 24 hours, pull out the tray, transfer the barley koji to a bowl, and mix to redistribute. Then return the barley koji to the tray and return the tray to the incubation chamber. After about 36 hours, the koji may begin to produce its own heat. At this point, unfold the cloth so that it does not cover the grain and adjust the temperature as necessary to maintain 85°F/29°C. Incubate the barley koji for a total of 72 hours, or until the layer of grains looks like a fuzzy white mat. It should be strong enough that you can pick up large pieces of it, if not the whole thing.

7. At least 6 hours and up to 24 hours before the koji is finished, begin soaking the peas in at least 2 quarts of water. If you are soaking for 6 hours, room temperature is fine, but if you are soaking for longer, soak the peas in the refrigerator.

8. Combine the peas with plenty of water in a 6-quart saucepan. Boil over high heat until soft, about 2 hours. Skim off any scum or hulls that rise to the surface and add water as necessary to replace what evaporates.

9. Pour the cooked beans into a strainer and let them drain for at least a few minutes. Then spread them out on a tray to cool slightly and steam off moisture.

10. When the beans have cooled to below 100°F/38°C, combine them in a bowl with the koji and ¾ cup of the salt, and mash together with a potato masher or meat grinder. The mixture should be chunky and dryish, like mashed potatoes.

11. Rinse the inside of your fermentation vessel with just-boiled water, making sure to coat the entire surface. Then sprinkle 1 tablespoon salt into the vessel, making sure to coat all the sides and the bottom of the vessel.

12. Spoon the mixture into your vessel, doing your best to pack it firmly and remove as many air bubbles as possible.

13. Salt the top of the paste well, then set a small piece of unbleached cotton cloth or parchment paper cut to fit the diameter of your vessel on top. It should sit right on the salt layer. Weight the paste as best you can; see page 247.

14. Cover the entire vessel with cloth or paper, securing it in place. Place it in a cool, dark place and follow the instructions for aging on page 248. It is ready after 6 to 9 months.

15. When you are ready to harvest your miso, open it up and follow the harvesting instructions on page 251. You may need to scrape off the top surface of the miso until you get to something that looks nice and rich in color. You can either strain off the tamari (the liquid pooling on the top of the miso) or you can mix it back into the miso and eat it as is. Your miso may be chunky; if you prefer a smoother paste, process it in a grinder or food processor. Store in an airtight container. The miso will keep indefinitely in the refrigerator.

Heirloom Cranberry Bean Miso

By Cheryl Paswater, Contraband Ferments

YIELD: 1 QUART

CHERYL (SEE PAGE 282) WRITES: This miso came about after I became obsessed with cranberry beans about two or three summers ago. I first encountered the cranberry bean at the Union Square Farmers Market, and at the time I was also mentally stalking Sean Brock after his book *Heritage* came out. It really expanded my understanding of heirloom beans as well as seed saving. I random-found some dried heirloom cranberry beans that winter and immediately thought of miso! I'd never seen a cranberry bean miso before.

PROCESS **miso (page 254)**	INCUBATION **10–12 months**

1½ cups (260 g) dry cranberry beans

1½ cups (260 g) dried or fresh rice koji

¼ cup (32 g) chopped leeks

2 garlic cloves, minced

1 strip seaweed (dulse, kombu, or other), chopped

1–2 teaspoons (3-6 g) red pepper flakes, depending on your desired heat level

4 tablespoons (65 g) sea salt, plus extra to prep the vessel and top the miso

1. Soak the beans for 8 to 24 hours. If you're making your own koji, time the soaking with the koji cycle; you will want your beans to be cooked when your koji is ready to come out of the tray.

2. Drain the beans and place them in a large pot. Cover with fresh water and cook over medium heat until al dente, 1 to 1½ hours. Or steam them in a pressure cooker for 30 minutes.

3. Drain the beans, reserving the bean cooking water.

4. If your koji is dried, place it in a bowl. Once the bean cooking water has cooled to below 100°F/38°C, add enough water, a tablespoon at a time, to the koji to moisten it. You will likely have to add more liquid once the first bit is soaked up. You want it to be wet enough that the koji feels well hydrated.

5. Transfer the cooked beans to a bowl and mash with a potato masher, being sure to break the hulls on each bean.

6. Add the koji to the mashed beans and mix well. Add the leeks, garlic, seaweed, pepper flakes, and 3 tablespoons of the salt, and mix again. The mixture should have a toothpastelike consistency.

7. Rinse the inside of your fermentation jar or crock with a bit of the bean cooking water, making sure to coat the sides well. Sprinkle 1 tablespoon salt into the vessel, making sure to coat all the sides and the bottom.

8. Spoon the miso mixture into your fermentation vessel, doing your best to pack well and remove as many air bubbles as possible.

9. Salt the top of the miso well, then set a small piece of unbleached cotton cloth or parchment paper cut to fit the diameter of your vessel on top. It should sit right on the salt layer. Weight the miso as best you can; see page 247.

10. Put a lid on your miso, then label and date it. Store it in a cool, dark place away from direct sunlight. We like to keep ours under our bed! Let it age for 10 to 12 months.

11. When you are ready to harvest your miso, open it up and follow the harvesting instructions on page 251. You may need to scrape off the top surface of the miso until you get to something that looks nice and rich in color. You can either strain off the tamari (the liquid pooling on the top of the miso) or you can mix it back into the miso and eat it is as is. Your miso may be chunky; if you prefer a smoother paste, process it in a grinder or food processor. Store in an airtight container. The miso will keep indefinitely in the refrigerator.

A GUIDE TO MAKING
NUT & SEED TASTY PASTES

Nut and seed tasty pastes are fun, fast, and gratifying, since they often take just a few days to ferment. Unlike miso, these pastes don't have an indefinite shelf life. They should be made and enjoyed relatively quickly. Refrigerated, their shelf life is similar to other nut butters. The reason is that nut oils are quick to go rancid. Unfortunately, as magic as koji is, it doesn't stop rancidity. That doesn't mean you will get sick from eating rancid oils, at least not in the short term; rancidity is more of a concern in terms of oxidation and free radicals that are damaging to our cells. Rancid oils don't taste good, but they are common in the processed foods of our modern world, and so people are less likely to know the off flavor. Rancid oil will smell and taste musty or somewhat like cardboard.

Since these nut and seed tasty pastes aren't aged, you can use a much lower salt ratio. We have had good results with 3 to 5 percent salt by weight (see how to calculate salt on facing page). For these tasty pastes, we generally use equal parts koji (any variety) to nuts or seeds (any variety). Add at least 1 teaspoon of white miso to get things started on the right foot and give you deeper flavor. You may use more miso and any variety you'd like; it really depends on what kind of flavor you are looking to build. Chef Josh Fratoni gives his tasty pastes a much bigger boost; they are usually 2 percent miso at the start.

Sometimes fermented nut and seed pastes can finish a little dull, by which we mean that the nut butter is too far forward and the umami is a bit hard to taste. This is the downside of balancing the breakdown of the protein without the breakdown of the oils. Josh shared that he has experimented a little bit with fermenting the pastes at higher temperatures in a water bath with an immersion circulator to speed up the action of the enzymes to see if he could get more flavor development in a shorter time.

Try pumpkin seeds, pistachios (Cheryl Paswater recommends pistachios with barley koji), tahini (this was good but didn't have a long shelf life), cashews, sunflower seeds, chestnuts (Josh says they came out like umami kettle corn), pine nuts, peanuts, macadamia nuts (see Cheryl's recipe, page 281), hazelnuts — the sky is the limit!

Soak nuts and seeds for 8 to 12 hours. Peanuts should be soaked and then boiled for about 20 minutes. Sesame seeds don't need to be soaked.

Coarsely grind the nuts or seeds in a food processor, then proceed as you would for miso: Mix in koji, white (or other) miso, and salt, and add enough boiled water (cooled to below 100°F/38°C) to get a toothpaste-thick texture. Pack tightly in a jar, leaving a little room for expansion. Sprinkle a thin layer of salt on top. Cover with unbleached cotton cloth or parchment paper and weight down as best you can. Secure a cloth or paper cover over the jar. Age for 5 to 21 days at room temperature.

Macadamia Tasty Paste

By Cheryl Paswater, Contraband Ferments

YIELD: 1 PINT

CHERYL WRITES: This recipe stems from my excitement around koji. After going bean crazy with my misos for two or three years, I started making nut- and seed-based misos after realizing they can be an amazing base, adding umami flavor, for dairy-free cheeses. I started with cashews and moved on to other nuts and seeds. I really love the flavor profile and nutritional benefits of macadamia nuts and decided I would give it a try.

INCUBATION **3–5 days**

---

1 cup (132 g) macadamia nuts

½ cup (90 g) fresh or dried rice koji

1–2 teaspoons (16–32 g) sea salt

---

1. Soak the macadamia nuts for 6 to 12 hours.

2. If your koji is dried, place it in a bowl and add ¼ cup boiled and cooled water. Mix well so that the koji is evenly moist.

3. Drain the nuts. Place the nuts and koji in a blender or food processor. Process, adding a little more water as needed, until you have a tooth-pastelike consistency.

4. Add the salt and mix well.

5. Transfer the blended mixture to a small jar, making sure to tap or press out any air pockets. Leave ½ to 1 inch of headspace for expansion.

6. Put a lid on the jar and label it with its contents and the date. Ferment at room temperature for 3 to 5 days.

7. When you are ready to harvest the paste, open it up and follow the harvesting instructions on page 251. You can either strain off the amino sauce (the liquid pooling on the top of the paste) or you can mix it back into the paste and eat it as is. Your paste may be chunky; if you prefer a smoother paste, process it in a grinder or food processor. Store in an airtight container.

Calculating Salt for Nut Tasty Pastes

To figure out how much salt you need to add to your tasty paste, add together the weight of koji and the weight of the dried nuts, then multiply that weight by the salt percentage you desire. For example, if you have 1 cup (132 grams) of nuts and ½ cup (90 grams) of koji and want a tasty paste that is 4 percent salt, you'll need about 9 grams of salt. Here's the math:

$$132 + 90 = 222$$

$$222 \times 0.04 = 8.88 \text{ grams of salt}$$

Meet the Maker

Cheryl Paswater
Contraband Ferments

HEIRLOOM CRANBERRY BEAN MISO (RECIPE ON PAGE 278)

Cheryl Paswater combines her career as an artist (and we want to stop and take note right here — how often do you hear that being an artist is someone's day job?) with a side fermentation business. She teaches, advocates, connects people, and makes ferments. She says, "I like to describe Contraband Ferments as a project. We started out with our fermentation CSA and then we progressed into fermentation education, private coaching to individuals and other fermenters, health coaching, and doing custom orders and work for people who have food intolerances or other needs that keep them out of the conventional pool. We really love helping spread fermentation love to everyone!"

When we asked Cheryl to pick one word to describe herself, she said, "Unicorn." But of course.

Q: What's your miso journey?

A: I started making miso a few years ago and kept it small and basic. I didn't really eat soy, so I started off by making black bean miso and chickpea miso, and then over the years I expanded into adding leeks, garlic, red chiles, and other ingredients. I kept things simple for a long time and then got in the idea of nut-based misos. I started with cashews and pistachios and then went deeper into that. As I've gotten more and more interested in food anthropology and seed saving, I began to play with heirloom beans for miso. Now I make misos with orca beans, cranberry beans, and more. The more obsessed I got with making miso, the more infatuated I became

with koji in general, as well as alternative ways to using miso in other ferments. I started making misozuke (miso pickles), miso-cured eggs, and miso turkey meatloaf, among other things.

Q: What is your advice for the home fermenter working in a small space in regard to making miso?

A: Start small and make it work. I live in Brooklyn, New York, where space is tight, and I run a fermentation CSA. I've had to get creative about where to stash all my miso (and I make A LOT of miso!). I have miso all over my apartment — in the kitchen, in hall closets, under my bed, and anywhere I can find a cool-ish dark space. I like to work in smaller batches for storage reasons, though on occasion I do larger 5-gallon buckets and tuck them away in a big closet.

Q: What are your five rules for making miso-like tasty pastes?

A:

5. **Magic.** Talking, singing, pet names . . . they all are part of the making experience. Do that.

4. **PLAY!** The best way to know what you like is to experiment. We like to make bigger batches (say a gallon) and then mix and match ingredients and pack them into quart jars.

3. **Add other flavoring agents.** Some of my favorite misos have garlic, chiles, leeks, dandelion greens, mustard greens, collard greens, and more.

2. **Buy good koji.** I love getting koji from the crew at Odon in Portland, Oregon, or from South River Miso in the Berkshires. There's nothing better than fresh, well-made koji!

1. **Source good beans.** Sometimes beans sit for a long time and are super-dry; even after you soak and cook them they're still rock hard. Buy good beans and heirloom varieties if you can . . . you won't regret it.

Q: If you were a miso, what kind would you be?

A: Cranberry bean with leeks, garlic, red chile, and seaweed. (See the recipe on page 278.)

MORE TASTY

Here are a few more ideas to take tasty beyond the bounds of convention. First we introduce you to Rich Shih, whose mantra could be "just the method," as he is continually exploring new ways to transform ingredients. You will see this in his recipe for a cheese-based miso (page 286).

We have been marrying our misos with dried fruit and putting them away for a secondary ferment, which is nothing new — makers have been doing variations of this theme for a while. For example, *The Book of Miso* describes a variation of farmhouse barley miso, which is a date miso from the island of Amami Ōshima in Japan.

You will also meet Korea's *jang*, starting with the mother paste, doenjang. From there we move on to Korea's soy sauce guk ganjang and the deeply delicious (and beautiful) gochujang — because peppers ferment so well.

MISO METHOD CHEESE, page 286

Meet the Maker
Rich Shih
Our Cook Quest

The banner on Rich Shih's website, Our Cook Quest (ourcookquest.com), has a very simple mantra: "Learn | Share | Exchange." This pretty much sums up Rich's vision for his work. Rich didn't follow a traditional culinary training path. He found his way through his genuine interest in food. He finds people doing innovative things, tries to figure out how to do it himself, and — here is the unique part of the process — contacts them to ask them to share, and he reciprocates by sharing his own creations. In a world where chefs survive by innovating in front of the pack, Rich's genuine interest in collaboration shines through, and people open up to him. Maybe it's because you can quickly tell from talking to him that nothing he does seems to be ego-driven in any respect. It's about sharing and exchange. It's about a better world.

We first met Rich over a meal that extended for several hours at the Commonwealth Market & Restaurant in Boston. Rich has held classes there and knows the chef, Nicco Muratore, and staff pretty well. In fact, seeing the way everyone greeted one another, it felt more like we had been invited to a low-key but excellent neighborhood restaurant owned by Rich's family. The meal was a fluid experience of enjoying what Chef Nicco was exploring that night and what Rich continued to pull out of his backpack to share with us and the staff. It reminded us of a magic act where more seemed to come out than could possibly fit in his bag. The flavors just kept appearing, starting with a full thermos of a delightful koji kombucha, a miso cheese that he has shared with us for this book (photo opposite; see page 286 for recipe), and other fermented (read: bursting with umami) tasty pastes. He'd brought these for us to try and take with us on our book tour, which we gladly did. The name of the restaurant where we met was very appropriate, because as we have gotten to know Rich better, it's clear to us that he is driven by a vision

of a world that is unified and shares a common wealth, and food plays a big part in helping us all get there.

That night Rich shared his dream project: an educational center grounded in food but with a much bigger vision that at its heart is very simple. As Rich said, people know how to do things — things that other people would like to learn. There was a time in our society when we formed guilds, sharing knowledge in our small communities for the betterment of everyone in that community. We helped each other, and without paywalls or Stripe-enabled cell phones to monetize that knowledge transfer. That's the heart of Rich's vision for this place: somewhere you can go that's awesome because the people there feel good about helping each other out. Food is center because it touches everyone, and it has a way of pulling people together communally.

Rich quickly and easily deflects any praise for his skill and innovation, saying he learned it from others doing it better. Still, his work speaks for itself. Scroll through his @ourcookquest Instagram feed for inspiration on what is possible.

Miso Method Cheese

By Rich Shih, Our Cook Quest

YIELD: ABOUT 1 PINT

RICH (SEE PAGE 285) WRITES: "When I first learned how to make miso, I wanted to explore the possibilities beyond what had already been done before.

"I started by looking at the base nutritional composition of soybeans to understand the protein to carbohydrate to fat loading. In other words, I used the percent composition of each major nutrient component (protein, carb, and fat) of soybeans as a starting point. The next logical step was to compare it to other foodstuffs and draw parallels to experiment with. Throughout the testing process, ricotta cheese miso was the clear standout. I discovered that the enzymes yielded aged cheese flavors in as few as 2 months, which is five times faster than with Parmesan. This was serious potential that I'd never seen in any other medium prior or since."

You can use miso cheese as you would miso or an aged cheese you'd add to a recipe. One simple and delicious application is to make a compound butter by mixing 1 tablespoon of miso cheese into a stick of butter. There's also the option of drying it, so it's more like a hard cheese.

PROCESS **miso (page 254)**　　　　FERMENTATION/AGING **at least 2 months**

---

1 cup (250 g) rice koji (preferably fresh; see note below)

2 tablespoons (35 g) kosher salt

1 cup (250 g) ricotta or any fresh cheese

Note: If you have to use dried koji, combine 1 cup (250 grams) of the dried koji with 1½ tablespoons (25 grams) of boiled and cooled-to-lukewarm water in a small bowl. Allow the koji to hydrate for a couple of hours at room temperature. If you don't want to wait, process the mixture into a rough paste.

---

1. Combine the koji and salt in a medium bowl. With clean hands, mix the koji and salt together so they're evenly distributed. Keep mixing and squeezing the koji and salt together to break

down the koji into a paste as much as possible. Don't worry too much about making it superfine or missing some grains. The pieces will have the opportunity to break down during the fermentation process.

2. Add the ricotta and mix thoroughly. Pour the contents into a pint jar. Cover the surface of the mixture with plastic wrap, then cover with a lid. Secure the lid just fingertip tight; do not seal it tightly.

3. Place the jar in your refrigerator and let the cheese age for at least 2 months. Taste the miso cheese as it evolves. When it has the flavor of an aged cheese, like Parmesan or Romano, it's done. If not, put it back in the refrigerator for another month, or until the flavor comes through.

Apricot Tasty Paste

Once we started experimenting with dried fruit tasty pastes, we couldn't stop trying different variations. They are incredibly easy and incredibly delicious. They are technically secondary ferments. If you can't wait to try these, they work with purchased miso as well. These can be used immediately; however, we encourage you to wait. Place the mixture in a jar, tighten the lid, and put it on top of your fridge for a few weeks. When you "rediscover it," you will be amazed at how the flavors have blossomed.

These tasty pastes age similarly to miso. If you are using a light young miso, it will be sweet and salty, and it's best stored in the fridge to keep those qualities intact, though we prefer it aged out of the fridge. It just gets better.

Apricot and sweet miso make an amazing pairing. The hint of salt from the miso amplifies the apricot flavor in a delicious way. Use this in salad dressing, as a spread on a grilled cheese sandwich, or as is on a cheese and charcuterie plate — who needs jam when you can have apricot tasty paste?

AGING **2 weeks or more**

½ cup (125 g) dried unsulfured apricots

½ cup (125 g) Light Miso (page 263), Sweet White Miso (page 262), or Sweet Miso Formula (page 269)

---

1. Place the apricots in a food processor and process to a paste.

2. Add the miso and pulse until well mixed. Scoop into a pint jar, taking care to press as you go to avoid creating air pockets. Use a butter knife to work out air pockets as needed. Press and smooth the top.

3. Tighten the lid on the jar and place on the counter at room temperature (around 70°F/21°C) for at least 2 weeks. Burp the jar after 5 or 6 days by quickly loosening and tightening the lid. It will keep changing as the enzymes work on the sugar. Continue to age, or store in fridge to slow down fermentation.

Variation: Date Tasty Paste

Use dates in place of apricots for an even sweeter fruit paste.

½ cup (125 g) dried unsulfured pitted dates

½ cup (125 g) Light Miso (page 263), Sweet White Miso (page 262), or Sweet Miso Formula (page 269)

Variation: Fig Tasty Paste

Imagine if the filling of a Fig Newton met a salty, rich companion. The sweet seedy texture is wonderful. Again, the process is the same as with the apricot tasty paste. The saltier miso brings out a deeper flavor that can age longer.

1 cup (250 g) dried unsulfured figs

½ cup (125 g) Red Miso (page 267) or Salty Miso Formula (page 269)

KOREAN JANG

Jang means "soybean sauce." *Doenjang* is "thick soybean sauce," *ganjang* is "salty soybean sauce," and *gochujang* is "pepper soybean sauce."

Doenjang is considered by many to be the root, and even the soul, of Korean cuisine. It is said that doenjang is made not with the hands but with the heart. Traditionally, households made their own doenjang every year. The paste and the iconic soup *doenjang jjigae* evoke the warmth of mothers and home. Traditionally made doenjang is fermented in onggi pots outside, with the changing of the seasons an essential factor in its evolution. The method is steeped in generations of tradition. We loved this piece of wisdom: the salinity of the brine is correct when you drop an egg in it and the "floating part of the egg is the size of a 500-won coin." Because doenjang is such an important part of the family's daily meal, it is said to be bad luck if something goes wrong with the fermentation.

When traditionally made, the beans for doenjang are cooked and crushed in a large wooden mortar to a texture that is slightly chunky and pressed into blocks (*meju*) in the fall when the air is dry and the sun is still warm. These blocks are hung from rice straw ropes and pick up wild aspergillus and other microbes. At the beginning of the lunar calendar, the dried meju blocks are rinsed, set in the onggi pots, and covered with brine. Before the lid is closed, a bit of charcoal, a few jujubes (Korean red dates), and some chile peppers are added. The charcoal is to collect dust and impurities, the jujubes for sweetness, and the chiles to keep out bad microbes.

As winter turns to spring, the ferment warms and enlivens. After 60 days, the meju blocks are removed from the brine and kneaded into a paste that is then pressed back into an onggi pot, where it will continue to ferment for months or years.

Doenjang

Doenjang is eaten daily and is said to help with digestion, especially with heavy meals. Like miso, it is a rich umami paste that gets deeper and darker as it ages. Unlike miso, it is often cooked or boiled when eaten. Also unlike miso, the process of inoculation is still completely wild, at least among the traditional makers.

   Doenjang can be consumed at a young golden state after about 1 year, at a coppery 3 years, or at a deep brown 10 years and beyond. This recipe has its roots in every doenjang video and post we could find, combined with Maangchi's doenjang recipe from her book *Maangchi's Real Korean Cooking: Authentic Dishes for the Home Cook* and many hours of trial and error.

PROCESS **miso (page 254)**	FERMENTATION/AGING **9–12 months**

5 cups (1,000 g) dry soybeans

1–2 cups rice hulls (to line the bottom of your inoculation tray)

2 chunks charcoal (optional)

1 teaspoon (7 g) honey (optional)

1½ gallons (5.7 L) water

9 cups (2.6 kg) salt, plus extra to top the final fermentation

2 whole chiles

3 jujubes (Korean red dates; optional)

1. Soak the soybeans for 8 to 24 hours.

2. Cook or steam the soybeans until they are soft enough to squeeze easily between your thumb and ring finger, following the instructions on page 54. Drain, reserving the bean cooking water.

3. Mash the beans with a large mortar and pestle, potato masher, or food processor, adding very small amounts of the bean cooking water if needed. You want a consistency that is fairly dry, but not so dry that the paste doesn't stick together smoothly in a block. It should feel almost like Play-Doh. You may need to mash the beans in a few batches because it is hard to get a consistent texture. The paste should be neither too chunky nor too smooth.

4. Divide the mash into three or four equal portions. Roughly form each piece into a rectangular brick. Work and pat each brick as needed to work out all of the cracks and air pockets. You want a nice even surface.

5. Let the bean blocks (meju) dry until they are solid, which should take 3 to 4 days in the open (don't use a dehydrator). We put them on a stainless-steel cooling rack set over a seed-starting pad or a heating pad set to low. Rotate the blocks regularly so that all the sides dry evenly. As they dry, the blocks will form cracks — that is okay.

Recipe continues on next page

Doenjang, *continued*

6. At this point the meju are traditionally hung with rope made of rice straw under the eaves to fully dry. If it is the fall, you can try hanging them like this using cotton twine. We have had the best success placing 1 to 2 inches of organic rice hulls in a casserole dish and placing the meju blocks on a wire rack directly above the hulls, rotating the blocks every so often. These hulls have many of the microbes you are looking to catch. You will likely see white fungus starting to grow in the cracks. Dry the blocks for 6 weeks.

7. Remove the blocks from the rope or casserole dish and give them some low heat to encourage the fungus to grow. You can put them in a box and cover it with an electric blanket set to low, or you can put them in a bread proofing box set to 90°F/32°C. The blocks should smell a little like the forest floor and may have brown, yellow, white, or greenish fungi growing on them. Keep them over low heat for 2 weeks.

8. Remove the blocks from their incubation chamber. Hang them again, this time inside, or place them back on the drying rack over the rice hulls.

9. Now it's time to prepare your fermentation vessel. If you are using a gallon jar, spritz the inside with some Everclear alcohol or no-rinse brewing sanitizer, or sanitize it by boiling it in a water bath for 5 minutes and air dry. We don't recommend the following charring procedure for glass. You can also spritz your crockery with Everclear if you do not have charcoal.

To sanitize an earthen onggi pot or crock using the traditional method, set a piece of charcoal on a flame until it is glowing. Use tongs to pick it up and drop it in the onggi pot or crock. Drizzle a teaspoon of honey over the hot charcoal and place the lid on top to trap the hot smoke. After 5 or so minutes, remove the lid and pluck out the charcoal with tongs. Wipe out the inside of the crock with a clean towel. Your vessel is ready.

10. Wash the fungi off each meju block with cold water. Set them aside to dry, then place them in the fermentation vessel. Combine the water and salt and mix well to form a brine solution, then pour the brine into the vessel. Add a fresh piece of charcoal, if you have one, along with a few dried chile peppers and a few jujubes, if you can find them.

11. Cover the vessel with cheesecloth and put on the lid. Let the blocks soak in the brine for 2 to 3 months. Traditionally, the onggi pots are set outside, and on clear sunny days the lid is removed (but not the cloth) to allow the ferment to interact with the sun, and then the lid is replaced at night. This diligence is in a way a meditation. You can put your pot in a window and do the same. Take care that the sunny spot isn't so sunny that it heats up the pot; the solar gain through a window can overheat this ferment. Do not stir.

13. Knead the remnants of the meju blocks with your hands until you have a nice moist paste. Pack the paste back into the fermentation vessel, pour ½ cup of the soy sauce over the top, and sprinkle a few tablespoons of salt on the surface. (At this point it is like packing miso into a fermentation vessel; see page 251.) Cover with a cloth and put the lid on. The doenjang is now ready for its final fermentation.

14. Place the fermentation vessel in a sunny window so that you can open it on a sunny day every few weeks to bring the sun's energy to the paste. Age for 5 to 6 months or longer.

12. After 2 to 3 months, open the crock. Discard the charcoal, chiles, and jujubes. Strain the contents of the fermentation vessel through cheesecloth, being sure to save the liquid. Squeeze the meju in the cloth to release the brine, but do not overly dry it out. The brine is now soy sauce. You can heat it to make guk ganjang (see the recipe on page 293).

A house full of good jang is a house full of blessings.

KOREAN PROVERB

MISO AND OTHER FERMENTED BEAN PASTES, PLUS TASTY SAUCES

GOCHUJANG, page 294

DOENJANG, page 289

GUK GANJANG,
opposite

Guk Ganjang

Korean soy sauce is having a bit of a moment, partially due to the popularity of Korean "temple food," a Zen Buddhist diet for a clear mind. At the he(art) of this food is the slow pace and patience required of fermentation.

Guk ganjang is an alluring reddish brown soy sauce that imparts sweet, sour, and earthy notes; the color is lighter than that of Japanese soy sauces. It differs from shoyu and tamari in that it is made from soybeans only — there are no grains. It is the sister act to doenjang: the same meju blocks give life to both of these foods. After the meju has soaked in its salty brine for 60 days, the solids are strained from the now amber liquid. The solids go in one onggi to become doenjang, and the liquid is boiled and poured into another onggi to become guk ganjang.

Not all Korean soy sauce is guk ganjang; it is specifically the by-product of doenjang. Guk ganjang is predominantly used as a flavoring for soups (*guk*).

Soy sauce from Doenjang (page 289)

1. After the doenjang is tucked in, strain the soy sauce again to catch any remaining bits of meju. You want it to be clear. Place the soy sauce in a small saucepan and bring to a boil over medium-high heat. Reduce the heat to low and simmer for 20 minutes.

2. Turn off the heat and let the sauce cool to body temperature. Pour into a sanitized jar. If you use only clean utensils to retrieve the sauce from the jar, the guk ganjang will keep indefinitely at room temperature.

It takes a lot of energy and time to make ganjang, *and the longer it ages, the more energy it takes. It's made from things in nature such as light and the wind. Starting from the beginning of its life, when it's just a little bean, to the energy it takes to create, sprout, and ferment to make the soy sauce out of it is what soy sauce is all about.*

JEONG KWAN, ZEN BUDDHIST NUN

MISO AND OTHER FERMENTED BEAN PASTES, PLUS TASTY SAUCES

Traditional Gochujang

YIELD: ABOUT ½ GALLON

Gochujang is a rich, velvety, deep-red Korean hot pepper paste with a wonderful mix of sweet, hot, sour, and salty flavors. It is used liberally as a condiment on just about everything. This recipe is also in our book *Fiery Ferments* and is based on a recipe by Maangchi in her book *Maangchi's Real Korean Cooking: Authentic Dishes for the Home Cook.* Our primary adjustment to all the recipes that we tried was to reduce the salt a bit. Despite this reduction, our pastes last for years unrefrigerated, and the final flavor still is the important salty element along with the rich, sweet heat.

For this recipe, you'll need powdered meju. You can use some of the meju for Doenjang (page 289), pulverizing it to a powder (after it is hung and is still dry), but traditional gochujang calls for *meju garu*, which is made from six parts soybeans and four parts rice. Instead of being dried in blocks, this mixture is shaped into balls with a hole in the middle, through which a straw rope is threaded so the balls can be hung. These meju balls collect the microbes from the open air — they are said to have wild *Aspergillus oryzae* as well as *Bacillus.* You can find powdered meju garu in Asian markets.

The malted barley in the recipe contains amylase, which helps break down the rice.

The gochugang is placed in an onggi pot and, like doenjang, aged with the influence of the seasons, usually for 6 months to a year, though it can age longer. Five-year-old gochujang has a deep flavor, having lost much of the sweetness, and is nearly black in color.

FERMENTATION/AGING **3 months or longer**

- 8 cups (1.9 L) tepid water (around 100°F/38°C)
- 2 cups (324 g) barley malt powder (see note at right)
- 5 cups (790 g) sweet rice flour (also called glutinous rice flour)
- 6 cups (509 g) Korean hot pepper powder (*gochujangyong gochugaru*)
- 1 cup (92 g) powdered meju or meju garu
- 3 cups (237 mL) brown rice syrup
- ¾–1 cup (205–275 g) salt

Note: *Barley malt comes in a couple of forms. For this recipe you want diastatic malt powder. This barley was sprouted and then dried and milled, retaining the diastase enzymes that were created by the sprouting process. We need these diastase enzymes to break down the starches in our gochujang into simpler sugars for the microbes to ferment. Barley malt powder can be found in Asian markets under the Korean name* yeotgireum. *The nondiastatic malt powder simply provides sugar and doesn't contain any of the enzymes we are looking for.*

1. Combine the tepid water and barley malt in a saucepan, whisking to blend. Then whisk in the rice flour. Allow to sit for 2 hours.

2. Set the saucepan on your stovetop over medium-low heat and cook until reduced by one-third, about 1 hour. Stir regularly to prevent sticking.

3. Let the mixture cool completely, and then stir in the hot pepper powder, powdered meju, rice syrup, and salt. Stir until thoroughly mixed. The mixture will be shiny and creamy.

4. Transfer to a gallon jar, an onggi pot, or your favorite fermenting vessel. Cover; for a more traditional method, cover with cheesecloth secured with a rubber band, followed by the lid.

5. Set in a sunny spot to ferment for 3 months. Maangchi recommends removing the lid on sunny days to allow some sunlight to shine through the cheesecloth onto the ferment. In midsummer or in a hot climate, move it to a cooler location in the afternoons. We live in Oregon and have many days with very little sunshine in the wintertime. We have found fermenting this paste in a sealed Weck jar in a window to be our most successful method.

6. After fermentation, transfer the paste to the refrigerator, where it will keep indefinitely. In Korea, this paste is not refrigerated.

Does Glutinous Rice Have Gluten?

The short answer is no, glutinous rice does not have gluten in it.

"Glutinous" in this case refers to the level of amylopectin, which is a kind of starch that is very common in plants. In fact, it makes up at least 75 percent of all the starch molecules in most plants. It is just waiting for a shower and massage to develop its distinctive elastic mesh. When cooked, like in the case of the glutinous rice flour in gochujang, the amylopectin helps thicken the sauce.

Glutinous grains (barley, wheat, and rye varieties) are different — they contain gluten, and it has nothing to do with starches. Instead, these grains contain two proteins, glutenin and gliadin, that when mixed with water produce gluten.

SOY SAUCES

People often call both shoyu and tamari soy sauce, not realizing there is a difference between the two. True tamari has a long history and is a by-product of miso — it's that little bit of liquid that pools on the top of each batch. The most traditional tamari was made in autumn with two parts soybeans to one part water. The soybeans were soaked, steamed, mashed, and formed into small balls. These were then hung outside to catch wild aspergillus, similar to Korean meju. Once the soybean balls were inoculated and koji had grown on them, they were dried for 2 weeks. The dried koji balls were placed in large cedar vats with salt and water. The resulting mash was then pressed under weights and allowed to ferment for at least 1 year. After fermentation, the liquid became tamari, and the paste, called moromi, was often used as miso or as a pickling medium. Like doenjang, this ferment became a two-for-one — have your sauce and eat it, too.

Shoyu in Japanese means soy sauce and it is one of the many variations on the theme of fermenting a liquidy mash of wheat and soy. Soy sauce has its roots in China, but it has long since been a mainstay throughout Asia — from Guk Ganjang (page 293) in Korea to *kecap manis*, a sweet and syrupy sauce from Indonesia.

Soy sauce, like so many of the world's traditional ferments, has been industrialized to find a shortcut around the long time it takes to truly get the rich flavor. Temperatures are often manipulated to hasten the process, stronger monocultured spores are used, and sometimes more salt is added. Lesser-quality ingredients are often used, and in the case of soy sauce, some are made of defatted soy grits — a by-product of soybean oil. The cheapest brands don't ferment at all and use a process known as "rapid hydrolysis" or "acid hydrolysis," which involves heating the defatted soy meal and sometimes cornstarch with hydrochloric acid to release the aminos.

Don't use these soy sauce imitations in your ferments. Instead, look for soy sauce that is unpasteurized and naturally brewed using traditional methods. Not only will the flavor be better, but the fermentation will be stronger with the soy sauce's contribution of enzymes.

Wheat-Free Tamari

YIELD: ABOUT 2 QUARTS, PLUS A FEW CUPS OF MOROMI PASTE

Originally, tamari was just a by-product of miso — that tiny bit of liquid gold umami that pools on top of a batch of miso. Because it is so delicious and there is a relatively small amount in each batch, some enterprising souls in central Japan in the fourteenth century set out to make just this sauce by intentionally making a wetter soy miso.

You will make this with soybean koji. To be completely wheat-free, be sure to use rice flour to disperse the spores when you make it.

This recipe requires daily stirring for at least a month or so. After that, stir once or twice a week. Keep it on your counter, where you will remember it, and it just becomes part of your routine.

PROCESS **miso (page 254)**	INCUBATION **at least 1 year**

3 pounds (1.3 kg) Soybean Koji, page 206

1 tablespoon (17 g) unpasteurized miso

1 cup (275 g) sea salt

10 cups (2.4 L) water that has been boiled and cooled

1. Mash the finished koji or run it through a meat grinder if you have one. Place the koji in a gallon jar, crock, or other vessel. Stir in the miso.

2. Dissolve the salt in the water, then stir the brine into the bean mixture. Cover your vessel with a lid. We use a crock with a water trough that allows air to flow out but prevents fruit flies from coming in. You can also secure a tight-weave cloth on the top of a crock and cover with a plate. The cloth is to discourage fruit flies and the plate is to discourage evaporation.

3. Let ferment at room temperature. Stir daily for the first month, then one or two times a week thereafter. As it progresses, feel free to taste whenever you stir. It will ferment for at least a year.

4. When the tamari is to your liking, strain out the solids using a press bag, a jelly bag, or several layers of cheesecloth in a colander set over a bowl. Pour the contents into the bag or cheesecloth, then press and squeeze to obtain as much of the liquid tamari as you can.

5. Store the tamari in a jar in a cool cabinet. If you find yeast or mold growing on the top of the liquid, just remove it — it is harmless to you and the sauce. To avoid yeast growth, store in a refrigerator. The moromi can be used just like miso. Store it in the fridge in an airtight container; it will keep indefinitely.

Shoyu

This is a classic soy sauce and worth the wait. This recipe is adapted from the recipe that comes with GEM Cultures' shoyu koji spores, which you will need to make shoyu. As with tamari, the mash left over from pressing the shoyu is called moromi and can be used like miso. In some areas of Japan, the moromi is left unpressed and salt-pickled vegetables are added to it.

The Book of Miso suggests that the shoyu fermentation time can be cut in half by keeping the vat between 68°F/20°C and 77°F/25°C, and we have used that temperature range in this recipe. However, keeping it in the cooler ambient temperature of your home is fine; it will just take a little longer.

PROCESS miso (page 254)	INCUBATION 40–50 hours for the koji, 12 months or more of aging for the shoyu

7 cups (1,225 g) soybeans

6½ cups (1,140 g) whole soft wheat berries or bulgur

2 teaspoons (2 g) shoyu koji starter

3½ cups (18 g) sea salt

1 gallon (3.8 L) boiled and cooled water

1. Soak the soybeans for 8 to 24 hours.

2. Rinse, then steam the beans until they are soft enough to easily crush between your thumb and ring finger, following the instructions on page 54.

3. While the beans are cooking, toast the wheat berries or bulgur: Preheat the oven to 350°F/180°C. Place the wheat on a baking sheet and toast in the oven for 20 to 25 minutes, or until evenly browned, stirring regularly. A slight charring is okay, as it will help develop the flavor. Alternatively, place the wheat in a skillet over medium-high heat and toast until slightly

brown with a nutty smell, stirring constantly, 5 to 15 minutes.

4. Coarsely chop the toasted whole berries in a grain mill, meat grinder, or high-powered blender (there's no need to chop if you're using bulgur).

5. Transfer the steamed beans to a colander to drain, then place in a large bowl and add the chopped toasted wheat. Mix thoroughly. When the mixture has cooled to body temperature (around 90°F/32°C), disperse the starter across the mixture. Stir to make sure that the starter is evenly distributed. Grow the koji on the beans following the instructions on page 206.

6. When the koji is ready, combine the salt and water in a 2-gallon crock (we like using a 10-liter water-seal crock) or other fermentation vessel. Mix until the salt is dissolved, and then stir in the koji. Cover with a lid, fingertip tight, or secure a cloth over the top.

7. Keep the crock on your counter where it is moderately warm, and remember to stir it daily for the first few weeks. After this time, place it in a warmer spot, like on top of a water heater — somewhere that stays in the mid- to upper 70s Fahrenheit/lower 20s Celsius. Continue to stir it regularly, about once a week. As it ages, the color will change first to a rust brown and then a deeper dark brown to black. The solids and liquid may separate or stay as a homogenous thick paste. Ferment for at least 12 months, or until the smell is pungent and the color is dark.

8. When the shoyu is to your liking, strain out the solids using a press bag, jelly bag, or several layers of cheesecloth placed in a colander set over a bowl. Pour the contents into the bag or cheesecloth, then press and squeeze the moromi to obtain as much of the liquid as you can.

9. Store the shoyu in a jar in a cool cabinet. If you find yeast or mold growing on the top of the liquid, just remove it — it is harmless to your or the sauce. To avoid yeast growth, store in a refrigerator. The moromi can be used just like miso. Store it in the fridge in an airtight container; it will keep indefinitely.

Cooking with All the Tasty Pastes

These tasty pastes — from the myriad misos and doenjang to your own creations — are going to become your flavor superpower in the kitchen. The complex flavors are ready for use. At times you'll want the flavor of a paste to be at the forefront of a dish. At other times, you'll use a paste to subtly enliven and heighten the taste of the dish, without being front and center, so the secret to the deliciousness is preserved.

We want you to discover what these pastes have to offer whether you make them yourself or not. All of the recipes in part III can be made with purchased pastes just as well as with homemade pastes. We have designed the recipes to use primarily miso, since miso is

widely available, but we hope that you will try any of the fermented bean pastes or other tasty pastes you create. Your own pastes may have a different texture from the smooth uniformity you find in tubs of purchased miso, but don't let that keep you from exploring the possibilities. After all, we cannot guess what you are going to dream up, but with some simple cooking techniques, you could be making sauces, soups, dressings, and marinades with your pastes.

When shopping for fermented bean pastes, look for unpasteurized versions so that you can truly enjoy the health benefits and flavor. Check labels. Gochujang, for example, is an amazing ferment, but the industrialized versions are not fermented and are full of corn syrup.

Making Miso Broth

The first rule of miso is to avoid boiling it. If miso gets boiled, you will lose some of the nutritional benefits as well as some of the subtler flavors. With any broth or soup, boil or simmer all of the other ingredients first and then take the pot off the heat to add the miso.

To help the miso dissolve in the hot broth, you can place the miso paste in a small bowl, add a ladle of your warm stock, whisk until smooth, and return to the main pot. Or you can place the miso in a long mesh strainer or a tea strainer, soften it in the broth, then use a spoon to work the miso through the strainer. How much will depend on the type of miso you use and the flavor you are trying to achieve. It's best to start small.

Miso broth can be as simple as adding miso to hot water. That said, a traditional miso broth is always made with dashi broth, as these two flavors are synergistic — umami + umami. The miso is tasked with bringing out the flavor of the dashi, and vice versa. To that same end, we often mix a little of our Shiitake Tasty Paste (page 271) with miso for similar deliciousness. Use 1 tablespoon of miso per cup of dashi, or to taste.

MAKING DASHI

Dashi is a stock base made from a combination of kombu (dried kelp) and bonito flakes (dried, fermented, and smoked bonito fish, which are in the same family as tuna). It's important that you don't boil the kombu. You can make the dashi ahead of time and store it in the refrigerator, where it will keep for about a week.

2 strips (20 g) kombu, broken into a few pieces

1 quart (1 L) water

1 cup (10 g) loosely packed bonito flakes (omit if vegetarian)

1. Place the kombu in the water in a saucepan and soak for about 30 minutes.
2. Place the saucepan over medium heat and slowly bring the water to a near boil. When you see the bubbles form on the edges of the pan and the kombu, remove the kombu.
3. Add the bonito flakes, if using, and bring the broth to a boil over medium-high heat. Skim off any foam that develops. Lower the heat and simmer gently for 5 minutes. Remove from the heat and allow to rest for 5 minutes. Strain out the bonito flakes.

Fun Fact: Aspergillus Fermented Fish

Fish were first fermented with *A. glaucus* in the 1700s to improve shelf life, which the fermentation did dramatically. These days, fish are fermented for flavor, fragrance, and mouthfeel.

Katsuobushi is a smoked skipjack tuna fermented with *A. glaucus*. There are two types of katsuobushi: one is made by smoking the fish after it has been boiled (arabushi), and the second ferments that arabushi for at least 2 months using *A. glaucus* (karebushi). Bonito flakes used in dashi are made using the same process but with a different fish.

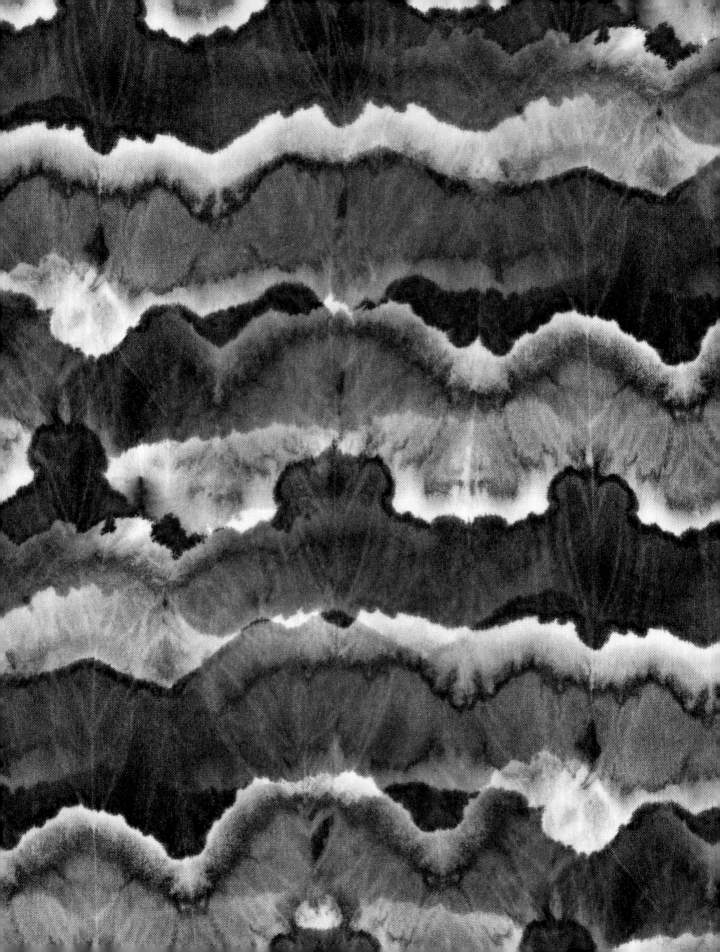

PART III

Eating

The ferments in this book lend themselves to flavoring just about anything and are all about the sides, or as the late anthropologist Sidney W. Mintz called them, the fringe. In this section, we swing the spotlight away from the center of the big plate (hearty main courses) to find inspired and scrumptious ideas for embellishing and enhancing our foods with these marvelous flavors. We start with sauces and condiments and then move on to mouthwatering ways to start your day, followed by ways to incorporate these powerhouse ferments into your day to sustain you through to the end. And when you get to that end, who doesn't want dessert? We have delicious recipes for that, too.

All of these recipes can be made using either homemade or purchased ferments. So please, dive in and enjoy them, whether or not you ever intend to make miso, natto, tempeh, or the others, or while you wait patiently for your own to be ready.

TASTY SAUCES, PICKLES & CONDIMENTS

I n this section, we will show you the easiest, tastiest ways to start integrating these ferments into your meals. We hope that this is where you begin your journey — where the rubber meets the road! The sauces, dressings, dips, intensely flavored pickles, and delicious and sublime condiments we feature here will add sparkle to any simple meal.

Velvety Peanut Sauce

YIELD: 1 CUP

We almost feel guilty calling this a sauce, as it is so creamy and thick. This condiment comes to the table regularly at our house — Christopher is a huge peanut butter fan. That said, this sauce is way more than peanut butter. All of the flavors play well together, so this sauce is not too hot, not too sweet, and not too salty — it is just right. We use it as a rich condiment alongside any dish that needs a little punch of flavor and protein.

¼ cup Doenjang (page 289)

¼ cup Gochujang (page 294)

¼ cup unsweetened peanut butter

¼ cup water

1. Combine the doenjang, gochujang, peanut butter, and water in a small bowl and stir until well mixed and creamy. We like the thick consistency of this sauce, but for a thinner sauce feel free to add more water.

2. Serve immediately. Store any leftover sauce in the fridge, where it will keep for 3 months.

Miso Mustard

YIELD: ABOUT 1½ CUPS

This recipe calls for a simple homemade mustard, but feel free to substitute prepared Dijon or hot Asian-style mustard (about 1 cup) to your miso/tasty paste instead. You can use any type of tasty paste you like. Use a dark miso for earthy, rich flavors or a fruit tasty paste for a fun and surprising sweet-salty combination. If it is too thick, thin it out with a bit of mirin.

FERMENTATION **3 days**

¼ cup brown mustard seeds

¼ cup white mustard seeds

2 teaspoons salt

1 cup water

1 tablespoon miso

4 tablespoons mirin

1. Combine the white and brown mustard seeds, salt, water, and miso in a blender and process. As the mustard seeds break down, they act as a thickening agent. Keep blending until the mixture reaches a paste consistency.

2. Ladle your mustard into a pint jar, pressing out any air pockets as you do. When you have placed it all in the jar, you may need to work out any remaining air pockets with a butter knife.

3. Place the lid on the jar and tighten. Set the jar on your counter to ferment for 3 days.

4. Open the lid (it may pop slightly as the CO_2 is released) and stir in the mirin. Your mustard is now ready to serve.

5. Store the mustard in the refrigerator. Technically this will last indefinitely, but it is so tasty that we are pretty sure it will all be gone rather quickly.

Miso as Dressing, Sauce, Dip, or Spread

When adding miso to any preparation, first thin it with another liquid, such as oil, vinegar, or water. This makes it much easier to incorporate. Simply place the miso in a small bowl, add the other liquid, and whisk until smooth.

If you have a favorite sauce, dip, or spread, you may love it even more if you add a bit of miso. Often just a dab will give it that extra deliciousness, without imparting much flavor. Of course, if you want the miso to be flavor-forward, simply add more.

Add miso (like the Fig Tasty Paste on page 287) to your butter to be used as a fancy toast spread or to be melted into mashed potatoes (or just put a little miso straight into the mashies). To make a miso butter, soften the butter and then mix in some miso. Start with a sweet white miso and then try others.

Banana Ketchup

Banana ketchup is a bright red condiment that was invented in the Philippines during World War II as a way to make ketchup without tomatoes, which had become scarce (bananas, however, were plentiful) — hence the red dye.

We skip the red dye and make this ketchup as a way to use overripe bananas. We think the beige color is more appealing, but if you don't, feel free to add turmeric to brighten it up. We also like that ours is fermented with koji. It can be used like ketchup and is especially fun as a dip for fried tempeh.

FERMENTATION **2 days**

¼ cup white, wine, or rice vinegar

1 teaspoon black peppercorns

5 cardamom pods, crushed lightly

5 whole cloves

1 star anise

1 cinnamon stick, broken into a few pieces

1 tablespoon peanut oil

2 small shallots, diced

2 garlic cloves, diced

3–4 ripe bananas (about 1 pound)

2 tablespoons Rice Koji (page 201) or other grain koji

½ teaspoon salt

½ teaspoon powdered turmeric (optional)

---

1. Combine the vinegar, peppercorns, cardamom pods, cloves, anise, and cinnamon stick in a small saucepan over high heat. Bring to a boil, cover, then lower to a very low simmer. Simmer for about 15 minutes. Set aside to cool.

2. While the spices simmer, warm the oil over medium-high heat in a small skillet. Add the shallots and garlic, and sauté until translucent and lightly caramelized, about 5 minutes.

3. Peel the bananas, break into pieces, and put in a bowl. Add the onion mixture, koji, salt, and turmeric, if using.

4. Strain the cooled infused vinegar into the bowl. Discard the spices. Using a hand blender, thoroughly purée the mixture until smooth.

5. Put the mixture in a pint jar. Tighten the lid and leave on the counter for a few days. It will begin to create CO_2 after about a day. Burp the jar to release any pressure as needed. Feel free to taste as it ferments.

6. The ketchup is done when you like the flavor. Store in your refrigerator, where it will keep for 2 to 3 months.

Miso Pesto

YIELD: ½ PINT

We made this pesto many times over the course of one summer, trying various additions. The Chickpea Miso (page 266) was our favorite, as it added just the right salty, creamy richness the Parmesan would normally provide, along with a light acidity normally brought by a lemon. Enjoy this simple pesto in all the ways you would regularly enjoy pesto, such as on pasta or pizza. We particularly liked it mixed in with Natto (page 118).

2 cups fresh basil leaves, loosely packed

3 garlic cloves, minced

1 tablespoon Chickpea Miso (page 266) or other tasty paste

3–4 tablespoons extra-virgin olive oil

Place the basil leaves, garlic, miso, and 3 tablespoons of the oil in a food processer. Process until smooth, scraping down the sides as needed. If you like a more fluid consistency, add more oil. Use immediately or store in the refrigerator in an airtight container. It will keep for at least 6 weeks.

The Mayos: A Base for Tasty Miso Sauces

Homemade mayonnaise is a basic pantry staple for us. Mayonnaise is so simple to make that we think it is one of the easiest places to replace a processed condiment with a nutrient-dense whole food. When we make anything with "mayo," we use homemade olive oil–based emulsions — there is no comparison in flavor, texture, and nutrition. We use both the egg-based mayonnaise (page 308) and an almond-based mayonnaise (page 309); for us the choice is purely culinary, depending on the meal, as each of these aiolis imparts different flavors and textures. You can use the recipes interchangeably when we call for mayonnaise in our recipes — both are a solid healthy food choice, and more importantly, delicious!

These mayos lend themselves to innumerable variations on a theme, especially with the fabulous umami of the ferments in this book. With a little bit of miso, they can become fantastic spreads or dips. Experiment with your favorite flavors. We add about 1 teaspoon of miso (any kind) to ½ cup of mayonnaise (or aioli). As is, it makes a great spread or fry sauce. Splash in a little lemon juice or other acid, and it is over the top. And if you want these spreads to be a little thicker, or to have a more neutral flavor, replace the olive oil in the mayo with your favorite oil — sunflower, grapeseed, almond, and light expeller-pressed sesame oil are some good choices.

Olive Oil Mayonnaise

YIELD: 1 PINT

Homemade mayonnaise is simple to prepare and worth the effort. It has the reputation of being tricky, as it can be sensitive and every once in a while will fail (instead of the oil and solids becoming one, they separate). It seems that ours fails about once every 20 or so batches. We used to wonder what we did wrong; now we know it can be a variation in the protein in the eggs. Usually adding another egg yolk will emulsify the mayo. The flavor of olive oil is wonderful, but some of the extra-virgin olive oils can be very bitter. Taste yours to make sure you like the flavor. If it is bitter, try using light or virgin olive oil (not extra-virgin). Add a little garlic (or better yet fermented garlic paste) to make a delicious aioli.

1 whole egg

1 egg yolk

1 cup olive oil

Pinch of salt

1 tablespoon lemon juice or raw sauerkraut brine (see note below)

Note: Raw, live sauerkraut brine not only will provide the needed acidity but will also add lactobacilli to your condiment. If you use sauerkraut brine, let the finished mayonnaise sit on the counter for 6 hours to allow the lactobacilli to multiply and continue to acidify your condiment, for a subtle sour flavor and to increase the probiotic content. The bacteria will slow down the oxidation of the oils, thereby giving the mayonnaise a longer shelf life in the fridge.

1. Combine the egg and yolk in a blender or food processor and blend for a few seconds.

2. With the motor running, drizzle in the oil in a thin stream. When the mixture reaches the desired consistency, stop adding the oil. Again with the blender running, add the salt and the lemon juice. As soon as the lemon juice has been incorporated and the mayonnaise has thickened, it's ready.

3. Store the mayo in the refrigerator, where it will keep for about 1 week; if you used sauerkraut brine and let the mayo ferment for 6 hours, it will keep for about 1 month in the fridge.

Almond Mayonnaise

YIELD: 1 PINT

We have made this almond mayonnaise for years; the exact number of years feels disconcerting and we are not going to share how many decades it has been. This is our updated version of a recipe for "almonaise" by Marilyn Diamond in *The American Vegetarian Cookbook*. Use it as you would any mayonnaise — as a sandwich spread, for example — or get creative and drizzle it on mashed potatoes or roasted vegetables.

To get all the benefits almonds have to offer, first soak and blanch them to make them easier to digest and break down the phytates (antinutrients that "lock up" the minerals and inhibit your body from absorbing them).

½ cup raw almonds

½–¾ cup water

½ teaspoon salt

1 generous teaspoon nutritional yeast (optional)

1–1¼ cups olive oil

3 tablespoons lemon juice or raw sauerkraut brine (see note below)

Note: Raw, live sauerkraut brine not only will provide the needed acidity but will also add lactobacilli to your condiment. If you use sauerkraut brine, let the finished mayonnaise sit on the counter for 6 hours to allow the lactobacilli to multiply and continue to acidify your condiment for a subtle sour flavor and to increase the probiotic content. The bacteria will slow down the oxidation of the oils, thereby giving the mayonnaise a longer shelf life in the fridge.

1. Soak the almonds for 1 day and then remove the skins. Alternatively, you can blanch them: Prepare a bowl of ice water and boil a separate small saucepan of water. Drop the almonds into the boiling water and boil for a few minutes. Remove them with a slotted spoon and plunge into the bowl of ice water. The skins will wrinkle and the almonds will slip easily from the skin as you pinch them between your thumb and forefinger.

2. Put the almonds in a food processor or high-powered blender. Grind them to a fine powder.

3. Add about half of the water, the salt, and the nutritional yeast, if using, and blend well. Then blend in enough of the remaining water to give the mixture a creamy consistency.

4. Remove the blender cap and run the blender on low. Add the oil in a thin, continuous stream until the mixture thickens.

5. Keep the machine running and add the lemon juice. Continue to blend for about 1 minute and it will thicken a bit more.

6. Store the mayo in the refrigerator, where it will keep for about 1 week; if you used sauerkraut brine and let the mayo ferment for 6 hours, it will keep for about 1 month in the fridge.

(Not) Blue Cheese Salad Dressing

YIELD: ABOUT ½ CUP

This salad dressing is a play on blue cheese dressing. It is a simple creamy dressing made with stinky-style fermented tofu in place of its pungent dairy cousin. This recipe yields enough dressing for one dinner salad. We generally double this recipe so that we will have some on hand to toss with some greens as a quick side salad. This also makes a nice dip.

You can buy the mayo for this dresssing or make it yourself. If you make this with the homemade mayonnaise on page 308, substitute a neutral-flavored oil for the olive oil, as the dressing will taste too strongly of olive oil otherwise. You could also use the almond-based mayo on page 309. We find that the fermented tofu salts the dressing nicely, so we don't add any more salt.

¼ cup sour cream

3 tablespoons mayonnaise

1 tablespoon Stinky-Style Fermented Tofu (page 89)

1 tablespoon lemon juice

Freshly ground black pepper

Combine the sour cream, mayonnaise, fermented tofu, lemon juice, and pepper to taste in a small bowl. Mix together with a whisk or hand blender. Serve immediately. It will keep in a tightly sealed container in the fridge for about 1 week.

Simple Miso Salad Dressing

YIELD: ABOUT ½ CUP

This salad dressing takes about a minute to make and is always a crowd-pleaser. This recipe uses chickpea miso, cilantro, and lime, but feel free to substitute the miso, the acid, and the herbs for others you prefer. For example, use red miso instead of chickpea miso, raw apple cider vinegar in place of the lime juice, and ginger in place of the cilantro for a warm and earthy dressing.

The miso acts as a thickener. If you want a thinner dressing, add a tablespoon or two of water. If you are using a saltier miso, you may want to reduce the amount of miso.

¼ cup olive oil

1½ tablespoons Chickpea Miso (page 266)

Juice of 1 or 2 fresh limes (about 2 tablespoons)

1 tablespoon finely chopped fresh cilantro

1 garlic clove, minced

Combine the oil, miso, lime juice, cilantro, and garlic in a small bowl with a whisk or a hand blender. Serve immediately. It will keep in a tightly sealed container in the fridge for about 3 weeks.

Miso Dijon Salad Dressing

YIELD: ABOUT ½ CUP

We are including this variation because we want to inspire you to never buy bottled dressing again.

¼ cup olive oil

1½ tablespoons Sweet White Miso (page 262)

¼–½ teaspoon Dijon mustard

2 tablespoons rice vinegar

1 garlic clove, minced

Freshly ground black pepper

Combine the oil, miso, mustard, vinegar, garlic, and black pepper to taste in a small bowl with a whisk or a hand blender. Serve immediately. It will keep in a tightly sealed container in the fridge for about 3 weeks.

Koji Cultured Cream (or Butter)

YIELD: ABOUT 1 PINT

Jeremy introduced this cultured cream to Kirsten. As makers of all things dairy — from yogurt and kefir to hard and soft cheeses — we have cultured a lot of cream and have found this one to be less finicky than the "traditional" methods. It has some subtle, sweet fruity notes from the koji and can be used any way you might use sour cream.

The beauty of koji cream is that it can be heated, even boiled, and it doesn't break like a standard sour cream. Yes, that means your stroganoff won't separate! Jeremy thinks this is because the koji enzymes denature and stabilize the proteins — pretty cool, right?

You can take this one step further and make your own koji cultured butter; see step 3.

FERMENTATION **24–36 hours**

2–3 tablespoons light Rice Koji (page 201) or Rice Amazake (page 217)

1 pint good-quality heavy cream

1. If you are using fresh koji, break up any clumps with clean hands. Place the koji or amazake in a quart jar. Add the cream and use a hand blender briefly to break up the lumps and incorporate the koji or amazake. Place the lid on the jar and let stand at room temperature for at least 24 hours.

2. The cream should now be thick and have a floral, fruity smell. It should taste mildly sour, like

a sweet, thick sour cream. If not, let it sit a few more hours, checking it regularly. After 36 hours, the fats will begin to separate from the whey. At this point you are headed to butter. The cultured cream will keep in the fridge for 2 weeks or more, but the flavors will continue to change.

3. If you want to make butter, we suggest that you stop the fermentation at the cream stage and then use a blender or food processor to whip the cream until it separates into the fat (the butter) and whey. The cultured butter will keep in the fridge for at least 2 months.

PICKLE MAGIC: FERMENTING VEGETABLES WITH MORE MICROBES

In our previous books *Fermented Vegetables* and *Fiery Ferments,* we explored lactic acid fermentation in depth, from bright, flavorful krauts to deeply complex fermented condiments. Here we will show you how to bring amazing flavors to your vegetables by employing enzymes — amylase to bring out the sweet flavors and protease for savory notes (converting those proteins to aminos) — as well as other microbes, yeasts, and some lactic acid bacteria. Most of these pickles are ready to use in just a few hours to a few days and should be eaten within a few days.

That said, with a little care, some of these pickling mediums can work (like kasuzuke) or even live (nukadoko) for years. In particular, we're talking here about pickling beds traditional to Japan, where pickling is an art form in and of itself. Like a sourdough starter, a pickling bed is a live medium that can be used over and over. Building the bed and sustaining it is no different than caring for a pet, a houseplant, or your garden's soil. They require regular love and some feeding to thrive, but they will reward you with incredible salty, deeply flavorful pickles.

In all of these recipes, you can radically change the mood and flavor of the pickles simply by changing how you prepare the vegetable.

It's all about surface area. If you are 30 minutes from serving dinner and you want to pickle some cucumbers to accompany the meal, you would slice them and drop them in the pickling medium. The greater surface area of the cucumbers will cause them to take on flavor more quickly. If you are thinking about a meal the next day, you might submerge the whole cucumbers and slice them after pickling. For an even deeper flavor, you could salt and press the cucumbers in order to get rid of extra water first and then leave them in the pickling bed for days or weeks. Each of these pickled cucumbers will have a different quality.

Sake Lees Pickles
(Kasuzuke or Sake Kasu)

The lees are the compressed yeasted rice mash left over after sake has been filtered. Sake lees usually come in a pressed cake form, called *kasu,* that is thick but slightly damp, like ricotta or soft crumbled tofu. Mechanically pressed lees from larger makers come in very dry sheets. You will need to tear these into small pieces and massage them in a bit of water until you have a paste consistency.

These pickles are thought to have come onto the food scene in the Kansai region of Japan around 1,200 years ago. First melon and then eggplant lees pickles were made by Buddhist monks and consumed by samurai in wartime.

Sake lees are said to be high in protein and vitamins B_6, B_{12}, and B_1. They can be used as is for pickling, but salt, sugar, and sometimes mirin are usually added to the medium. Vegetables, after being pressed with salt for 2 days to release excess liquid, can be fermented in a bed of sake lees for anywhere from a few hours to several years. For added flavor, try adding dried chile, hot mustard powder, ginger, or shiso leaves.

FERMENTATION **a few days to 18 months**

- 2 pounds whole vegetables
- 3 tablespoons sea salt, plus extra for pressing vegetables
- 3 cups sake kasu (lees; see headnote)
- 1 tablespoon sugar
- Your favorite spices or herbs (optional)

1. Place the whole vegetables in a container, such as a small plastic tub or casserole dish, that you will be able to apply a lot of weight to. Sprinkle liberally with salt.

2. Cover the dish with plastic wrap or clean cotton cloth. On top of this place something heavy, such as a pot of water, a crock, or a tub of flour. Allow the vegetables to weep for 2 days. At that point, they should be limp. If they are not, let them sit for a few more hours.

3. Combine the kasu, 3 tablespoons salt, sugar, and spices, if using, in a bowl and mix well. Cover the bottom of your fermentation vessel with some of the kasu mixture, place the vegetables on top, and cover with another layer of kasu mixture, making sure that the vegetables are all tucked in and submerged. If your vessel is narrow, you may need to have several layers of the kasu mixture and vegetables. Cover the container with plastic wrap or a clean cotton cloth, then weight it down.

4. Let the vegetables ferment for a few day to 18 months. Check the fermentation regularly over the first weeks; there will be a lot of CO_2 action, and you'll want to make sure that the vegetables stay submerged in the kasu mixture.

5. Once the vegetables are cured to your liking, you can enjoy them. Save the rest of the pickles in an airtight container in the refrigerator for up to 2 weeks. Keep the kasu mixture in an airtight container in the refrigerator. You can use it once more. It will last for a few weeks. There are, however, many creative uses, including dehydrating it and using it as a flavoring powder.

Miso Pickled Vegetables

YIELD: A LITTLE LESS THAN 1 POUND OF PICKLES

This simple, light, and sweet miso pickling bed (*miso doko*) can be made in just a few minutes. After a few hours in the bed, the vegetables will have mellow flavor. After a few days, they become boldly flavored small bites to be used in small servings or as garnish.

This miso bed will last for a number of pickle batches. Its vibrant, salty sweetness will eventually dissipate. You will know the bed is getting "tired" when it becomes watery and the flavors become thin. At this point, you can use the miso mixture to flavor another dish that will be eaten immediately, such as a soup. It will also do well as a marinade for meat. When we consign our pickling mixture to a meat marinade, we sometimes add a little mirin or vinegar to revive it a bit.

This recipe makes a little over 1 cup of miso for pickling, which can usually accommodate about 1 pound of vegetables. Use any vegetable. Daikon is one of the most common, and for very good reason — the intense flavor and crunch are addicting. But any root — carrots, turnips, celeriac, parsley root, parsnip, rutabaga — is delicious. We even pickled our kale and collard stems (for 2 days), with delicious results.

FERMENTATION **a few hours to a few days**

---

1 pound vegetables of choice, sliced into ⅓-inch slices or spears

Salt, if needed (see step 1)

⅔ cup younger mellow miso, like a white rice (page 262) or chickpea (page 266)

⅓ cup dark aged miso, such as Red Miso (page 267) or your own salty miso creation (page 269)

3 tablespoons mirin

3 tablespoons sake

---

1. If you are using vegetables with a high water content, like radishes, eggplant, summer squash, cucumbers, onions, and the like, toss them with a little salt in a bowl and let sit for an hour or so. This will pull out the excess moisture. Rinse off the salt water and pat dry.

2. Combine the misos, mirin, and sake in a small bowl and mix well. Spread half of the miso mixture in a small square glass container and cover with a layer of thin porous cotton cloth; the butter muslin used for cheesemaking is great, or you can use three layers of regular cheesecloth, as it is a little gauzy and porous. Place the cut vegetables in a single layer on the cloth, then put another layer of cloth over the veggies and spread the rest of the miso mixture on top. The cloth will allow you refrain from rinsing the vegetables after pickling, which will preserve more of your pickling bed. However, you can, of course, just poke your veggies down into your miso bed and fish them out when you are ready.

3. Let the vegetables sit in the pickling bed for at least a few hours and up to 3 or more days. You can taste the vegetables frequently to evaluate their flavor.

4. Once the vegetables are cured to your liking, lift off the top layer of cloth and remove the vegetables. Eat immediately or place in a container in the refrigerator, where they will keep for up to a week. Refrigerate the pickling bed between uses.

SOY SAUCE CURED
EGG YOLKS, page 323

MISO MARINATED
EGGS, page 322

MISO PICKLED
VEGETABLES

NUKA PICKLED
DAIKON, page 317

AMAZAKE PICKLED
VEGETABLES, page 316

Koji Pickles

There are three common ways to use koji to pickle vegetables: amazake (*ama-koji*) to enhance the sweetness in vegetables; shio koji, which is similar to ama-koji except that the salt plays an important role in bringing out more of the savory notes in vegetables; and sour amazake, which you can use to make a fermented lactic acid pickle without the salt (tasting a crunchy, lactic-fermented vegetable with all of the funk, some of the sweet, and none of the salt can be mind-blowing). All three need to be raw and unpasteurized for the enzymes to do their work.

We have not yet found a vegetable that did not come out uniquely tasty using a koji treatment. And we have tried a lot of them: asparagus, all the brassicas — broccoli, Brussels sprouts (cut in half), cabbage, cauliflower, the hardy greens and Asian greens, turnips, and radishes — beets, carrots, cucumbers, eggplant, garlic, green beans, leeks (premarinate them for a potato-leek soup), onions, summer squash, turmeric root, winter squash . . . We even tried melon slices — yum. You get the picture.

When working with watery vegetables, like celery, cucumber, daikon, eggplant, summer squash, and turnip, it's best to salt and press the vegetables first in order to release some of the juices and concentrate the flavors.

All of these koji-based sauces have a similar consistency and are used in the same way. The important thing is to make sure all of the surfaces of the cut vegetables are covered with the sauce. Simply place the veggies in a bowl and cover with the amazake (page 217), shio koji (page 223), or sour amazake (page 218). The thinner you cut the vegetables, the sooner they will be finished and the more deeply the flavors will penetrate.

Another method is to place the vegetables in a ziplock plastic bag and spoon in a few teaspoons of the koji sauce of your choice. Seal the bag, removing as much air as possible, and move the vegetables around a bit to make sure their surfaces are coated. Then press the vegetables and the sauce together at the bottom of the bag. Similarly, you can use a vacuum-seal bag. For a super-quick pickle, you can place the bag in a sous vide setup and set the temperature to 110°F/43°C; in an hour you will have a tasty pickle.

Ama-koji pickles should be served with the amazake they were pickled in — don't throw it out, because it is tasty stuff. Shio koji and sour amazake pickles tend to be the quickest and easiest, as they can be ready to eat within 15 to 30 minutes. If you're short on time, you can ferment cucumber slices, daikon or turnip spears, and other veggies in shio koji for a tasty snack or side dish (and shine as the cook — nobody will know how easy it is). You can also use this method to make quick vegetable salads, such as Javanese carrot salad (page 350).

Making and Using a Nuka Pot (Nukadoko)

The nuka pot (or *nukadoko*) is a long-living pickling bed made of rice bran and salt. It is not uncommon in Japan for this bed to become a family heirloom, passed down through the generations. Like fine cheeses and wines, the nuka pot will have the unique flavor of place. Also like wine and cheese, the flavor will mature over time, becoming smoother and more complex. But unlike wine, this medium requires almost daily care. To sustain the microbial equilibrium, the bed must be used frequently.

Vegetables will come out limp but still very crunchy, with a tangy, salty, earthy flavor — they are now pickled, or *nukazuke*. You can add these highly seasoned vegetables to dishes at will. They also make great tempura. If you like fried okra, then sink okra whole in the pot, remove when cured to your liking (we pickle them for 24 hours), batter, and fry.

You can buy starter for a nuka pot online (see source guide, page 388), and in some well-stocked Asian markets you can find kits of inoculated rice bran that you just add water to. When building your own pot, the bed should look and feel like moist sand, the kind you would use to build a sand castle. Nuka beds are traditionally kept in cedar tubs, but many people choose to keep them in a ceramic crock with a lid. Any lidded vessel will do. The lid shouldn't be airtight.

FERMENTATION **6–8 weeks for the nuka pot; 8–24 hours for nukazuke**

---

  2 pounds (900 grams) dry rice bran

  ½ cup (135 grams) sea salt

  1 tablespoon (5 grams) dry yellow mustard

  ½ ounce (14 grams) kombu (dried kelp), broken into pieces

  8 garlic cloves, sliced

  1 (1-inch) piece fresh ginger, minced

  1 tablespoon (7 grams) cayenne pepper

1–2 dried shiitake mushrooms, run through a blender to make a powder (optional)

  6 cups water, or as needed

     Vegetables, as needed

1. Place the rice bran in a large skillet and toast over medium heat until lightly browned, 5 to 10 minutes. Stir frequently to keep the bran from scorching.

2. Combine the toasted rice bran, salt, mustard, kombu, garlic, ginger, cayenne, and shiitake mushroom powder, if using, in a 1- to 2-gallon glass or ceramic lidded vessel. Mix well. Add the water in a few batches, mixing it in with your hands. Stop adding water when the mixture has the consistency of wet beach sand. You want it to be damp and clumping but not a wet slurry. Make sure you get any dry spots out of the corners. Cover with a lid that is not airtight.

Recipe continues on next page

3. Bury ½ cup of vegetables (you can use scraps, since you won't be eating them) in the mash. If the vegetables are small, you may want to gather them up in a bit of cheesecloth so that you can find and remove them easily later. Place the vessel in a spot that is on the cool side of ambient home temperatures (in the lower to mid-60s Fahrenheit/15°C–19°C) and let sit for 1 day.

4. After 1 day, remove the veggies, scraping off as much of the paste back into the pot as you can. Discard these veggies and then add ½ cup of fresh veggies. Repeat this process daily for 1 month. Stir regularly from top to bottom to aerate (which discourages mold) and manage temperature, moisture, and flavor. If you have questions, turn to troubleshooting on page 380.

5. After the first month, you may want to start using veggies that you want eat, like peeled carrot or daikon sticks. (The vegetables are edible all along, but they will bland, and they will still be a little bland in this young pot.) Vegetables are ready to eat in 8 to 24 hours. Watery veggies like celery, eggplant, and cucumbers should be pressed with salt and allowed to sit for an hour or two before being put in the pickling bed to remove excess water.

6. After 6 to 8 weeks, the nuka flavors become more mature. At this point, the pot should have a pleasant earthy smell. Refresh the pot monthly with salt, bran (use 13 percent salt by weight of the rice bran), and a few flavoring spices.

PRO TIP

Zero Waste,
Super Taste

We found that when we needed to feed the pot but didn't have anything we wanted to pickle, we could put in vegetable scraps, like carrot tops or cucumber peels, for a day. Some scraps, like kale stems, were reborn as food — when chopped finely, they became a delicious garnish (like capers but different). Vegetables will come out limp but very crunchy and tangy. They will be salty and earthy.

Keeping Up with the Nuka Pot

The thing is, to sustain the microbial equilibrium the bed must be used frequently. This means regularly adding fresh vegetables as well as continued frequent stirring from top to bottom to aerate. If you can't keep up with the nuka pot's demands, you can put an established pot in the refrigerator. This is also good if you will be traveling or the weather is very hot. Another stategy is to leave it out and keep some vegetables in it for weeks. This alleviates a constant turnover but you still have to aeriate it every few days. You will find it will also need fresh rice bran and salt (use 13 percent salt by weight of the rice bran) every few months.

Tofu Misozuke

YIELD: 1 POUND

This recipe makes a block of tofu fermented in miso. Different misos will give your tofu completely different flavors. For a mellow tofu, use Chickpea Miso (page 266) or brown rice miso, which we prefer over the sweet miso. For a rich, deeply flavored tofu, use Red Miso (page 267), Barley Miso (page 265), or a salty tasty paste (page 269). We have even mixed gochujang (page 294) with miso for a fun spicy flavor. We cannot speak to how fully the enzymes left in the miso are able to ferment the tofu, yet it is quite likely that the live microbes in the miso do transform the tofu over time, at least on the edges.

Tofu misozuke is very satisfying on a cheese board with crackers and cucumbers. We also use it with pasta dishes and love it in Misozuke Malfatti (page 360). Get creative with this base recipe, both during the fermentation and in its use. We love using different fresh or fermented herbs, like sweet basil, Thai basil, or shiso leaves, with the miso.

CURING TIME **2 months**

1 (16-ounce) block extra firm tofu

1 cup unpasteurized miso of your choice

2 tablespoons sake

Herbs (fresh or fermented), red pepper flakes, or other spices (optional)

A few sheets of nori, strips of kombu (soaked in water for 30 minutes), or fermented or dried herbs, for wrapping the tofu (optional)

1. Press the tofu: Place a cutting board on top of a dish towel to catch any liquid. Place a couple of paper towels on the cutting board and set the tofu block on the paper towels. Place another paper towel on top of the tofu, then place a second cutting board or casserole dish on top, followed by something heavy, like a sack of rice. Allow the tofu to drain at room temperature until it is just moist and feels quite firm, about 2 hours.

2. Meanwhile, mix the miso and sake in a small bowl. The sake makes the miso creamy enough to spread.

3. Coat the whole block of tofu with the miso mixture (see the note, page 320). Sprinkle on herbs and spices, if using.

4. Wrap the tofu in nori, soaked kombu, or fermented or dried herbs. (For a very simple misozuke, you can skip this step.)

Recipe continues on next page

Tofu Misozuke, *continued*

5. Wrap the entire block in cheesecloth. This will make it easy to manage.

6. Place four paper towels in the bottom of a sealable container large enough to fit the tofu. The towels will soak up more moisture as the tofu ages and the enzymes break down the protein.

7. Put your tofu block in the container and seal. Label the container with the date and contents and set it in the refrigerator.

8. Let the tofu ferment for 2 months. Change the paper towels once a week, but otherwise keep the container sealed. If you notice surface mold, remove it.

9. Unwrap the tofu. If you used kombu, discard it. If you used nori or herbs, they make a tasty rind, so there's no need to remove them. Store the tofu in a sealed container in the fridge, where it will keep for about 1 month.

Note: *If you would like a cleaner, tidier block of tofu misozuke, wrap the tofu in a neat layer of cheese-cloth, then spread the miso mixture and apply the wrappings on the outside. Then the infused tofu can be cleanly removed from the miso coating.*

Miso Marinated Eggs

YIELD: 6 EGGS

You may have met these eggs on a ramen bowl or bento plate as *miso tamago* or *tamago misozuke*. The flavor changes with the type of miso you use. A sweet white type miso (page 262) will render a mellow, almost sweet egg. Chickpea Miso (page 266), or mellow miso, is a little saltier but still has some sweetness tinged with a heartier umami and is one of our favorite choices. Barley Miso (page 265) is darker in color and has a heartier flavor — the eggs will soak up more color, giving the white pretty brown edges when cut open. Red Miso (page 267) is the most pungent of the group and will impart the most flavor and the most impressive color. Any miso will work, but unpasteurized miso will immediately begin to transform the flavor of the egg. The miso can be reused a number of times for many batches of eggs. The longer the eggs cure, the stronger the flavor will be. In general, the saltier and darker the miso, the shorter the time needed.

CURING TIME **7 hours to 4 days**

---

6 eggs

1–1½ cups miso of your choice

---

1. Fill a pot with enough water to cover the eggs by about 1 inch of water. Bring the water to a boil over high heat.

2. While you are waiting for the water to boil, fill a bowl with warm water from the faucet (it should feel about body temperature) and place the eggs in the water. This will warm them slightly, which reduces the likelihood that they will crack when they hit the boiling water (if your eggs are already at room temperature, you can skip this step).

3. When the water is boiling, gently lower the eggs into the water and set a timer for 7 minutes. During this time, prepare a bowl of ice water.

4. Drain the eggs in a colander, then transfer them to the ice water bath. When cool, crack the eggs carefully and peel, taking care as the whites will be a bit more tender than a fully cooked hard-boiled egg.

5. Place 2 to 3 tablespoons of the miso in the palm of your hand. Pat the miso into a pancake, then nestle the peeled egg in the center and wrap the miso around it. Place in a sealable storage container. Repeat with remaining eggs. Don't worry if your miso is moist and you won't allow you to wrap it neatly; do the best you can to get coverage on the whole egg. Alternatively (this may take more miso), place a layer of miso on the bottom of a container, add the peeled eggs, then cover with the remaining miso. Place the container in the fridge and age for at least 7 hours and up to 4 days.

6. When eggs are ready, gently remove and continue to wipe off most of the miso. Press the excess miso into a resealable container, removing any air pockets. Store the eggs in an airtight container in the refrigerator for up to 1 week.

Soy Sauce Cured Egg Yolks

YIELD: 2 YOLKS

Cured eggs yolks are packed with protein, flavor, and fun. We make these when we have egg yolks left over from recipes that call for lots of egg whites — like macaroons.

Soy sauce works quickly on the yolks — the salt draws out the moisture while the amino acids in the sauce change the flavor. In as little as 6 hours, the yolk is transformed into an ultrarich treat that has developed a light skin and a creamy center. It can be the protein jewel atop a bowl of grain and veggies or the perfect topping to buttered toast. When allowed to age for 3 days, that same yolk becomes translucent and is no longer liquid. In this case, however, longer is not better. After 2½ days, the yolk begins to shrink until it finally disappears in the soy sauce. However, if you allow it to cure until hard you can dry it, then grate the dried yolk on foods as a garnish — much in the same way you would use a bit of Parmesan cheese.

CURING TIME **6 hours to 3 days**

¼ cup soy sauce, Shoyu (page 298), or Wheat-Free Tamari (page 297)

1½ tablespoons mirin

A few chile flakes, a small piece of kombu, or a few sprigs of cilantro (optional)

2 eggs

1. Place the soy sauce, mirin, and chile flakes, if using, in a quarter-pint jar, then stir to combine.

2. Crack the eggs into two separate small containers, separating the yolk from the white and taking care not to break the yolk while removing as much white as possible. Set the whites aside for use in another recipe or freeze for use later. Gently lower the yolks into the soy sauce mixture.

The yolks will float. Allow to cure in the refrigerator for 6 hours to 3 days (yolks cured for at least 12 hours or until hard will preserve longer). For the first 6 hours, spoon a little mixture over the top of the yolks every hour or two.

3. After curing, remove the yolks from the mixture and serve immediately or refrigerate in an airtight container for up to 1 day.

4. If curing eggs, preheat the oven to 180°F/82°C.

5. Place the yolks on a wire rack on a baking sheet. Bake for 35 minutes, then allow yolks to cool completely on the rack. Wrap individually in waxed paper and place in an airtight container. They will keep in the refrigerator for up to 3 months.

Miso Cured Egg Yolks

YIELD: 2 YOLKS

Enzymes from the miso break down proteins into amino acids, creating lusciously flavorful yolks. Use two small quarter-pint jars for nesting the yolks; you will use much less miso that way. You won't reuse this miso for curing yolks, but you can use it in soup for a pleasant egg flavor. This recipe specifies two yolks and jars, but you can make as many as you would like.

Many recipes say to wrap the yolks in cheesecloth before submerging in the miso, but in our trials, we found the cheesecloth wasn't necessary.

You can also dry these yolks, then grate on food as a garnish.

CURING TIME **8 hours to 3 days**

3–4 tablespoons miso of choice

2 tablespoons mirin

1 tablespoon sake (optional)

2 eggs

---

1. Combine the miso, mirin, and sake, if using, in a small bowl. Divide the mixture evenly into two quarter-pint jars.

2. Using a spoon, form a hole in the miso for the yolk. You want the hole to be as deep as possible without exposing the glass underneath.

3. Crack the eggs into two separate small containers, separating the yolk from the white and taking care not to break the yolk while removing as much white as possible. Set the whites aside for use in another recipe or freeze for use later. Carefully drop one yolk into the depression of a jar, then repeat. Smooth the edges around the yolks so they have as much contact with the miso as possible. The top will be exposed. This is fine. Place the lids on the jars and allow to cure, in the refrigerator, for 8 hours to 3 days.

4. When ready, remove the yolks from the mixture and serve immediately. You can carefully rinse off excess miso or leave for more flavor. Or, refrigerate in an airtight container for up to 1 day; don't rinse off the miso until ready to serve.

5. To preserve yolks for drying, cure until they are a bit hard. When ready to dry, preheat the oven to 180°F/82°C.

6. Place the rinsed yolks on a wire rack on a baking sheet. Bake for 35 minutes. Allow the yolks to cool completely on the rack. Wrap individually in waxed paper and place in an airtight container. They will keep in the refrigerator for up to 3 months.

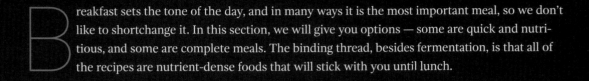

BREAKFAST

reakfast sets the tone of the day, and in many ways it is the most important meal, so we don't like to shortchange it. In this section, we will give you options — some are quick and nutritious, and some are complete meals. The binding thread, besides fermentation, is that all of the recipes are nutrient-dense foods that will stick with you until lunch.

Basic Warm Amazake

YIELD: ABOUT 1 PINT

This warm smoothie can be enjoyed as is, but a little freshly grated ginger or Fermented Ginger Honey (page 327) on top is also delicious. Different spices can completely change the mood. Try a drop of vanilla extract. Or add to the vanilla a drop of almond extract and cinnamon, and you are on the road to a warm fermented horchata. We also like a sprinkle of cardamom or freshly grated nutmeg.

1 cup Rice Amazake (page 217)

1 cup just-boiled water

Freshly grated ginger or ginger honey, for topping (optional)

Combine the amazake, hot water, and ginger, if using, in a blender. Blend until smooth. Enjoy immediately.

Amazake Milk

YIELD: 1 QUART OR MORE

Blended amazake is a great milk substitute, and it's both healthful and easy to make.

2 cups Rice Amazake (page 217)

2–3 cups water

Combine the amazake and 2 cups of water in a blender and blend until smooth. If you want a thinner amazake, add more water. Refrigerate and use within 1 week.

Amazake Mocha (or Latte)

YIELD: ABOUT 1 PINT

You cannot go wrong with coffee and chocolate, separately or together. This recipe has all the ingredients for a mocha drink, but leave out the chocolate and you have a wonderful latte. Substitute hot water for the coffee and you have a simple drinking chocolate. For a richer mocha, add 1 teaspoon coconut manna. Drink warm in the morning or pour over ice for a refreshing afternoon beverage.

1 cup brewed coffee (see note below)

½ cup Rice Amazake (page 217)

1 tablespoon unsweetened cocoa powder

Combine the coffee, amazake, and cocoa powder in a blender. Blend until smooth. Enjoy immediately.

Note: We like to brew our coffee with a tablespoon or two of chicory. This gives the mocha a smooth, bitter note that balances the sweetness of the amazake.

Amazake Broth

YIELD: ABOUT 1 PINT

If you are looking for a savory warm beverage, the combination of broth (bone broth, vegetable broth, or miso soup) and amazake makes a delicious breakfast on the go. If you use miso soup, you also have an easy-to-eat healing food perfect for anyone who's down with a cold or the flu. To make the creamy miso even richer, add 1 tablespoon sesame oil (we prefer the mellowness of untoasted sesame oil, but feel free to use toasted) or other healthy oil.

When it's made with chicken broth, we find that this drink tastes remarkably like cream of chicken soup. It reminds Christopher of his childhood and canned cream of chicken soup. He likes this in a thermos to drink on the road.

½ cup Rice Amazake (page 217)

1½ cups hot broth or miso soup (page 301)

Combine the amazake and broth in a blender. Blend until smooth. Enjoy immediately.

Amazake Drinks

All of the amazake drink recipes in this section are based on our Rice Amazake recipe on page 217. But feel free to experiment with other grain amazake ferments. For all of these recipes, we use a blender or stick blender to make a smooth, creamy consistency. You can think of these drinks as warm smoothies. We hope these recipes get you started and you use them as a launching point for all kinds of delicious drinks and smoothies.

Fermented Ginger Honey

YIELD: ABOUT 1 PINT

Part culinary, part medicinal, this simple but powerful ferment might become a staple in your pantry. You can drizzle it on anything that needs a little of ginger's warmth, and it has become one of our favorite ferments for flavoring sweet amazake drinks. Just a small amount gives the whole drink deeper sweet flavor. The potent medicinal properties of the honey and ginger are said to be synchronistic, and even more so when paired with the power of fermentation.

It is important to use raw honey for fermentation, as it is alive with all of the necessary microbes. Honey is a naturally antimicrobial environment, but a little bit of liquid (in this case, from the ginger) disrupts the natural balance and causes the microbes to begin to ferment.

INCUBATION **1 month**

1¼ cups thinly sliced fresh ginger (no need to peel)

1½ cups raw honey

1. Place the sliced ginger in a pint jar with a tight-fitting lid. Pour in the honey, leaving about 1½ inches of headspace. This space is important, as the honey will become bubbly and active and will need room to keep from overflowing. Place the lid on the jar and tighten.

2. Place the jar on your counter and let ferment for 1 month. Every other day, or more often as needed, crack open the lid and allow the CO_2 to escape. Then retighten the lid and turn the jar over to allow the honey to coat any ginger that may have floated to the top.

3. When the bubbles start to settle down and the honey becomes runny, with a warm earthy ginger flavor, it is done. Enjoy. The ginger will have soaked up the honey and taken on a wonderful gentle candied flavor. Store this stable ferment in your cabinet or refrigerator. This will last a year or more.

Amazake Chai

Smooth and rich, this chai is creamy, even though it has no milk, and sweet, though it has just a drop of added honey. For a thicker drink, use more amazake or less tea. Use your favorite bagged chai to make the tea, or make your own chai mix (see note below).

If you haven't yet made fermented ginger honey, you can substitute ½ to 1 teaspoon raw honey with ½ teaspoon grated fresh ginger.

1½ cups hot chai tea

½ cup Rice Amazake (page 217)

½–1 teaspoon Fermented Ginger Honey (page 327), with 2 to 3 slices of the ginger

Combine the tea, amazake, honey, and ginger slices in a blender. Blend until smooth. Enjoy immediately.

Note: *For homemade chai, combine 2 cups water with a 1-inch slice of fresh ginger, 3 lightly crushed cardamom pods, 3 whole cloves, 1 teaspoon black peppercorns, 3 whole allspice pods, and a broken-up cinnamon stick in a medium saucepan. Simmer for 10 minutes, then remove from the heat and add a heaping tablespoon of loose black tea or two black tea bags. Steep for a few minutes, pour through a strainer, and it's ready to use with the amazake.*

Amazake Golden Milk

Golden milk has received a lot of attention in the last few years for its anti-inflammatory properties. We thought this traditional warm drink was a perfect candidate to become an amazake drink, and we weren't disappointed — it was even better than we'd hoped. Traditionally, golden milk is made with whole cow's milk, but we substitute coconut oil, which gives the milk rich mouthfeel and deep flavor. The fat is necessary to help your body absorb the anti-inflammatory curcumin (from turmeric) and piperine (from black pepper), so don't skip it.

If you haven't yet made Fermented Ginger Honey, you can substitute ½ to 1 teaspoon raw honey with ½ teaspoon grated fresh ginger.

1¼ cups water

1 tablespoon coconut oil or ghee

1 cinnamon stick

½ teaspoon black peppercorns

½ teaspoon powdered turmeric

½ cup Rice Amazake (page 217)

½–1 teaspoon Fermented Ginger Honey (page 327), with 2 to 3 slices of the ginger

Combine the water, oil, cinnamon, peppercorns, and turmeric in a small saucepan over medium-high heat. Simmer for 10 minutes. Pour through a strainer into a blender and add the amazake, honey, and ginger slices. Blend until smooth. Enjoy immediately.

AMAZAKE GOLDEN MILK

Oat Koji Granola

YIELD: ABOUT 2 QUARTS

This granola smells like cinnamon raisin oatmeal cookies. It is wonderfully sweet and crunchy, with no added sugar or oil. We came up with this recipe for two reasons: we wanted a way to capture the tastiness of oat amazake, and we wanted to make a granola that our neighbor, who loves granola but is sensitive to oats, could eat. This is by no means a health claim, but he can enjoy this fermented version without ill effects.

1½ quarts Oat Amazake (page 218), made with oat groats

1½ tablespoons ground cinnamon

¼ teaspoon ground cloves

1½ tablespoons vanilla extract

½ teaspoon almond extract

1–2 cups chopped dried fruits and/or nuts

1. Combine the amazake, cinnamon, cloves, vanilla, and almond extract in a bowl and mix thoroughly.

2. Preheat the oven to 170°F/77°C, or set a dehydrator to 105°F/41°C.

3. Spread the granola mixture out on baking sheets (for the oven) or on dehydrator trays lined with silicone or Teflon sheets (for the dehydrator). Don't worry that the oat groats will tend to clump; they will break apart later for a good granola texture.

4. Bake the granola in the oven for about 1 hour 15 minutes, or until it is just a bit a moist (you don't want it to get too crispy). If using a dehydrator, dehydrate for 10 to 12 hours, or until thoroughly dry. The finished granola will be lightly crispy with a chewy center.

5. Mix in the chopped fruit and nuts. Store the granola in an airtight container. It will keep for about 1 month.

Oat Amazake Currant Scones

YIELD: 6–8 SCONES

This is a modification of the currant cream scones recipe from *The Tassajara Recipe Book* by Edward Espe Brown. Our copy of the book is so beaten and battered that you can hardly identify it. The cover gave up protection many years ago, then the spine lost its yoga suppleness and eventually broke. Two generations on the farm have turned to this scone recipe time and again, making modifications over the years. It only made sense to try oat amazake in the recipe, and the resulting scones are truly wonderful.

---

2 cups all-purpose flour (or substitute up to ½ cup whole-wheat pastry flour)

2½ teaspoons baking powder

½ teaspoon salt

6 tablespoons chilled butter

1 egg

½ cup Oat Amazake (page 218)

1 tablespoon grated lemon zest

1 teaspoon vanilla extract

½ cup dried currants

1 tablespoon cream or milk

1 teaspoon sugar

---

1. Sift together the flour, baking powder, and salt into a large bowl. Mix in the butter with a pastry cutter, two forks, or your fingers (our preferred method) until the butter is in little lumps the size of small peas.

2. Beat the egg in a separate bowl, then add the oat amazake, lemon zest, and vanilla to the egg and stir to combine.

3. Add the currants to the flour mixture and stir to combine, then add the egg mixture and lightly toss until it comes together. With your hands, fold the dough in the bowl until it pulls away from the sides of the bowl and you have a shaggy ball. It's important to keep a light touch here because overworking it excites the gluten.

4. Let the dough rest on the counter for 15 minutes to relax the gluten.

5. Preheat the oven to 400°F/200°C.

6. Using a rolling pin or your hands, flatten the dough to form a circle about ¾ inch thick, tapering the edges to round them. Brush the cream on top, then sprinkle the sugar over the top. Cut into six or eight wedges and place them on an ungreased baking sheet about 2 inches apart.

7. Bake for 15 minutes, or until the scones are puffed up a bit and lightly browned. Transfer to a wire rack or cutting board to cool.

Breakfast Scrambled Tofu Squared

YIELD: 2 SERVINGS

Christopher and I discovered tofu scramblers as an alternative to scrambled eggs in college. We love them. We use sprouted tofu, which is made with sprouted soybeans. Sprouting releases the phytates and makes the tofu easier to digest. It also improves the nutrient value. Sprouted tofu is richer in protein, calcium, and iron, slightly lower in carbohydrates, and slightly higher in omega-3 fatty acids than regular tofu. We also discovered that adding "stinky" tofu elevates the flavor profile beyond measure — especially the cheesy quality.

We make a basic scramble with the veggies that we have on hand. Kale, thinly sliced, is one of our favorites.

1 (16-ounce) block firm or extra firm sprouted tofu

1 tablespoon medium- to high-heat oil (avocado and sunflower are good choices)

1½ teaspoons fermented garlic paste; 1 garlic clove, grated; or ½ teaspoon garlic powder

2 tablespoons finely diced onion

½–1 cup mixed seasonal vegetables, such as thinly sliced kale and diced green bell pepper

1 teaspoon grated fresh turmeric or ½ teaspoon powdered turmeric

¼ teaspoon freshly ground black pepper

1–2 tablespoons Stinky-Style Fermented Tofu or more to taste (page 89)

Nutritional yeast

1. Line a plate with a few paper towels and set the tofu on the plate to drain for 10 minutes. Pat the tofu dry with fresh paper towels, then crumble it.

2. Place the oil in a large skillet over high heat and allow it to warm. Add the crumbled tofu and fry until cooked and warmed thoroughly, about 10 minutes.

3. Reduce the heat to medium-high. Add the garlic paste, onion, mixed vegetables, turmeric, and black pepper to the pan and sauté (turning down the heat as needed) until the vegetables are cooked and the onions are translucent, about 5 minutes.

4. Remove the pan from the heat. Add the stinky tofu, mix thoroughly, then sprinkle on the nutritional yeast to taste (we use about 3 tablespoons). The yeast will soak up any moisture and give the scramble a creamy texture. If the mixture is too dry, add a little bit of water to achieve a good consistency. Serve immediately.

South River Porridge

YIELD: 2 SERVINGS

This recipe comes from *The Little Book of Miso Recipes* by Christian and Gaella Elwell at South River Miso (see page 252). Not only is the porridge delicious, but the recipe is clever, using miso to presoak (or lightly ferment) the oats in order to break down the starches and make them more accessible to your body. Without imparting a noticeable taste of its own, the miso will liquefy the cereal, unlocking its essential nutrition and creating a wholesome sweet taste as it ferments overnight.

You can use other whole, rolled, cracked, or ground cereal grains, although cooking times will vary. Cook the grains in the evening, and your breakfast will be ready for you in the morning. Breakfast hot cereals could also be "processed" overnight with amazake.

1 cup rolled oats

¼ cup dried mulberries (optional)

2 cups water

2 teaspoons unpasteurized Light Miso (page 263)

1. Combine the oats, mulberries (if using), and water in a medium saucepan and bring to a boil over high heat. Reduce the heat to low and let simmer, uncovered, until the water is absorbed, 5 to 10 minutes. Let the oatmeal cool to body temperature (100°F/38°C or below).

2. Add the miso and stir thoroughly. Cover and let sit overnight at room temperature (about 70°F/21°C). Reheat in the morning (without boiling) and serve.

Smoky Bacon-ish Tempeh

YIELD: ½–1 POUND

Fried, crispy, salty, and smoky — this marinated tempeh hits all the same happy places as bacon. We have served this to die-hard bacon lovers who are chagrined to admit that while it's not bacon, it is just as good.

A nice long 8-hour marinade will give tastier results than a quick one, but if you are low on time the shio koji works wonders in as little as 2 hours. We use peanut or coconut oil to fry the strips, but any high-heat oil will do. And do fry them in a generous amount of oil. We tried baking and broiling the strips, but they dried out, weren't sufficiently crispy, and lost the richness of the flavor.

2–3 tablespoons Shio Koji (page 223)

2 teaspoons liquid smoke

1 (8- to 16-ounce) cake tempeh of your choice

3–4 tablespoons peanut, coconut, or other high-heat oil

---

1. Combine the shio koji and liquid smoke in a quart-sized ziplock bag and mix well. Place the tempeh in the bag and massage to throughly coat it in the marinade. Let sit in the fridge for at least 2 hours and up to overnight.

2. Remove the tempeh from the marinade and slice thinly into strips. Heat the oil in a skillet over medium-high heat. Place the strips in the oil and fry until golden on the bottom, about 3 to 4 minutes. Turn over and fry the second side until golden, about 4 minutes. Serve hot.

Infusing Tempeh and Oncom with Flavor

Tempeh has a nutty flavor on its own, but it is also a blank canvas for many flavors. The same holds true for oncom. It is similar to tempeh but much milder, with a softer texture to boot. Oncom has become our favorite base for chorizo and breakfast sausage; see our recipe on page 163.

For years, our go-to method to bring tempeh's flavor to the forefront was to simmer it in a mixture of soy sauce and water for about 10 minutes per side. Then it was ready to use as is or to sauté or fry. Don't get us wrong, that is still a fantastic way to prep your tempeh, but let's talk about shio koji for a moment.

Shio koji is our preferred marinade for tempeh or oncom. Its enzymes develop the flavors that are already in the tempeh and oncom, pulling out more sweetness and umami while salting these unsalted ferments. We find that it leaves tempeh and oncom ready to take on the flavors of any cuisine without the distinctive undertones of soy sauce. You will also see enzymatic flavoring action at work action as you try the following three breakfast "sausage" recipes.

When we marinate tempeh and oncom, we find that you don't need to use all of it at once. We often marinate a bunch and take out just as much as we want for a particular dish. Tempeh and oncom will keep for about 4 days in the marinade in the refrigerator.

Oncom Maple Breakfast Sausage

YIELD: ½–1 POUND

These breakfast "sausage" links or patties are a delightful alternative to a traditional maple breakfast sausage. The maple syrup infuses the oncom with a subtle sweetness, making these as appropriate in a savory egg breakfast as they are with a stack of pancakes. We also find that they make a delicious breakfast sandwich on a biscuit, bagel, or English muffin. A normal cake of oncom or tempeh can be cut into links. For patties, grow the oncom or tempeh (as pictured at right) in round silicone molds.

We get the best results from marinating the oncom in a ziplock bag, as the air can be removed, which will fully submerge the oncom. That said, the oncom can also be marinated in a covered dish as long as you rotate it a few times over the course of the marination.

These links or patties are baked, which gives them a crisp outer crust and a soft juicy interior. If you want a fattier sausage, fry them in a few tablespoons of oil in a nonstick pan. This recipe can also be made with tempeh, which will give it a chunkier texture and a nuttier flavor.

8 ounces Black Oncom (page 163) or tempeh of your choice

¼ cup Shio Koji (page 223)

3 tablespoons maple syrup

1 teaspoon dried sage

1 teaspoon dry mustard

¼ teaspoon freshly ground black pepper

½ teaspoon onion powder

1. Slice the oncom into strips that are about ½ to ¾ inch wide and 3 inches long, or form into patties.

2. Combine the shio koji, maple syrup, sage, mustard, pepper, and onion powder in a quart-size ziplock bag and mix well. Place the strips in the bag and massage to thoroughly coat with the marinade. Allow the strips to marinate in the fridge for 2 to 8 hours.

3. Place your oven rack in the bottom third of your oven. Preheat the broiler.

4. Grease a baking sheet. Place the strips on the baking sheet, leaving space between each strip. Broil for about 5 minutes, or until nicely browned on top. Turn each piece with tongs, then broil for another 3 minutes or until evenly bronzed. Serve hot.

Oncom Mexican-Style Chorizo

Mexican chorizo is a spicy, crumbly sausage with a very distinctive look (it's bright red) and flavor that is owed to the combination of spices and the mild dried ancho chile. Our oncom version is so close to the original that you could fool some people — okay, admittedly there is hardly any grease, but the flavor and crumbly texture are pretty fantastic. Use in any recipe where you might use chorizo, from breakfast burritos to nachos or taco salad. It also makes a delicious topping for a baked potato.

The two spices that give this chorizo the "right" flavor are ancho chile powder and finely powdered Mexican oregano. If you can't source ancho chile powder, pasilla chile powder also works quite nicely. If you can't find Mexican oregano, simply pulverize regular oregano in a spice grinder.

This chorizo is fried after it is marinated. For the best results, use a nonstick pan and stir fairly frequently while watching it. It will brown quite readily, which gives it a wonderful flavor and mouthfeel, but this also makes it susceptible to burning.

8 ounces Black Oncom (page 163) or tempeh of your choice

2–3 tablespoons Shio Koji (page 223)

2–3 tablespoons ancho or pasilla chile powder

1 teaspoon garlic powder

½ teaspoon powdered Mexican oregano

¼ teaspoon freshly ground black pepper

Pinch of ground cloves

Pinch of ground cinnamon

2 tablespoons avocado oil

1. Crumble the oncom into chunks ranging from ¼ inch to 1 inch.

2. Combine the shio koji, chile powder, garlic powder, oregano, black pepper, cloves, and cinnamon in a lidded container and mix well. The marinade will be quite thick and sticky. Place the chunks in the marinade and stir to thoroughly coat. Allow the oncom to marinate in the fridge for at least 2 hours or overnight. The marinade will coat the oncom and there won't be any liquid, so there is no need to stir.

3. Warm the oil in a nonstick pan over medium heat. When the oil is hot, add the entire container of marinated oncom. Let it cook without disturbing it until you see browning, which will happen in 5 to 6 minutes; then immediately stir and lower the heat. Continue to cook, now stirring frequently, until the oncom is evenly cooked, 5 to 10 minutes. Serve immediately.

Traditional Natto Breakfast Bowl

YIELD: 1 SERVING

Natto is a breakfast food and is typically seasoned and eaten with rice. The seasoning varies by taste, but its simplest form is some soy sauce and sesame oil. Natto packages usually include karashi, a hot Japanese yellow mustard, that can be added to the mixture. In our experience, the soy sauce packet is not a high-quality soy sauce but a sugary chemical approximation; use your own instead. We also add some kimchi or other fresh or fermented vegetables for garnish.

1 cup warm rice

¼–½ cup Natto (page 118) or your favorite natto

1 teaspoon soy sauce (Shoyu, page 298, or Wheat-Free Tamari, page 297)

1 egg

Sliced scallions and nori, for garnish

1 teaspoon sesame oil

Grated daikon

Toasted sesame seeds, for garnish

1. Place the warm rice in a serving bowl.

2. Place the natto into another bowl. Add the soy sauce and egg and stir together vigorously.

3. Place the natto mixture on the rice, then top with scallions and a few slices of nori. Drizzle the oil on top, add the grated daikon, and then sprinkle with toasted sesame seeds.

Whipping Natto

Traditionally, natto is stirred vigorously before being eaten to increase its viscosity. There are very distinct ideas (and they vary widely) on how many times this innocuous-looking pile of shiny fermented soybeans needs to be whipped. We have read anywhere from 10 to 424 times, the latter being reported by *Japan Times* as the optimal number. Any more whipping and you lose the goo. When we visited Heidi Nestler (see page 125) for "natto boot camp," for every recipe we ate, the natto was stirred 200 times with two chopsticks. Very soon after the stirring commences, the natto resembles a foamy slurry that has the still-whole soybeans suspended in it. This gooey foam indicates happy, healthy natto.

We know you are thinking, YUM! Well actually, yes, yum is part of it. The Japanese say that this agitation awakens the umami (delicious) and increases the nattokinase (natto's superpower), enabling the nutrients to be digested better. However, there is no scientific evidence that whipping increases nattokinase activity. In fact, science would say that adding air to the surfaces is bad for protein activity. It's really up to you whether you like to eat your natto stirred or unstirred.

Natto Eggs Benedict

YIELD: 4 SERVINGS

In this recipe, we wanted to lose the anxiety of the careful orchestration it takes to make a warm, creamy stack of traditional eggs Benedict for everyone. Instead, we wanted these eggs to be a doable, powerful breakfast for any day of the week. We nixed the fickle hollandaise sauce and replaced it with a simple homemade mayonnaise, which can be made ahead of time. We replaced the Canadian bacon, which needs to be crispy and warm, with natto, which is prepared ahead of time, leaving you more time for the toast and eggs. We've also added a bed of greens to bring some freshness to the plate. We love the chewy texture and sour taste of a true levain sourdough bread, but feel free to substitute English muffins.

A fun variation is to whisk a tablespoon of gochujang (page 294) into the mayonnaise.

4 tablespoons Natto (page 118)

½ teaspoon sesame oil

½ teaspoon soy sauce (Shoyu, page 298, or Wheat-Free Tamari, page 297)

1–2 scallions, sliced thinly

1 tablespoon white or cider vinegar (optional)

4 slices sourdough bread

Butter, for spreading on toast

½ pound baby greens

4 eggs (see note below)

1 cup mayonnaise, preferably homemade (page 308)

Note: The fresher the eggs, the tighter the whites. This helps the eggs retain their shape, as the stronger white is less likely to break into stringy threads. To determine the freshness of your eggs, put them in a bowl of water. If the eggs lie horizontally on the bottom, they are fresh. As eggs age, the air pocket increases, making them stand upright in water; use these older eggs for something else.

1. Combine the natto, oil, soy sauce, and scallions in a small bowl and stir together.

2. Fill a shallow pot or pan with 2 to 3 inches of water and set it over medium heat. Add the vinegar, if using, to help speed up the setting of the white. Bring the water to about 180°F/82°C; at this temperature the water is very hot but no bubbles are disturbing the surface.

3. While you are waiting for the water to heat, toast the sourdough and butter each piece. Set each piece of toast on a plate and top with the greens.

4. Line a plate with a paper towel. Crack an egg into a small bowl, then slide it into the water. If you want to cook your eggs simultaneously, crack each one into a separate bowl and add them to the hot water one after another. Cook for 1 minute, then stir gently so that the eggs don't stick. For a soft egg, cook for about 3 minutes; for a firmer egg, cook for about 5 minutes. When done, remove the eggs with a slotted spoon and set them on the towel-lined plate briefly to drain the excess water.

5. Place the eggs on the toasts. Dollop the mayonnaise on top of the eggs, dividing it equally among them. Garnish each egg with a tablespoon of the natto mixture.

SMALLER BITES

Humans have been using bowls for a very long time. This section focuses on the sustenance we might serve in bowls — salads, tapas, dips, and spreads.

Natto Polenta

By Ann Yonetani, NYrture

YIELD: 4 SERVINGS

ANN WRITES: The classic way to eat natto in Japan is on a bowl of warm rice, seasoned with a bit of soy sauce, spicy mustard, and perhaps some chopped scallions or a raw egg yolk. Mix it all up and put it down the hatch for a quick Japanese power breakfast. For those who may not have fresh steamed rice in an automatic cooker at all times and don't want to try natto first thing in the morning, there is natto polenta, which pairs well with a wide range of savory foods for lunch or dinner.

4 cups water

1 cup polenta cornmeal

1 tablespoon butter (optional)

Salt

½ cup Natto (page 118)

1 cup diced tomatoes

4 tablespoons diced figs

Handful of basil leaves

1. Bring the water to a boil in a medium saucepan over medium-high heat. While whisking gently, add the cornmeal in a thin stream. Add the butter, if using, and salt to taste. Cook, stirring constantly, for 2 minutes, then reduce the heat to low. Cook, stirring occasionally, for 30 to 45 minutes, depending on the consistency you desire.

2. Divide the polenta between four bowls, then top each bowl with 2 tablespoons natto, ¼ cup diced tomatoes, 1 tablespoon diced figs, and a few basil leaves. Serve immediately.

Note: *We tried this recipe with Betty Stechmeyer's Koji Polenta (page 221) instead of traditional polenta; it was sweeter but also quite tasty.*

Turn Your Rice Bowl Inside Out

Lunch bowls are fantastic, but they're not very portable. Everyone loves a sandwich not just because it's versatile and yummy but because it's easy to take on the go. We have found that a wonderful (gluten-free) way to have your bowl and have your sandwich is by making a Japanese rice sandwich, called *onigirazu*. It is almost like sandwich meets sushi, or *onigiri* (Japanese rice balls wrapped in sushi) supersized.

To make one, lay a piece of plastic wrap on your counter, and lay a sheet of nori on it diagonally, so that one corner points at you. Place ½ cup of warm rice in the middle of the nori sheet and spread it evenly into a square, leaving space around the edges. Place fun fillings (see the lists below) in single thin layers across the rice. We usually pick three or four fillings. Use small amounts so that you don't overfill. Place another ½ cup of rice on the top of the fun stuff, then fold each corner of the nori tightly toward the center of the filled rice. Wrap tightly in plastic wrap. The moisture from the warm rice will bind everything together. If you intend to eat the sandwich right away, let it rest for a few minutes so it will set. When you're ready to eat, cut the onigirazu in half with a sharp knife.

Vegetables

▶ Avocado slices

▶ Any fresh vegetables, sliced thinly

▶ Any greens, cut to lay flat

▶ Any fermented pickles, sliced thinly

▶ Any fermented vegetables (from kraut to kimchi)

Proteins

▶ Natto (page 118)

▶ Strips of flavor-infused tempeh (page 335)

▶ Fried egg

▶ Miso marinated egg (page 322)

▶ Cured egg yolks (page 323)

▶ Tofu Misozuke (page 319)

▶ Sardines or other small fish, spread thinly

Condiments

▶ Pone Yay Gyi Shallot Salad (page 351)

▶ Velvety Peanut Sauce (page 304)

▶ Miso mayonnaise (page 307)

▶ Thin layer of fermented stinky tofu (page 89)

Hand Rolls
(Temaki Zushi)

By Heidi Nestler, Wanpaku Foods

YIELD: 20–32 ROLLS

This is how Heidi Nestler (see page 125) introduces folks to natto. She marvels at how her conversations about natto are starting to change. She is spending less time talking about how to hide the goo. One woman stopped her midsentence to say, "What is it that people don't like about this again?" Nonetheless, it is important to present it well to those who are new to natto.

These cheery little rolls can make a wonderful starter or hors d'oeuvre for a potluck, a simple meal, or a nice dinner party. If your friends have never had natto, these are sure to get the conversation going.

2¼ cups short-grain rice

1 cup Natto (page 118)

Dash of soy sauce (Wheat-Free Tamari, page 297, or Shoyu, page 298), plus extra for serving

Dash of sesame oil

2 drops rice wine vinegar

Grated wasabi, for serving

10–16 sheets nori, cut in half

2 avocados, sliced lengthwise

10–20 green shiso leaves

Fresh or fermented cucumbers, carrots, or daikon, sliced into long, thin spears (optional)

Sushi-grade fish (we like tuna or cuttle-fish), cut into strips (optional)

1. Cook the rice, following the instructions on page 67.

2. Stir or chop the natto, then combine in a bowl with the soy sauce, oil, and vinegar. Mix well.

3. Set out some soy sauce and wasabi in small serving dishes.

4. To make a hand roll, place one of the half sheets of nori in the palm of your hand. Place a small amount of rice on the nori, add a bit of the natto mixture, and top with avocado, shiso leaves, vegetables, and fish, as desired. Be careful not to add too much! Roll into the shape of an ice cream cone. Enjoy! If you like, dip the roll in soy sauce or wasabi. If you've used fermented vegetables, you may not need any extra seasoning at all.

Natto Toast Is the New Avocado Toast

In recent years, toast has gone beyond butter and jam and become the substrate for all manner of ingredients from simple to indulgent, including the now ubiquitous avocado toast. Well, we ask the avocado to move aside — it's time for natto toast. We thought: Why not take everyone's favorite comfort food and make it the happy, easy backdrop for a food many are not comfortable with yet?

Below we share our favorite natto spreads, but consider this as a launching point for you to come up with your own combinations. We use 2 tablespoons of natto, the amount in a container of purchased natto. It makes one fully loaded slice or two pieces of normal spread thickness.

BASIC GUIDELINES

First, start with great bread. It doesn't matter if your favorite is rye, sourdough, or gluten-free; it all makes yummy toast. To your natto, add:

▸ Something fatty (think sesame oil, olive oil, or mayonnaise [page 308])

▸ Something salty (olives, feta, nukazuke [page 317], or soy sauce fill that roll)

▸ A little something with a bright flavor, usually that has some acidity (such as a splash of citrus, chopped fermented veggies, or mustard)

▸ Something fresh (like scallions, garlic chives, a bit of freshly grated turmeric or ginger, or grated carrots)

SOME COMBINATIONS

The idea here is to get you thinking about mixing and matching flavors that wouldn't naturally go together. You may notice that we keep coming back to Mediterranean flavors. Perhaps that's because those salty-savory flavors work nicely with the alkaline umami of natto. The three natto recipes that follow — Tapenatto, Natto Miso Hummus, and Pimento Cheese (Meets Natto) — are excellent on toast but can be so much more.

▸ Natto prepared for Traditional Natto Breakfast Bowl (page 339) or any similar classic combination of flavors

▸ 2 tablespoons natto mixed with 1 teaspoon Miso Pesto (page 307) atop a layer of fresh greens

▸ 2 tablespoons natto mixed with 1 tablespoon crumbled feta, 5 or 6 chopped Kalamata olives, and 2 teaspoons olive oil

▸ 2 tablespoons natto with 1 tablespoon blue cheese and 2 teaspoons olive oil, topped with slices of pear

▸ 2 tablespoons natto with 3 tablespoons chopped lemon-dill kraut, plain kraut, or kimchi, on top of (someone had to do it) mashed avocado

Tapenatto

YIELD: ABOUT 1½ CUPS

This spread is a play on a traditional tapenade. The salty-sour flavor of olives pairs nicely with the natto and makes a delicious spread. Conveniently, if you are a little natto shy, these stronger flavors offer a good disguise. If you use canned green olives, you will get a much subtler flavor than if you use the brine-cured variety.

The recipe calls for thick aged balsamic vinegar. The regular balsamic vinegar is too thin and strongly flavored for this dish.

After the tapenade has been chopped in the food processor, chop the natto by hand. This is important, as one doesn't want to anger the biofilm. In other words, putting natto through a food processor gives a whole new meaning to the stringiness that is *neba neba*.

Serve on crusty bread or use as a dip.

1 cup green olives, pitted, canned or cured

½ cup Kalamata olives, pitted

2–3 garlic cloves, fresh or fermented

2 tablespoons capers

4–5 dried tomatoes, soaked in hot water for a minute to soften if not packed in oil

½ teaspoon thick aged balsamic vinegar (optional)

2–3 tablespoons Natto (page 118)

1. Combine the green and Kalamata olives, garlic, capers, tomatoes, and vinegar, if using, in a food processor and pulse a few times until the mixture is finely chopped. Alternatively, finely chop all the ingredients by hand. Put the mixture in a bowl.

2. Chop the natto on a cutting board until all the beans are at least halved or smaller. Add the chopped natto to the olive mixture and stir to incorporate. Serve immediately.

Pimento Cheese (Meets Natto)

YIELD: ABOUT 1 CUP, ENOUGH FOR 2–4 SANDWICHES

When faced with fermented mash-ups of his childhood foods, Christopher frequently comments, tongue in cheek, "Just like mama used to make." It is usually a sign that he likes it. During the development of this book, he found natto hidden in all sorts of surprising foods (admittedly, some were better than others). When Kirsten hid natto in a grilled cheese sandwich the first time (nothing like gooey cheese to hide gooey natto), she was accused of putting tiny, beany paratroopers in the sandwich that dove out at each bite. The sharp cheese was tasty with natto, though, so she took it a step further by adding cream cheese, mayonnaise, and pimentos. All that is to say that this is not your mama's pimento cheese, and this pimento cheese is darn good in a melted sandwich as well.

If you want a denser, less gooey spread, use more cream cheese. Conversely, for a creamier, goopier spread, use a little more mayonnaise.

¾ cup grated sharp cheddar cheese

2–4 tablespoons cream cheese, softened

1–2 tablespoons mayonnaise (page 308)

1–2 tablespoons chopped pimento (jarred or fresh roasted; if you use fresh roasted be sure to add an extra tablespoon of rice or white wine vinegar)

1 teaspoon rice or white wine vinegar

2 tablespoons Natto (page 118)

Dash of salt or Shio Koji (page 223)

1. Combine the cheddar cheese, cream cheese, mayonnaise, pimento, and vinegar in a medium bowl. Mix well with a fork to incorporate the cream cheese evenly.

2. Gently stir in the natto, add salt to taste, and serve. Store any unused portion in a sealed container in the refrigerator for up to 1 week.

Natto Miso Hummus

YIELD: 2–3 CUPS

KIRSTEN WRITES: When I was in seventh grade (read: middle school angst), my mom discovered hummus and would make me hummus sandwiches to take to school. The thing is, the average American had never heard of hummus, especially in rural Arizona. By the time lunch came around, the massive quantities of fresh garlic all but announced my "strange" lunch, which I was known for. As you can imagine, my early relationship with hummus was on shaky ground. Luckily for me and everyone else, that has changed, but here I am pushing the envelope on this tried-and-true spread. This is delicious, but if you have preteens in your home, you might refrain from putting it in their lunch.

A bean spread seemed like the natural place to hide natto, and it was one of the first things we tried making with natto, before we knew anything about it. We did not know that vigorous stirring of natto was a thing (see page 339). Naturally we used a food processor to mix the natto in the garlicky tahini-garbanzo mixture. The biofilm came to life and infused the garbanzo beans with vigor, and the whole thing all but grew out of the bowl. We soon realized that topping the hummus with the natto is the way to bring these two beans together.

The salt in this recipe comes from the miso. We generally find it to be enough, but sometimes we add a just a tiny bit of salt to the natto topping.

2 cups cooked chickpeas

⅓ cup tahini

3 tablespoons Chickpea Miso (page 266)

1–2 tablespoons lime juice

2 garlic cloves

2–3 tablespoons Natto (page 118)

1 tablespoon toasted sesame oil, plus more to garnish

1 tablespoon thinly sliced scallions

1–2 tablespoons capers (or finely chopped nukazuke stems, page 317), for garnish

1. Combine the chickpeas, tahini, miso, lime juice, and garlic in the bowl of a food processor and blend until smooth. You may need to add a little water as it is processing to achieve a creamy smoothness. Spoon the hummus into a bowl, and use the spoon to create a small crater in the center.

2. Combine the natto, oil, and scallions in a small bowl and mix well. Scoop the mixture into the depression in the hummus. Garnish with the capers and an extra drizzle of oil.

Spicy Javanese Carrot and Cucumber Pickle Salad
(*Acar Timun*)

YIELD: ABOUT 3 CUPS

This classic Javanese pickle salad is served often alongside such standards as *nasi goreng* (Indonesian fried rice). Its bright, salty acidity pairs well with anything fried, including tempeh (page 157) and Myanmar-Style Shan Soup (page 94).

Because it is only slightly acidic and mildly spicy, we eat this as a side salad with many dishes. It is traditionally made with vinegar, but we found it a wonderful place to showcase the power of shio koji. The shio koji does all the work and infuses the vegetables with dynamic flavor, bringing out sweet notes as well as some acidity. We like a hint of sweetness, but feel free to add more sweetener if you want a more typical Southeast Asian flavor.

This is one example of how to make quick salads in a bag. We use this technique with many vegetable combinations (though we don't always add sugar) and encourage you to do the same.

SMALLER BITES

FERMENTATION **about 24 hours**

---

2 medium carrots, julienned

1 medium to large cucumber, julienned

2 large shallots, sliced into thin rings

1 large jalapeño, sliced into thin rings

¼ cup Shio Koji (page 223)

1 tablespoon sugar (granulated, coconut, or palm work well)

1. Combine the carrots, cucumber, shallots, and jalapeño in a gallon ziplock bag and toss around a bit to make sure they are evenly mixed.

2. Add the shio koji to the bag and toss again to make sure the vegetables are evenly coated. Press as much air out of the bag as possible, then seal.

3. Leave the bag on the counter for 1 hour, then place in the fridge and let ferment until the next day.

4. When you are ready to serve, open the bag and pour the contents into a bowl. Stir in the sugar. Any unused portions can be stored in the refrigerator for up to 1 week.

Pone Yay Gyi Shallot Salad

YIELD: 2–4 SERVINGS

Burmese salads are wonderfully complex, with various flavors and textures. They include crunchy deep-fried garlic, chickpeas, or peanuts; they might have tomatoes, mango, or fish; and many have the intense flavors of ferments — like fermented tea leaves, pickled ginger, or, in this case, a dark brown bean paste.

Kirsten ordered this salad at a restaurant below Mount Popa the day after we'd visited a pone yay gyi factory. It had been a bumpy, mountainous road, and we were all a little jostled, so when the dark, oily plate of thinly sliced shallots came, the rest of the family opted out. She admitted that the oil looked difficult to swallow given her queasiness, but she found it delicious and spent the remainder of the meal trying to convince the rest of the table to taste it. Christopher did, despite the teenagers teasing him about it, and he liked it. All this to say, think of this side dish as more of a condiment. It is a perfect "fringe" — its deep, salty, pungent flavor is a wonderful accompaniment to a meal. We like it alongside rice, noodles, or beans and decorated with delicious sides. (Crazy fusion tip: we've even added a teaspoon of this salad to homemade aioli for a great sandwich dressing.)

In Myanmar, shallots are usually referred to as onions, and here they are tiny bulbs with reddish purple outer skin and a pink blush inside. The markets are overflowing with baskets of them, which makes sense, as they are used chopped, sliced, as a paste, raw, cooked, or fried in just about every dish. The European shallots that we tend to find in the United States work perfectly well. A little goes a long way with this punchy raw "onion" salad. As a condiment, it adds a punch of flavor that makes a statement with even the most bland of starches.

---

1 tablespoon Pone Yay Gyi (page 95)

2 tablespoons peanut oil

1 medium to large shallot, sliced very fine

Red pepper flakes

1. Combine the pone yay gyi and oil in a medium bowl and mix well.

2. Add the shallot slices to the dressing and stir, then garnish with pepper flakes to taste.

Miso-Glazed Vegetables

YIELD: 2–4 SERVINGS

This is a basic miso glaze. Use a sweet white miso for a sweet caramelized flavor, or a darker, older miso for a much deeper caramelized flavor. It depends on your mood or your meal — or what you happen to have in your fridge. This recipe will glaze about 1 pound of vegetables. This is one of our favorite ways to eat winter veggies like kohlrabi, carrots, turnips, onions, and rutabaga, but feel free to use it with any vegetables.

1 pound vegetables, cubed or thickly sliced

2 tablespoons miso of your choice

2 tablespoons butter or coconut oil

1 teaspoon maple syrup

Juice of 1 lemon

Salt and freshly ground black pepper

1. Combine the vegetables, miso, butter, and maple syrup in a large skillet. Add enough water to just cover the vegetables. Bring this mixture to a boil over high heat, then lower the heat to simmer and cook, stirring as needed, until the vegetables are tender and the liquid is thick (timing will vary depending on type of vegetable). If the vegetables are not tender when the liquid is thick, add more water, a little at a time, until they are.

2. Cook for a few more minutes to further thicken the sauce and caramelize the vegetables; you want them golden brown. The glaze will start to stick a bit.

3. Add the lemon juice and stir to coat the vegetables with the glaze. Season with salt and pepper to taste and serve.

Miso as Marinade or Glaze

Perhaps the most approachable way for someone to get to know miso is through its incredible ability to transform meat. It's like barbecue sauce with a superpower: its enzymes soften a tough cut while penetrating the meat with flavor. Our simplest marinade is just miso, mirin, and a splash of doburoku (page 232).

For cooking success, it is important to note that miso marinades act a little differently than barbecue sauce. The flavor, with the help of the enzymes, penetrates the surface of the meat or vegetable. However, the excess must be blotted off. Since miso doesn't melt, it can create a charred effect that is not the yummy kind — instead, it can become a bitter, burned-tasting crust. Simply blot off the excess with a paper towel, leaving a thin layer that barely coats the food.

LARGER BITES

The ferments in this book aren't just lovely accents — they really bring a main meal together, and sometimes are even the main attraction. In this section we want to give you a peek into the versatility of these ferments. As you know by now, miso is not just the soup that comes with your meal. We hope these recipes act as a springboard for your ideas. Enjoy.

:::::::::::: Thai Marinated Tempeh :::::::::::::

By Smiling Hara Hempeh

YIELD: 2–4 SERVINGS AS A SALAD OR BOWL TOPPING

This quick and easy marinade will work with any tempeh. The tempeh is as delicious on fresh salad as it is on a bowl of flat, wide Thai-style drunken noodles — or on anything that needs a little zip.

This recipe is the result of a collaboration between Chad and Sarah from Smiling Hara Hempeh (page 154) and their friend Ashley, who is a dietitian. Chad and Sarah have worked hard to introduce tempeh to the sometimes-unadventurous American eater by collaborating with bloggers and chefs to push the boundary on how yummy it can be. Now, they say, more and more people love it.

---

14 ounces tempeh of your choice

1 tablespoon chili garlic sauce

¼ cup lime juice

¼ cup maple syrup

¼ cup peanut butter (or any nut butter)

¼ cup Wheat-Free Tamari (page 297), Shoyu (page 298), or coconut aminos

---

1. Cut the tempeh into thin, bite-size pieces (we like cutting them into small triangles).

2. Combine the chili garlic sauce, lime juice, maple syrup, peanut butter, and tamari in a small bowl and whisk to combine. Taste and adjust the flavor to your liking.

3. Add the pieces of tempeh to the marinade and toss to coat. Cover and refrigerate for at least 2 hours or overnight.

4. Preheat the oven to 350°F/180°C.

5. Line a baking sheet with parchment paper. Arrange the tempeh pieces on the baking sheet. Reserve any leftover marinade to brush over the tempeh once baked.

6. Bake for 25 to 30 minutes, or until the tempeh is caramelized and golden brown. Remove from the oven and brush with any remaining marinade. Serve immediately. Store leftovers in the refrigerator for up to 3 days.

Betsy's Tempeh Panini

YIELD: 2 SANDWICHES

This recipe was contributed by Betsy Shipley, cofounder of Betsy's Tempeh (you can read her story on page 132). These sandwiches are crispy, chewy, messy, and comforting — really everything you want in a panini. We think the keys are in her sauce, which has a well-balanced flavor and nice creaminess from the avocado, and the thinner "panini tempeh" that Gunter made in the steel trays. However, any tempeh or oncom will be delicious. Use your favorite type of bread.

1 avocado

4 black olives, finely chopped

1 scallion, chopped

1½ teaspoons Dijon mustard

½ teaspoon lemon juice

1 basil leaf, finely chopped

6–8 ounces tempeh of your choice, cut into two patties

Grapeseed or avocado oil

4 slices bread

Tomato slices

1. To make the panini sauce, mash the avocado in a bowl, then add the olives and mash together. Add the scallion, mustard, lemon juice, and basil, and mix well. Taste and adjust for seasonings as desired.

2. Using a panini grill, grill the tempeh until you see grill marks, about 5 minutes. You will have to check it by lifting the top of the grill. If you do not have a panini grill, sauté the tempeh in 1 to 2 tablespoons of oil in a skillet over medium-high heat for 3 or 4 minutes, flipping once.

3. To assemble the sandwich, slather the panini sauce on all four slices of bread. Layer the tempeh and tomato slices on two pieces of bread and top with the other pieces. Brush a little grapeseed oil on the panini grill, set the sandwiches on the grill, and pull the top down. Grill until both sides are nicely browned and the sandwich is warmed through.

No Panini Maker? Try the Cast-Iron Hack

We don't have a panini maker, but we do have a few cast-iron skillets. Our brother-in-law Ted shared this technique with us and it works like a charm. Use two skillets (only one needs to be cast iron). The cast-iron pan needs to be the same size or smaller than the pan in which you will fry your sandwiches. Heat both skillets. Put your favorite fat in one skillet and add the sandwich, then place the hot, heavy cast-iron skillet on top of the sandwich. Allow the sandwich to cook for a few minutes. Remove the top skillet and reheat it on the burner. Turn the sandwich over, adding more fat if necessary.

Place the hot, heavy skillet on the sandwich and cook again for another few minutes. Take the skillet off and enjoy!

Tempeh "Fish" Tacos

YIELD: 2–4 SERVINGS

These "fish" sticks are crispy on the outside and tender on the inside. Fish tacos are a go-to family favorite in our house, so the fact that these tempeh alternatives satisfy everyone is saying something. The batter below is gluten-free and vegan but we have found that for extra fishy umami you can add a teaspoon of fermented fish sauce. You could also serve these "fish" sticks with fries for a tempeh "fish-and-chips."

The accompanying slaw is quick and easy. Depending on the season, we make it with fresh or dried tomatoes. While the two versions are a little different, both are absolutely tasty.

8 ounces Basic Soybean Tempeh (page 157) or tempeh of your choice

High-heat oil for frying, such as coconut or peanut

4–6 corn tortillas

TEMPEH BATTER

½ cup chickpea/garbanzo flour

¼ cup potato starch

2 teaspoons seafood seasoning, such as Old Bay

1 teaspoon dulse flakes or kelp powder

½ teaspoon baking soda

1 teaspoon apple cider vinegar

¼ cup seltzer water

SLAW

1 green cabbage (1½–2 pounds), finely chopped

1 large fresh tomato or 6–7 sun-dried tomatoes, diced

½ red onion, finely diced

½ bunch cilantro, chopped

Juice of 1 lemon

½–1 cup almond-based mayonnaise or regular, preferably homemade (pages 308 and 309)

Salt and freshly ground black pepper

1. To make the slaw, combine the cabbage, tomato, onion, cilantro, and lemon juice in a large bowl. Add the mayonnaise, using however much you like in your slaw. Season with salt and pepper to taste. Toss until evenly coated, then taste and adjust seasonings.

2. Slice the tempeh into strips.

3. To make the batter, combine the flour, potato starch, seafood seasoning, dulse flakes, and baking soda in a bowl and mix well. Add the vinegar and seltzer, and stir until just mixed. The batter will have the consistency of a fluffy pancake batter.

4. Heat about 1 inch of oil in a pan. Dip the tempeh pieces into the batter to coat. Fry the tempeh in the pan, turning once, until crispy and golden, about 3 to 5 minutes per side (see note).

5. While the tempeh is cooking, warm the tortillas in a skillet over medium heat to soften. Stuff the tortillas with the tempeh and slaw and serve hot with all the usual taco fixings.

Note: The pieces may soak up a lot of oil. If the pan dries out, add a little more oil when you turn the tempeh.

Tempeh Kimchi "Reuben" Sandwiches

YIELD: 6 SANDWICHES

This recipe is a spicy fusion facelift of the traditional Reuben, using tempeh in place of corned beef. Here we double up on Korean spice with gochujang and kimchi.

While this open-faced sandwich is now far from traditional, it is certainly a wonderful amalgamation of many traditional ferments. With so many strong and flavorful ingredients, it can be easily adapted. You can even make a vegan version with vegan mayo and cheese, or a gluten-free version with gluten-free bread.

½ cup water

1 rounded teaspoon Traditional Gochujang (page 294)

1 (16-ounce) block Basic Soybean Tempeh (page 157) or tempeh of your choice

2 tablespoons peanut oil

1 medium onion, diced

½ cup mayonnaise, preferably homemade (page 308)

6 pieces sourdough bread

2 cups kimchi

2 cups grated sharp cheddar cheese

1. Mix the water and gochujang in a skillet and bring to a boil over medium heat. Add the tempeh block and simmer for 1 to 2 minutes. Flip the block and simmer for 1 to 2 minutes longer. Remove from the pan and set aside to cool. Rinse the cooking liquid from the pan.

2. When the tempeh is cool, slice it thinly. Return the pan to the stove, add the oil, and warm over medium heat. Place the tempeh slices in the pan with the onion and sauté until the onion is translucent and the tempeh is lightly browned, about 5 minutes. Remove from the heat.

3. Slather the mayonnaise on all the slices of bread.

4. Adjust the oven rack to the middle position and turn on the broiler.

5. Put the bread slices mayo side up on a baking sheet. Divide the tempeh mixture between the six slices. Top each open-faced sandwich with a generous helping of kimchi, followed by the cheese.

6. Broil the sandwiches for about 5 minutes, until the cheese is bubbling. Serve hot.

Misozuke Malfatti

YIELD: 12–15 DUMPLINGS

Sometimes the path to recipe creation takes a circuitous route and lands in a more interesting destination than the one you set out to find. This particular recipe started out as a ravioli filling (just for the record, unadorned tofu misozuke makes a wonderful simple ravioli filling). We cut and rolled pasta into ravioli and had all the leftover pieces of dough. Our brother-in-law, a chef who spent years making pasta, explained that the Italian grandmothers would use these bits to make quick "malformed" pasta. But when Kirsten wanted to learn more, she found references to *malfatti* — "malformed" ricotta dumplings. So we veered that way and found that tofu misozuke stands in beautifully for ricotta.

These little dumplings are quite filling — just three or four, with a side of roasted asparagus and braised radicchio, makes a quick (impressive) meal.

7½ cups loosely packed spinach

¾ cup Tofu Misozuke (page 319)

¼ cup flour (einkorn, unbleached all-purpose, or a gluten-free mix)

⅓ cup plus 1 tablespoon semolina or rice semolina flour

1 small egg, beaten

2 tablespoons butter

1 sprig fresh sage (about 10 leaves)

1. Combine the spinach in a pan with a tablespoon of water over medium heat and cook until wilted, 2 to 3 minutes. Spread out the spinach in a colander to drain and cool, then squeeze out all the water. Place the squeezed spinach on a cutting board and chop finely.

2. Fill a medium pot with about 2 quarts of water and bring to a boil over high heat.

3. While you wait for the water to boil, place the tofu misozuke in a medium bowl and break apart lightly with a fork. Add the einkorn flour and toss to coat (you will get a crumbly consistency similar to a piecrust, and as you would with a biscuit dough, you should only mix minimally). Add the chopped spinach, 1 tablespoon of the semolina, and the beaten egg, and stir until everything just holds together and is barely mixed.

4. Scatter the remaining ⅓ cup semolina over a plate or small baking sheet. Using a teaspoon, scoop out walnut-sized portions of the dough onto the plate. Roll to coat in the semolina, pressing into a ball. Repeat until you have used up all of the dough.

5. By this time your water should be boiling. Add the malfatti dough balls (leaving the semolina on the tray behind) and simmer, reducing heat if necessary, until they float to the surface, about 5 minutes. Once they are floating, simmer for 1 more minute, then drain and return them to the pot to keep warm.

6. While the malfatti are simmering, melt the butter in a small skillet over medium heat. Add the sage leaves, gently cook for 30 seconds, then remove from the heat.

7. Place the malfatti on plates, pour the sage butter sauce over the top, and serve.

Springtime Variation

Try replacing the spinach with 1 cup packed, chopped dandelion leaves and ⅓ cup onion tops. The dandelion greens started out as a placeholder for spinach when we happened to run out of it. We quickly learned that the tougher dandelion greens don't replace spinach in a one-to-one ratio, as these greens do not wilt the way spinach does. We loved the light bitter flavor this spring green added but found the texture overwhelming on the first pass. We made it a few more times to find a pleasing consistency. We share this so that you know that if you want to experiment with other greens in malfatti, keep the texture and body of the greens you will be using in mind.

DESSERT

Sweet umami is the juxtaposition, or maybe the collision, of sweet and savory. In the following recipes, we bring a deeper aroma and flavor to some familiar sweets.

CocoNatto Bites

YIELD: 12 BITES

These beauties were created after an all-day natto tasting event in Portland, Oregon. We were staying with our good friends Claudia and Jeff. After the tasting, Claudia was convinced that she needed to incorporate natto into her diet and she wanted to do it through an enjoyable bite of something, preferably on the sweet side. The idea for these bites began late that night as a riff on her favorite coconut bliss balls, and over several iterations they morphed into these tasty treats. If you want to consume some natto every day and don't know where to begin, start here.

4 tablespoons Natto (page 118)

6 ounces coconut butter

¾ cup cacao nibs

½ cup coconut flakes

¼ cup unsweetened cocoa powder

4 tablespoons tahini

4 tablespoons honey

¼ cup plus 3 tablespoons white or black sesame seeds (see note below)

¼ teaspoon salt

Note: *Other topping options include grated coconut, cocoa powder, cocoa nibs, matcha green tea, and hemp seeds.*

1. Place the natto on a cutting board and coarsely chop.

2. Combine the coconut butter, cacao nibs, coconut flakes, cocoa powder, tahini, honey, 3 tablespoons of the sesame seeds, and the salt in the bowl of a food processor. Pulse several times until well combined.

3. Add the natto and pulse three or four times to incorporate well.

4. Place the remaining ¼ cup sesame seeds in a shallow bowl. Scoop out rounded tablespoons of the coconut mixture and form into balls. Roll each ball in the sesame seeds to coat. Place the bites in an airtight container and set in the refrigerator to harden for at least 1 hour. Enjoy. If you don't eat them first, they will keep for 2 weeks in the fridge.

Coconut Dulce de Leche con Miso

YIELD: 1–1½ CUPS

We have included two versions of dulce de leche with miso: one with coconut and one without. In either case, we give options for using white or hatcho miso. If we had to choose one version, it would be this coconut dulce de leche with hatcho miso. Its rich, dark, almost smoky flavor is complex and delicious, and keeps you coming back for more. The dark color and thick consistency could almost fool you into thinking it was a dark chocolate sauce.

When made with white miso, this sauce is hands-down a salted caramel sauce with a little something special — umami.

The coconut version is a bit thinner than the non-coconut one, but both syrupy sauces can be drizzled on everything. Don't limit yourself to the obvious choices like ice cream or Miso Cheesecake (page 366). Instead, break the boundaries and try it on the Myanmar-Style Fried Tofu (page 93).

1 (12.2-ounce) can evaporated coconut milk (see note below)

1½ cups coconut sugar

FOR WHITE MISO VERSION

3 teaspoons Sweet White Miso (page 262), or to taste (see note below)

FOR HATCHO MISO VERSION

1 teaspoon Soybean/All-Bean Miso (page 270), or to taste (see note below)

Note: Don't get the sweetened condensed canned coconut milk. It is much higher in fat than evaporated coconut milk and causes the sauce to break when the miso is added, leaving a puddle of coconut oil.

If your miso is at all chunky, blend until smooth before adding it to the sauce.

1. Place the coconut milk in small saucepan and stir in the coconut sugar. Bring to a low boil over medium-low heat, stirring regularly. Simmer for 20 to 30 minutes, keeping an eye on it and continuing to stir regularly. It will thicken but not as much or as quickly as many sugar syrups. It is done when you are done watching over it. The longer it goes, the thicker it gets, though it will never be as thick as a milk-based dulce de leche. It will be a deep, rich brown.

2. Remove the pan from the heat and allow to cool for 10 minutes.

3. Place the miso in a bowl and break it up a bit with a whisk. While whisking, pour in the dulce de leche. The white miso will incorporate quickly, while the hatcho will take a bit more whisking to fully integrate.

4. The sauce is now ready to use. Store in the refrigerator, where it will keep for 2 to 3 days. For a longer shelf life, keep the coconut dulce de leche plain and add miso to portions as you're ready to use them; the plain coconut dulce de leche will keep in the fridge for a few weeks.

Dulce de Leche con Miso

YIELD: 1–1½ CUPS

While the Coconut Dulce de Leche con Miso (page 364) won the taste test, not one person didn't approve of this thick, creamy, oh-so-sweet (and now a touch savory) dulce de leche.

This is wonderfully simple to make — no stirring for hours and instead just the gentle sound of a boiling can. It's true, you boil the can for 2 to 3 hours. The most important thing is that you keep the can fully submerged so that the can, and its contents, don't overheat, which could cause an exploding can. Don't worry; just fill a deep pot with plenty of water, simmer gently, and check every half hour (adding water as needed), and it will be fine.

How do you use this sweet condiment? Again, have fun with it — put it between two cookies, or use it as a center in a cupcake or as frosting for anything (including that Miso Cheesecake on page 366). Or play up its savory side. The hatcho version reminded us of black garlic, so we made a sourdough grilled cheese sandwich with sharp cheese, black garlic, and a slather of this sauce. Just sayin'.

1 (14-ounce) can sweetened condensed milk, label removed

FOR WHITE MISO VERSION

2 teaspoons Sweet White Miso (page 262), or to taste (see note below)

FOR HATCHO MISO VERSION

½ teaspoon Soybean/All-Bean Miso (page 270), or to taste (see note below)

Note: If your miso is at all chunky, blend until smooth before adding it to the sauce.

1. Place the unopened can in a large pot and fill with enough room-temperature water to cover the can by 2 to 3 inches.

2. Place the pot on the stove and bring to a boil over high heat, then reduce the heat and let simmer for 2 to 3 hours. Check the water level every half hour and add water if the water level gets within an inch of the top of the can. The longer it cooks, the darker the sauce will be.

3. Remove the pot from the heat. Allow the can and the water to cool for 30 minutes, then carefully remove the can using tongs. Place on a cooling rack and let cool to room temperature, 30 to 40 minutes. Don't attempt to open the hot can, as the pressure inside the can will cause the hot dulce de leche to spurt out.

4. Open the cooled can. The milk will be firm. If you would like it to be easier to work with, heat gently in a double boiler.

5. Place the miso in a bowl and break it up a bit with a whisk. While whisking, pour the dulce de leche into the bowl. The white miso will incorporate quickly, while the hatcho will take a bit more whisking to fully integrate.

6. The sauce is now ready to use. Store in the refrigerator, where it will keep for 2 to 3 weeks, provided that you use a clean utensil when serving and the container is well sealed.

Miso Cheesecake

YIELD: 12–16 SERVINGS

In our first book, *Fermented Vegetables*, Christopher was in charge of the desserts section and wanted so badly to include Kirsten's grandma's cheesecake recipe. In his family, a "cheesecake" came in a glass casserole dish and featured canned cherry filling, Cool Whip, and a thick layer of graham cracker crust. So, when Kirsten's grandma Irene made him a birthday cheesecake early on in our relationship and it didn't include any of these ingredients, he was skeptical. She called it a Philadelphia cheesecake, and while Christopher knew that she had lived in Philadelphia, and the cream cheese she used had that name on its label, still, it was round and naked on top and bottom. Then came that first, life-changing bite.

Decades later, working cheesecake into a book about fermented vegetables was a challenge. Christopher's idea, which failed our family taste testing miserably, was to add finely chopped and washed basic sauerkraut. He was thinking coconut flakes, but everyone agreed it registered in the tasters' minds as multiple hairs in their otherwise perfect cheesecake. It was the only cheesecake that was never finished in our house.

This time around, we think we have a winner. Miso adds a subtle, salty caramel flavor to this family favorite, rendering it a little lighter and creamier as well. We hope Kirsten's grandma Irene would approve.

---

4 eggs, at room temperature

1½ cups sugar

2 teaspoons arrowroot powder

4 tablespoons Sweet White Miso (page 262)

2 teaspoons vanilla extract

3 (8-ounce) packages cream cheese, at room temperature

1½ cups sour cream

1 cup Dulce de Leche con Miso (page 365) (optional)

---

1. Preheat the oven to 325°F/170°C. Lightly butter a 10-inch springform pan.

2. Crack the eggs into a large bowl and beat until somewhat thick and lemon-colored, about 3 minutes. Add the sugar, arrowroot, miso, and vanilla, and beat until well combined.

3. Place the cream cheese in another large bowl and beat until smooth. Gradually add the egg mixture, stirring well to combine.

4. Fold in the sour cream.

5. Pour the batter into the springform pan and bake for 1 hour, or until the sides have puffed up. Turn off the oven, open the door a crack (inserting a butter knife between the door and the oven works great), and let rest in the oven for 1 hour.

6. Remove the cake from the oven and let cool at room temperature for 1 hour. Then remove the outer ring and refrigerate for at least 4 hours before serving. Drizzle Dulce de Leche con Miso on top, if using.

Chocolate Miso Babka

YIELD: 12–15 SERVINGS

Challah, which is a soft and buttery egg bread, was a staple when our house was full of hungry kids and tired grownups. After a trip to New York, Christopher found and quickly fell in love with chocolate babka. By the time he got home, the loaf he had brought back to share was mostly devoured, leading to what the kids remember as the best culinary experiment ever. After about a dozen versions or so, Christopher came down to using his go-to challah with modifications. Now the evolution continues with the addition of miso, which lends a caramel note to this wonderful creation.

1¼ cups warm water (at body temperature)

2 teaspoons active dry yeast

¼ cup honey

2 tablespoons butter, melted

2 eggs

1 teaspoon salt

5 cups unbleached all-purpose white flour

6 ounces bittersweet chocolate

⅓ cup granulated sugar

3 tablespoons Sweet White Miso (page 262)

1½ teaspoons ground cinnamon

4 tablespoons (½ stick) unsalted butter, melted

Coarse granulated sugar, for sprinkling

1. Pour the warm water into a large bowl. Whisk in the yeast and honey. Let stand until the yeast begins to foam, 5 to 10 minutes.

2. Add the 2 tablespoons melted butter, 1 egg, and salt to the yeast mixture and whisk together.

3. Add the flour 1 cup at a time, stirring after each addition. You will probably want to start stirring with your whisk but abandon it for a sturdy wooden spoon, which you will abandon for your hands in the end. Knead the dough until all the flour is incorporated and the dough feels elastic, 4 to 5 minutes. If it is still sticky, add a bit more flour and work it in. Cover the bowl with a clean towel and set in a warm spot to rise until doubled in size, 90 to 120 minutes, depending upon the activity of the yeast, your kneading, and the temperature of the room.

4. Butter a 10-inch bread pan and line it with parchment paper.

5. Punch down the risen dough, dump it onto a floured work surface, and knead for 5 minutes, sprinkling on some additional flour if your dough begins to stick to the surface or to you. Divide the dough into three equal balls. Roll each ball into a ropelike shape about 1½ inches in diameter. Make sure all three are of equal length and relatively the same thickness. With a rolling pin, flatten each rope to about ½ inch to form rectangles.

Recipe continues on next page

369

DESSERT

Chocolate Miso Babka, *continued*

6. Combine the chocolate, granulated sugar, miso, and cinnamon in the bowl of a food processor. Process until they are a consistent crumble.

7. Brush each dough rectangle with the unsalted melted butter and sprinkle one-third of the chocolate-miso mixture onto each. Press the mixture into the dough with your fingers. Roll each dough rectangle up into a rope again and pinch the seam and ends to keep the chocolate-miso mixture safely inside.

8. Braid the three ropes together, but don't worry if they aren't perfect because you are going to twist them and cram them into the bread pan.

9. Place your braided dough in the pan, twisting it a bit if you like. Cover the pan with plastic wrap and let rise for 1 hour at room temperature or until the dough has risen enough to fill the pan.

10. Preheat the oven to 350°F/180°C.

11. Beat the remaining egg with 1 tablespoon of water in a small bowl. Brush the top of the risen dough with the egg wash, then sprinkle the coarse granulated sugar on top.

12. Bake for 35 to 45 minutes, or until the top is brown. Remove the bread from the pan by lifting out the parchment paper. Cool on a wire rack. No really, wait until it's cooled before you slice into this beauty so that it keeps its shape. Enjoy!

Why Bake with Miso?

The answer is flavor. Miso, as far as we can tell, does not impart a secret fermentation power to baked goods. What it does do very well is impart its bold, salty-nutty umami flavor that can make sweeter recipes, like the chocolate babka, much more complex. Also, miso doesn't melt, which is helpful because it doesn't affect your moisture ratios. It's like baking with a high-quality salty butter without the cholesterol, and in the end you get the same result as you would for vegan recipes.

Our suggestion is to bake with a white miso first, and if you really like the flavor profile, try the recipe again, this time using the same proportion of red miso.

Peanut Butter Oat Natto Bar

YIELD: 9 SERVINGS

Imagine if a Cliff Bar and a Rice Krispie treat had a baby, in Japan. It would be this energy-packed bar. If you think you would never try natto, try this bar first — you will be surprised. The natto provides a lightness and that marshmallow gooeyness, except it's extremely good for you.

1 cup rolled oats

½ cup crisp rice cereal

¼ cup coarsely chopped nuts or seeds (pecans, walnuts, almonds, and sunflower seeds all work wonderfully)

¼ cup coarsely chopped dried fruit (apricot, cherry, pear, apple, and raisins work well)

6 Medjool dates, pitted and coarsely chopped

⅓ cup rice flour

2 tablespoons ground flaxseed

1 tablespoon chia seeds

1 tablespoon sesame seeds

½ cup peanut butter

½ cup brown rice syrup

½ teaspoon ground cinnamon

1 teaspoon vanilla extract

1 (2-ounce) package natto or 4 tablespoons homemade natto (page 118)

1. Combine the oats, rice cereal, nuts, dried fruit, dates, and rice flour in the bowl of a food processor. Pulse several times until everything is chopped and combined. Add the flaxseed, chia seeds, and sesame seeds. Stir to combine. Transfer to a large bowl.

2. Combine the peanut butter, rice syrup, and cinnamon in a small saucepan over medium-low heat. Stir until well combined, until the mixture has a uniform color and a shiny, almost taffy-looking consistency, about 2 to 3 minutes. Remove from the heat and stir in the vanilla.

3. Add the natto to the oat mixture, then immediately pour the peanut butter–rice syrup over the oat-natto mixture and stir until everything is combined. We typically start with a wooden spoon, but once everything starts to clump together we abandon the spoon for our hands. The better you mix it now, the better your bars will hold together later.

4. Oil an 8-inch square pan and line it with parchment paper. Press the mixture firmly into the pan. Again, we usually start with the wooden spoon or rubber spatula but finish with our fingers, wetting them as we go. Chill in the refrigerator for at least 30 minutes.

5. Cut into nine bars. Wrap them individually in plastic wrap or waxed paper and store in the refrigerator. They will keep for up to 2 weeks.

Troubleshooting

Troubleshooting Natto and Its Cousins

Natto may smell slightly like ammonia, but it shouldn't smell "off." If the ferment does smell "off," then it probably is. If the strings are weak, then the ferment is weak. You will want to compost it and try again.

Not enough goo? My beans have hardly any threads.

This means there was not enough bacterial growth. The thickness and viscosity of the biofilm are signs of how happy the bacteria are. Scanty threads can be a result of various things being off, such as:

- There was not enough culture (spores), or it was old and inactive.

- The incubation temperature got too hot and the bacteria died, or the temperature was never warm enough for healthy growth.

- The incubation humidity was too low, allowing the beans to dry out.

- The beans were undercooked, not allowing the bacteria good access to the food/nutrient substrate.

- The beans were not hot enough to shock the spores into action.

- There was not enough moisture in the beans to support bacterial growth.

Texture is mushy.

The beans may have been overcooked. Overcooked beans often have a problem with flavor and texture. Broken-down beans have more sugar and amino acids available for the bacteria to eat, and they can easily overferment. In many cases, even when the flavor or texture is not what you want, if there are strings and the flavor isn't off (as in foul), then it is more likely a disappointing batch, not a bad batch.

My natto was fine when I put it in the fridge, but then it turned.

Sometimes natto will last in the fridge for months and months, and other times it only lasts a few weeks, likely due to some kind of contamination that got past the bascillus web of goo. Sometimes there is mold that you can see (send that one to the compost), or the strings will weaken and disappear (we send that to the compost, too). Contamination can also show itself through odors that are wrong — think sour milk or rotten potatoes (again into the compost). Keeping your natto in smaller containers makes a big difference in extending its life. We find that in these smaller containers (we use half-pint and pint jars), the biofilm stays more evenly dispersed instead of settling to the bottom as it does in large containers. Remember: the biofilm is the superpower that keeps the contaminants out.

The beans are chewy and a bit dry, like raisins, or they have the consistency of partially cooked beans.

This could happen with beans that were not sufficiently cooked before being fermented. More often, chewy beans are the result of too much airflow, which can cause them to dry out. This situation is more common at the edges of a batch, where the beans don't have as much mass to keep them evenly moist. It has also happened to us when we used steamer baskets to hold the beans for fermentation, which is sometimes recommended in spore instructions (for keeping airflow).

My natto is slimy, not the telltale stringy neba neba.

Sliminess can result from too much moisture and not enough oxygen and airflow. Make sure that whatever you use to cover your fermentation vessel (we usually call for plastic wrap or aluminum foil) has plenty of holes. Sliminess may also reflect poor bacterial growth that allowed other contaminants to grow.

My ferment has an off odor.

Natto, cheonggukjang, and other *B. subtilis* ferments do have very pungent odors. They will be stinky, especially to the uninitiated. However, there should still be an underlying earthy quality to the aroma. It might smell like ammonia — in some cases the ammonia is quite strong — but it shouldn't smell rotten or foul. Really, there is a difference. If your ferment does smell off, do not eat it; send it on to the compost.

My ferment is pink, orange, or some other abnormal color.

This is a sign that some kind of spoilage has moved in. It's time to move the ferment to the compost pile.

My natto has white dots.

Sometimes you will see small white dots form in the natto as it ages. They might look a little like mold at first, but upon closer inspection you will see that they have a crystalline form instead of a mold fuzziness. These are amino acid crystals, which can be a little crunchy but are delicious — little bursts of pure umami.

Troubleshooting Tempeh and Oncom

With tempeh, as with all ferments, remember to trust your senses and trust your gut. If your instinct says don't eat it, then don't.

The mycelium did not grow or is weak.

Check the temperature in your incubator. If it is too cold, the mycelium won't grow, and if it is too hot, it also won't grow, or it will stop mid-cycle. If you suspect it is chilly but warming it up doesn't work, it could be that the spores are inactive or weak.

Salting the water of your cooking beans can cause mycelial growth to come on much more slowly in the incubation phase, and in some cases not at all. We have found salting not to be worth the added trouble.

It may also be the case that the beans or other substrate have tough skins and the mycelium couldn't break through to feed itself. In the same vein, when the beans or nuts are too large, the gaps that the mycelium has to fill are too wide for it to develop a strong mat. This problem can be remedied in the future by chopping the beans smaller, and/or by adding small grains to help fill the spaces.

If the growth isn't as tight and the tempeh is crumbly, you can still use it. During the experiment phase for this book, we had a lot of crumbly tempeh. Try using the crumbly tempeh as you might ground meat. We often soaked it overnight in a small amount of shio koji and some spices and fried it up the next morning as part of a scramble.

The mold growth is patchy.

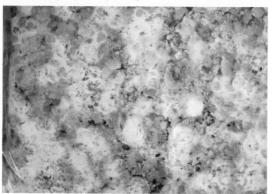

Sometimes tempeh will have spots of strong mycelium growth and then patches of sparse or no growth. This can indicate uneven starter distribution, a lack of good aeration in those spots, uneven heat, or colonies of other microbes having the upper hand in those portions. If the white growth is solid in some areas and all other signs point to normal, you can remove and discard the bad areas and eat the portion that is fine.

This can be common when fermenting in a plastic bag. When there are not enough holes poked in the bags, you will see weak growth with strong growth right around the holes. If you see this, you can remedy it by simply poking more holes in the weak areas and continuing to incubate.

My tempeh has black spots.

This is a sign that it's time to remove your tempeh from the incubator. After the fungus has grown and continues to mature, it begins to develop darker gray or black patches. This is the beginning of sporulation. If you are fermenting in a plastic bag, you will first see it around the holes, and it may happen as soon as 24 hours after inoculation. In an open tray, we have found that the temperature is easier to control, and sporulation happens later and isn't restricted to the holes. Once sporulation has started, it will likely continue in the refrigerator. If you want to stop the process, you can bake the tempeh to pasteurize it, but it is not necessary. As we mentioned earlier, overripe (strongly flavored) tempeh is a delicacy in Java (see page 153), which brings us to the next issue . . .

TROUBLESHOOTING

My tempeh has a heavy ammonia odor.
Tempeh smells nutty and earthy when it is young and growing. When it matures or is over-ripe, the flavors get stronger, as does the smell, which is ammonia-like. Overripe tempeh is not harmful; it just hits a different point on the flavor profile. However, it should not smell rank. If it does, it is a sign of spoilage.

My tempeh is sour.
Sour tempeh can be the result of many things, or a combination of imperfect conditions. The first thing you want to observe is whether it is just sour like pickles, or whether it is sour along with some other signs of spoilage, like poor texture or off smells. If it's just sour, this can be the result of the vinegar used to soak raw substrate (though this isn't very frequent, it has happened to us). If the tempeh is sour and shows other signs of being off, it is often the result of lactic acid bacteria, which is an indicator that other bacteria have had the opportunity to move in, so you don't want to eat it. This can happen if the tempeh started out too cold for too long and the mycelium didn't take hold before the opportunistic bacteria came on the scene. This can also happen with old or weak starter culture.

My tempeh is mushy or wet.
This can indicate a number of possibilities. Perhaps the beans were overcooked, or perhaps they were not dry enough when inoculated. If you fermented them in a bag, the perforations in the bag may have been insufficient (too small or too far apart). Or the humidity may have been too high within the incubation chamber. A mushy ferment often also lacks a good mycelial structure and the mold isn't able to breathe as well.

My tempeh is crunchy.
The substrate was undercooked.

My tempeh is slimy.
It's likely that your bean cake overheated and the opportunistic heat-loving *B. subtilis* (think natto) moved in. Does it have a distinctive natto aroma? You can think of it as a fail or a fusion. In theory, if *B. subtilis* is the only invading bacteria, it is not harmful, but we find it inedible. We generally compost these cakes, but they must smell yummy because our dog finds them and digs them out — she is definitely getting her *B. subtilis* probiotic benefits. Regardless, if your tempeh cake is slimy, pass it along to either the compost pile or another creature.

My tempeh is pink, orange, or some other abnormal color.
This is a sign of spoilage — time to move it to the compost pile. In the future, be sure to monitor the temperature. It may have gotten too high and the mycelium died, allowing other bacteria to move in. Also make sure your incubator allows plenty of fresh air. This can be as simple as opening the lid or door to check it, so that you are exchanging the CO_2 for fresh oxygen.

Troubleshooting Koji

There is no visible sign of mold on the substrate.
When the mold doesn't develop, most often the cause is inconsistent temperatures in the incubation chamber; temps that are too hot or too cold will deter the mold from forming. Or perhaps the grains were too hot when they were inoculated, killing the spores.

This can also happen if the grain is too dry because it wasn't cooked enough (the raw side of al dente); the mycelium needs some moisture in order to grow. In a similar vein, if the grain has a hard husk and it hasn't been broken by soaking or cooking, the mycelium will not be able to access the starches and feed itself to grow.

If none of these seems plausible, you may have an old starter and the spores are no longer viable.

There is irregular or poor mold growth on the substrate.

How long has your koji been growing? In the early stages, the mold can look inconsistent. If you carefully pick up some of the substrate with clean hands, you will often see that the mycelia are beginning to bind, even when the mold isn't yet apparent. If your koji has had ample time to grow and the substrate is not bound by the mycelia, then it's likely that there was not enough starter, it was dispersed unevenly through the substrate, or the starter is old and weak.

Irregular or poor growth can also happen when the koji has gotten cold during incubation.

My koji is wet and has an alcoholic flavor.

In this case, the koji substrate feels quite moist, like each grain is holding a plump little bit of moisture and the soft starches have liquefied under the skin of the grain. A wet koji mat comes from wet rice (or other grain). It's not uncommon with boiled rice, which is why most koji makers steam their rice and other grains.

Wet koji can be used; the enzymes are alive and well, though it is best used in a few days. In fact, wet koji is grown on purpose for certain ferments. This wet koji metabolizes quickly; the liquid will lose its sweetness and begin to taste slightly alcoholic. In a sense, it is creating sake (in rice koji, at least). If you can catch this syrup before it gets too alcoholic, you have a wonderful, sweet liquid very much like mirin. You can

squeeze this liquid out and use it where you would mirin or syrup — Jeremy Umansky says it is delicious on waffles, for example (see his mirin hack on page 222).

Sometimes water forms at the bottom of the fermentation dish or tray, and this can be a result of overheating.

My koji is slimy.

When the temperature in the incubation chamber is off and the koji mold doesn't have a firm hold on the substrate, other microbes have a chance to move in. Slimy suggests *Bacillus subtilis,* which is our friend in natto, but not in koji. Remember that the bacillus spores are fairly indestructible, so they are likely present on your grains and beans even after you cook them, and in the flour you use to disperse the koji spores even after you toast it. With the proper incubation temperatures, your koji spores will be happy and conditions will not be ideal for bacilli, so your koji can outcompete them. However, if the temperature gets too hot, the aspergillus will die off and the bacilli, which like those warmer climes, will do what they do best: make a biofilm. If this is the case, you will likely smell the telltale bacillus ammonia odor instead of the sweet, floral scent of a koji bloom.

A bacillus-infected koji will show up as a bad miso later. In short, you don't want to use a slimy, bacteria-infected koji.

My koji is yellow or greenish.

This koji is at the next phase of its life cycle. As the aspergillus ages, the white mold will begin to show fuzzy patches of light yellow and then olive green. This signifies the beginning of sporulation. Once this begins, the mold will bloom quickly.

If the koji has just begun to show signs of sporulation, remove it from the incubation chamber and cool it down by putting it in the fridge to slow the cycle. If the patches are small and isolated, you can remove them. If they are starting to spread all over, it's best to use the koji to make miso; you'll add salt to make miso,

which will kill the mold and stop sporulation in its tracks. If the koji has many green patches — if more of it is green than not — don't use it for other ferments, as it will taint the flavor.

My koji has black or very dark green patches.

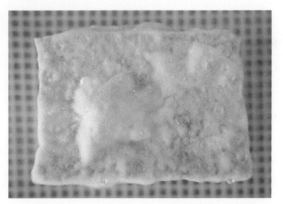

This indicates spoilage. Other microbes have moved in and taken up residence. If these spots are small and isolated, cut them out and discard them, keeping the rest. But first make sure that the rest of the koji is not somehow "off" — like sticky or slimy. If you do see other indications of spoilage, discard the entire batch.

My koji has brownish spots or coloring on the bottom.
If the koji otherwise smells and looks right, it is probably just staining. This can happen with metal trays. Unless the trays are stainless steel, they can develop rust while incubating and stain the koji. This can also happen with wooden trays; depending on the type of wood used, they may leach some color. To remedy this, place a piece of muslin in the tray to grow the koji on.

Troubleshooting Amazake

Despite amazake generally being considered a classic gateway ferment for koji, our first few batches of amazake were troublesome. If you have trouble at the beginning, don't give up. Here is what may have happened and what to do next time.

Amazake did not sweeten or soften.
It is possible that you did not have enough koji to work through the fresh grain. Remember that the koji is bringing in the enzymes to break down and sweeten the grain. You can add more koji, if you have it, and try incubating the amazake for a little longer. Amazake is also temperature sensitive. Make sure your grain is not too hot when you add your koji, or the enzymes will denature. The ideal temperature is 138°F/59°C, but at 140°F/60°C it's game over. If you think that your ferment got too cold, try making your incubation space a little warmer and give it more time.

Amazake is sour.
If your amazake sours or has a bitter flavor, the grain fermented too long. This can be a good thing if you're going to use the amazake for pickling or marinating.

Troubleshooting Miso (and Tasty Pastes)

My miso tastes lactic sour or smells alcoholic, or like acetone or ethanol.
In the early stages of fermentation, as the microbes are working things out, the miso may taste sour because the lactic acid bacteria are hard at work. Or it may taste like alcohol because the yeasts are processing the sugars.

Your ferment likely needs to age longer; time may mellow out the acidity. That said, not enough salt can cause the miso to have an abundance of lactic bacteria, creating a sour miso that doesn't right itself.

Add more weight to your miso as well. Weight is important in making sure that excess moisture (which in the anaerobic environment can put things out of balance) gets pushed to the top (becoming that yummy tamari) and gases are allowed to escape.

If the extra time and weight don't do the trick, there may be more factors at play, and you may not be able to correct them for this particular batch. If your miso is overly moist and/or low in salt, you may experience a sour/lightly alcoholic miso. Add less liquid in the future, or increase your salt ratio to at least 7 or 8 percent.

Acetone or ethanol smells most typically happen in small-jar batches and can signal that the miso was unable to breathe properly. To avoid this problem, make sure your ferment is well weighted to push out gases, and use a small crock, if you have one, as your fermentation vessel to take advantage of the ventilation offered by its micropores. If you smell acetone or ethanol during the fermentation, stir the paste thoroughly to release gases and then add more weight. You may have to do this a few times during the aging period. Often this stirring and extra weight will help resolve the problem; if they do not, discard the batch.

My miso has an off flavor and/or unusual dark color.
Chances are, before you get to flavor, you will notice an off smell. If something is amiss, your nose knows — trust it. As far as color, we mean here a dark color that is not the rich amber to red to shiny black of the Maillard reaction, but

instead an off tone, especially if the darkening happens in a light miso ferment where it cannot be attributed to aging. An unexpected or off dark color can be a result of koji that was contaminated with *B. subtilis*. This miso should be put out as an offering to the compost pile.

My miso has mold and other funky things on its surface.

As is the case for all ferments, the surface — that edge between oxygen and ferment — is susceptible to invaders, so it's likely that something will end up growing on your longer-aging miso. This is where your upbringing will have trained you to toss it and run. But don't! Oxidation, mold, and yeasts are normal. You can simply remove any such growth. Start slowly and work your way down, wiping away any mold or yeast growth off the edges of the crock. Then scrape off the happy microbial jungle layer. It may smell, but don't let that deter you. Keep going. Underneath all of this you will find the anaerobic layer of miso. It will look, smell, and taste good.

The mold has moved beyond the surface.
If you find mold growing anywhere other than on the surface of your miso, oxygen has likely penetrated the batch through air pockets. If you find these pockets in the upper layers, you may be able to remove the affected areas and reach "clean" miso. However, if you find mold in pockets throughout, it is best to take a deep sigh and send this miso to the compost. This happens when the miso hasn't been packed tightly or weighted enough to bring all of the liquid to the surface and push any CO_2 out of the ferment. Another thing, and this is a little less common, is that if the miso hasn't been mixed thoroughly and there are clumps of koji within the structure of the miso, they too can mold.

The miso is dry.
If your miso has been fermenting for a long time in a warm, dry space (most of our modern heating systems tend to dry the air dramatically), it is possible that the miso will dehydrate a little bit or even so much that it becomes a bit of a hockey puck. This is especially true if the ferment is in a ceramic crock — these tend to breathe out the sides more than an impermeable glass jar. To avoid dryness in your miso, make sure that you find a cool, unheated — but not cold — incubation space. If you know your space is challenging (perhaps the best you can do is under your bed), that is okay. Just check it regularly and perhaps use a glass jar as your fermentation vessel. If the miso is dry, you can often rehydrate it by adding water a little bit at a time and stirring it well to give it a nice paste texture. (Think miso bouillon cube . . .)

Troubleshooting a Nuka Bed

The vegetables taste too salty or not sour enough.

If your nuka is a new pot, it might not be ready. Keep priming the bed with vegetable scraps until the vegetables have a nice acidic flavor. Vegetables in a young pot will take longer to ferment than in a well-established pot.

The nuka bed is wet.

Vegetables with a high moisture content will lose water in the bed and make it watery. Soak up or pour off any excess water that is sitting on top of the bed, or add more rice bran and salt. Do this as soon as you see it, as too much moisture will smother the bed and it will sour.

Something is off.

If your pot smells sour or alcoholic, or if it has white mold on top, something is out of balance. Yellow mustard powder can be used to bring it back into balance. In the case of mold, scoop it off first. Add 1 teaspoon of yellow mustard powder and turn the nukadoko two times a day. Uncover it and set it in a sunny spot if you can. It should start smelling earthy and yeasty in a couple of days. If not, you can try a little more mustard, but if this doesn't help after about a week, the bed might be beyond saving. If the bed does return to normal, try aerating it a little less frequently; you don't want to overoxygenate the bacteria.

The pot developed a green or pink mold.

Discard immediately and start again.

ENDNOTES

[1] de Clercq, Nicolien C, Albert K Groen, Johannes A Romijn, and Max Nieuwdorp. "Gut Microbiota in Obesity and Undernutrition." *Advances in Nutrition* 7.6 (2016): 1080–89.

Luna, Ruth Ann, and Jane A Foster. "Gut brain axis: diet microbiota interactions and implications for modulation of anxiety and depression." *Current Opinion in Biotechnology* 32 (2015): 35-41.

Shechter, Ari, and Gary J. Schwartz. "Gut–brain nutrient sensing in food reward." *Appetite* 122 (2018): 32-35.

[2] Belleme, John, Jan Belleme, and Christina Pirello. *Japanese Foods that Heal: Using Traditional Japanese Ingredients to Promote Health, Longevity & Well-being.* North Clarendon, Tuttle Publishing, 2007.

Frias, Juana, Cristina Martinez-Villaluenga, and Elana Penas. *Fermented Foods in Health and Disease Prevention.* London, Academic Press, 2017.

Kim, Eun Kyoung, So-Yeon An, Min-Seok Lee, Tae Ho Kim, Hye-Kyoung Lee, Won Sun Hwang, Sun Jung Choe, Tae-Young Kim, Seung Jin Han, Hae Jin Kim, Dae Jung Kim, and Kwan-Woo Lee. "Fermented kimchi reduces body weight and improves metabolic parameters in overweight and obese patients." *Nutrition Research* 31.6 (2011): 436–43.

[3] David, Lawrence A., Corinne F. Maurice, Rachel N. Carmody, David B. Gootenberg, Julie E. Button, Benjamin E. Wolfe, Alisha V. Ling, A. Sloan Devlin, Yug Varma, Michael A. Fischbach, Sudha B. Biddinger, Rachel J. Dutton, and Peter J. Turnbaugh. "Diet rapidly and reproducibly alters the human gut microbiome." *Nature* 505.7484 (2014): 559–63.

Holscher, Hannah D. "Dietary fiber and prebiotics and the gastrointestinal microbiota." *Gut Microbes*, vol. 8, no. 2, 2017.

[4] Herian, Anne M., Steve L. Taylor, and Robert K. Bush. "Allergenic Reactivity of Various Soybean Products as Determined by RAST Inhibition." *Journal of Food Science*, vol. 58, no. 2, 1993.

[5] Sena, Cristina M., and Raquel M. Seica. "Soybean: Friend or Foe." *Soybean and Nutrition*, 2011.

[6] Khalil, Abdul W., et al. "Comparison of sprout quality characteristics of design and kabuli type chickpea cultivars (Cicer arietinum L.)." *LWT — Food Science and Technology*, vol. 40, no. 6, 2007, pp. 937-45.

[7] Chandra-Hioe, Maria, et al. "The Potential Use of Fermented Chickpea and Faba Bean Flour as Food Ingredients." *Plant Foods for Human Nutrition*, vol. 71, no. 1, 2016, pp. 90-95.

[8] Madode, Yann E., Martinus J. Nout, Evert-Jan Bakker, Anita R. Linnemann, and Djidjoho J. Hounhouigan. "Enhancing the digestibility of cowpea (*Vigna unguiculata*) by traditional processing and fermentation." *LWT — Food Science and Technology*, vol. 54, no. 1, 2013, pp. 186-93.

[9] Granito, Marisela, and Glenda Alvarez. "Lactic acid fermentation of black beans (*Phaseolus vulgaris*): microbiological and chemical characterization." *Journal of the Science of Food and Agriculture*, vol. 86, no. 8, 2006, pp. 1164–71.

[10] Biesiekierski, Jessica R., Simone L. Peters, Evan D. Newnham, Ourania Rosella, and Jane G. Muir. "No Effects of Gluten in Patients with Self-Reported Non-Celiac Gluten Sensitivity After Dietary Reduction of Fermentable, Poorly Absorbed, Short-Chain Carbohydrates." *Gastroenterology*, vol. 145, no. 2, 2013, pp. 320–28.

[11] Elyas, Selma H., Abdullahi H. El Tinay, Nabila E. Yousif, and Elsiddig A. Elsheikh. "Effect of natural fermentation on nutritive value and in vitro protein digestibility of pearl millet." *Food Chemistry*, vol. 78, no. 1, 1 July 2002, pp. 75–79.

[12] Starzynska-Janiszewska, Anna, Robert Dulinski, Bozena Stodolak, Barbara Mickowska, and Agnieszka Wikiera. "Prolonged tempe-type fermentation in order to improve bioactive potential and nutritional parameters of quinoa seeds." *Journal of Cereal Science*, vol. 71, 2016, pp. 116–21.

[13] Marco, Maria L., et al. "Health benefits of fermented foods: microbiota and beyond." *Current Opinion in Biotechnology*, vol. 44, 2017, pp. 94–102.

[14] Gobbetti, Marco, et al. "Sourdough lactobacilli and celiac disease." *Food Microbiology*, vol. 24, 2007, pp. 187–96.

[15] Di Cagno, Raffaella, et al. "Gluten-free Sourdough Wheat Baked Goods Appear Safe for Young Celiac Patients: A Pilot Study." *Journal of Pediatric Gastroenterology and Nutrition*, vol. 51, no. 6, Dec. 2010, pp. 773–83.

[16] Juckett, Gregory, et al. "The Microbiology of Salt Rising Bread." *West Virginia Medical Journal*, vol. 104, 2008, pp. 26–27.

[17] Saberi, Helen. *Cured, Fermented and Smoked Foods: Proceedings of the Oxford Symposium on Food and Cookery 2010.* Totnes, England, Prospect Books, 2011, pp. 84–87.

[18] Sumi, Hiroyuki, et al. "A novel fibrinolytic enzyme (Nattokinase) in the vegetable cheese Natto: a typical and popular soybean food in the Japanese diet." *Experientia*, vol. 43, 1987, pp. 1110–11.

[19] Iwamoto, Jun, et al. "Menatetrenone (Vitamin K2) and Bone Quality in the Treatment of Postmenopausal Osteoporosis." *Nutrition Reviews*, vol. 64, no. 12, 2006, pp. 509–17.

[20] Yanagisawa, Yashhide, and Hiroyuki Sumi. "Natto Bacillus contains a large amount of water-soluble vitamin K (Melaquinone-7)." *Journal of Food Biochemistry*, vol. 29, no. 3, 2005, pp. 267–77.

[21] Zhao, Xin, et al. "Comparisons of Shuidouchi, Natto, and Cheonggukjang in their physicochemical properties, and antimutagenic and anticancer effects." *Food Science and Biotechnology*, vol. 22, no. 4, Aug. 2013, pp. 1077–84.

[22] Cantabrana, Igor, Ramon Perise, and Igor Hernandez. "Use of *Rhizopus oryzae* in the kitchen." *International Journal of Gastronomy and Food Science*, vol. 2, 2015, pp. 103–11.

[23] Frias, Juana, Cristina Martinez-Villaluenga, and Elana Penas. *Fermented Foods in Health and Disease Prevention.* London, Academic Press, 2017.

Paredes-Lopez, O, and G I. Harry. "Changes in Selected Chemical and Antinutritional Components during Tempeh Preparation Using Fresh and Hardened Common Beans." *Journal of Food Science*, vol. 54, no. 5, 1989, pp. 968–70.

Jiménez-Martinez, Cristian, et al. "Diminution of quinolizidine alkaloids, oligosaccharides and phenolic compounds from two species of *Lupinus* and soybean seeds by the effect of *Rhizopus oligosporus*." *Journal of the Science of Food and Agriculture*, vol. 87, no. 7, 2007, pp. 1315–22.

[24] Ahmad, A, et al. "Enhancement of beta-secretase inhibition and antioxidant activities of tempeh, a fermented soybean cake through enrichment of bioactive aglycones." *Pharmaceutical Biology*, vol. 53, 2015, pp. 758–66.

Hong, Go-Eun, et al. "Fermentation Increases Isoflavone Aglycone Contents in Black Soybean Pulp." *Asian Journal of Animal and Veterinary Advances*, vol. 7, no. 6, 2012, pp. 502-11, doi:10.3923/ajava.2012.502.511. Accessed 25 Aug. 2018.

Nout, M J., and J L. Kiers. "Tempe fermentation, innovation and functionality: update into the third millenium." *Journal of Applied Microbiology*, vol. 98, no. 4, 2005, pp. 789–805.

[25] Onda, T., F. Yanagida, M. Tsuji, T. Shinohara, and K. Yokotsuka. "Time series analysis of aerobic bacterial flora during Miso fermentation." *Letters in Applied Microbiology*, vol. 37, no. 2, 2003, pp. 162–68.

[26] Onda, T., F. Yanagida, M. Tsuji, T. Shinohara, and K. Yokotsuka. "Time series analysis of aerobic bacterial flora during Miso fermentation." *Letters in Applied Microbiology*, vol. 37, no. 2, 2003, pp. 162–68.

[27] Belleme, John, Jan Belleme, and Christina Pirello. *Japanese Foods that Heal: Using Traditional Japanese Ingredients to Promote Health, Longevity & Well-being.* North Clarendon, Tuttle Publishing, 2007.

[28] Ito, A, et al. "Effects of Soy Products in Reducing Risk of Spontaneous and Neutron-induced Liver-tumors in Mice." *International Journal of Oncology*, vol. 2, no. 5, 1993, pp. 773-76.

[29] Shurtleff, William, and Akiko Aoyagi. *The Book of Miso: Savory, High-Protein Seasoning.* 2nd ed., vol. I, Berkeley, Ten Speed Press, 2001.

[30] Ogasawara, Masashi, Yuki Yamada, and Makoto Egi. "Taste enhancer from the long-term ripening of miso (soybean paste)." *Food Chemistry*, vol. 99, no. 4, 2006, pp. 736–41.

BIBLIOGRAPHY

Ahmad, A, et al. "Enhancement of beta-secretase inhibition and antioxidant activities of tempeh, a fermented soybean cake through enrichment of bioactive aglycones." *Pharmaceutical Biology*, vol. 53, 2015, pp. 758-66.

Albala, Ken. *Beans: A History*. New York, Berg, 2007.

Allen, Tony, and Charles Phillips. *Ancient China's Myths and Beliefs*. Rosen Publishing Group, 2012.

Belderok, B. "Developments in bread-making processes." *Plant Foods for Human Nutrition*, vol. 55, no. 1, pp. 1-14.

Belleme, John, Jan Belleme, and Christina Pirello. *Japanese Foods that Heal: Using Traditional Japanese Ingredients to Promote Health, Longevity & Well-being*. North Clarendon, Tuttle Publishing, 2007.

Bi, Hua, Haizhen Zhao, Fengxia Lu, Chong Zhang, and Xiaomei Bie. "Improvement of the Nutritional Quality and Fibrinolytic Enzyme Activity of Soybean Meal by Fermentation of Bacillus subtillis." *Journal of Food Processing and Preservation*, vol. 39, no. 6, 2015, pp. 1235–42.

Biesiekierski, Jessica R., Simone L. Peters, Evan D. Newnham, Ourania Rosella, and Jane G. Muir. "No Effects of Gluten in Patients with Self-Reported Non-Celiac Gluten Sensitivity After Dietary Reduction of Fermentable, Poorly Absorbed, Short-Chain Carbohydrates." *Gastroenterology*, vol. 145, no. 2, 2013, pp. 320-28.

Blake, Michael. *Maize for the Gods*. University of California Press, 2015.

Butt, Masood S., Muhammad Tahir-Nadeem, Muhammad K. Iqbal Khan, Rabia Shabir, and Mehmood S. Butt. "Oat: unique among the cereals." *European Journal of Nutrition*, vol. 47, no. 2, 26 Feb. 2008, pp. 68-79, doi:10.1007/s00394-008-0698-7. Accessed 20 Dec. 2017.

Campbell, Kristina. "Studying how gluten reacts with bacteria in the gut helps advance treatments for celiac disease." *Gut Microbiota News Watch*, 15 Feb. 2017, www.gutmicrobiotaforhealth.com/en/studying-gluten-reacts-bacteria-gut-helps-advance-treatments-celiac-disease/. Accessed 6 Aug. 2017.

Cantabrana, Igor, Ramon Perise, and Igor Hernandez. "Use of Rhizopus oryzae in the kitchen." *International Journal of Gastronomy and Food Science*, vol. 2, 2015, pp. 103-11.

Carvalho, Marcia, Teresa Lino-Neto, Eduardo Rosa, and Valdemar Carnide. "Cowpea: a legume crop for a challenging environment." *Journal of Science of Food and Agriculture*, vol. 97, no. 13, 1 Oct. 2017, pp. 4273-84, doi:10.1002/jsfa.8250. Accessed 11 Dec. 2017.

Chandra-Hioe, Maria, et al. "The Potential Use of Fermented Chickpea and Faba Bean Flour as Food Ingredients." *Plant Foods for Human Nutrition*, vol. 71, no. 1, 2016, pp. 90-95.

Chompreeda, Penkwan, and M. L. Fields. "Effects of Heat and Natural Fermentation on Amino Acids, Flatus Producing Compounds, Lipid Oxidation and Trypsin Inhibitor in Blends of Soybean and Cornmeal." *Journal of Food Science*, vol. 49, no. 2, 1984.

Chung, Bonnie. *Miso Tasty: The Cookbook*. London, Pavilion, 2016.

Clayton, Bernard. *Bernard Clayton's New Complete Book of Breads*. Simon and Schuster, 2006, pp. 265-70.

Cuadrado, Carmen, Gemma Ayet, Luz Robredo, Javier Tabera, and Rosa Villa. "Effect of natural fermentation on the content of inositol phosphates in lentils." *European Food Research and Technology*, vol. 203, no. 3, 1996, pp. 268-71.

Curran, Julianne. "The nutritional value and health benefits of pulses in relation to obesity, diabetes, heart disease and cancer." *British Journal of Nutrition*, vol. 108, no. S1, Aug. 2012.

David, Lawrence A., et al. "Diet rapidly and reproducibly alters the human gut microbiome." *Nature*, vol. 505, no. 7484, 2014, pp. 559-63.

de Clercq, Nicolien C., et al. "Gut Microbiota in Obesity and Undernutrition." *Advances in Nutrition*, vol. 7, no. 6, 2016, pp. 1080-89.

De Vuyst, Luc, and Patricia Neysens. "The sourdough microflora: biodiversity and metabolic interactions." *Trends in Food Science & Technology*, vol. 16, no. 1, 2005, pp. 43-56.

Di Cagno, Raffaella, et al. "Gluten-free Sourdough Wheat Baked Goods Appear Safe for Young Celiac Patients: A Pilot Study." *Journal of Pediatric Gastroenterology and Nutrition*, vol. 51, no. 6, Dec. 2010, pp. 773-83.

Ditzler, Joseph. "OSU study grows new respect for barley." *The Bulletin*, 28 Nov. 2017, www.bendbulletin.com/home/5792886-151/osu-study-grows-new-respect-for-barley. Accessed 29 Nov. 2017.

Du Bois, Christine M., Chee-Bent Tan, and Sidney Mints, editors. *The World of Soy*. 1st ed., Champaign, University of Illinois Press, 2008.

Duetsch, Jonathan, and Natalya Murakhver. *They Eat That?: A Cultural Encyclopedia of Weird and Exotic Food from Around the World*. ABC-CLIO, 2012, pp. 182-84.

Dufossè, L. "16 ñ Current and Potential Natural Pigments From Microorganisms (Bacteria, Yeasts, Fungi, Microalgae)." *Handbook on Natural Pigments in Food and Beverages*. Amsterdam, Woodhead Publishing, 2016, pp. 337-54. *Science Direct*. Accessed 22 Dec. 2017.

Duguid, Naomi, and Jeffrey Alford. *Burma: Rivers of Flavor*. New York, Artisan, 2012.

Duodu, K G., J R. Taylor, P S. Belton, and B R. Hamaker. "Factors affecting sorghum protein digestibility." *Journal of Cereal Science*, vol. 38, no. 2, 2003, pp. 117-31.

Dwivedi, Minakshee, K Vasantha, Y Sreerama, Devendra Waware, and R Singh. "Kaulath, a new fungal fermented food from horse gram." *Journal of Food Science and Technology*, vol. 52, no. 12, 2015, pp. 8371-76.

Elkhalifa, A E., B Schiffler, and R Bernhardt. "Effect of fermentation on the functional properties of sorghum flour." *Food Chemistry*, vol. 92, no. 1, 2005, pp. 1-5.

Elyas, Selma H., Abdullahi H. El Tinay, Nabila E. Yousif, and Elsiddig A. Elsheikh. "Effect of natural fermentation on nutritive value and in vitro protein digestibility of pearl millet." *Food Chemistry*, vol. 78, no. 1, 1 July 2002, pp. 75-79.

Flynn, Sharon. *Ferment for Good: Ancient Foods for the Modern Gut*. 1st ed., Hardie Grant, 2017.

Frias, Juana, Cristina Martinez-Villaluenga, and Elana Penas. *Fermented Foods in Health and Disease Prevention*. London, Academic Press, 2017.

Gobbetti, Marco, et al. "Sourdough lactobacilli and celiac disease." *Food Microbiology*, vol. 24, 2007, pp. 187-96.

Granito, Marisela, and Glenda Alvarez. "Lactic acid fermentation of black beans (*Phaseolus vulgaris*): microbiological and chemical characterization." *Journal of the Science of Food and Agriculture*, vol. 86, no. 8, 2006, pp. 1164-71.

Granito, Marisela, Juana Frias, Rosa Doblado, Marisa Guerra, and Martine Champ. "Nutritional improvement of beans (*Phaseolus vulgaris*) by natural fermentation." *European Food Research and Technology*, vol. 214, no. 3, 5 Dec. 2001, pp. 226-31, doi:10.1007/s00217-001-0450-5. Accessed 28 Dec. 2017.

Greiner, R, and U Konieczny. "Improving enzymatic reduction of myo-inositol phosphorus with inhibitory effects on mineral absorption in black beans (*Phaseolus vulgarisms* var Preto)." *Journal of Food Processing and Preservation*, vol. 23, pp. 249-61.

Hachisu, Nancy Singleton. *Preserving the Japanese Way*. Kansas City, Andrew McMeel Publishing, 2015.

Hall, Kenneth R. *A History of Early Southeast Asia: Maritime Trade and Societal Development, 100–1500*. Rowman & Littlefield Publishers, 2010.

Hashisu, Nancy S. *Preserving the Japanese Way: Traditions of Salting, Fermenting, and Pickling for the Modern Kitchen*. Kansas City, Andrews McMeel Publishing, 2015.

Herian, Anne M., Steve L. Taylor, and Robert K. Bush. "Allergenic Reactivity of Various Soybean Products as Determined by RAST Inhibition." *Journal of Food Science*, vol. 58, no. 2, 1993.

Herman, William H., and Paul Zimmet. "Type 2 Diabetes: An Epidemic Requiring Global Attention and Urgent Action." *Diabetes Care*, vol. 35, May 2012, pp. 943-44, doi:https://doi.org/10.2337/dc12-0298. Accessed 8 Dec. 2017.

Hitosugi, Masahito, Katsuo Hamada, and Kazutaka Misaka. "Effects of *Bacillus subtillis* var. *natto* products on symptoms caused by blood flow disturbance in female patients with lifestyle diseases." *International Journal of General Medicine*, vol. 8, pp. 41-46.

Holscher, Hannah D. "Dietary fiber and prebiotics and the gastrointestinal microbiota." *Gut Microbes*, vol. 8, no. 2, 2017.

Honda, Tetsuya, Mikiko Michigami, Yoshiki Miyachi, and Kenji Kabashima. "A case of late-onset anaphylaxis to fermented soybeans (natto)." *The Journal of Dermatology*, vol. 41, no. 10, 2014, pp. 940-41.

Hong, Go-Eun, et al. "Fermentation Increases Isoflavone Aglycone Contents in Black Soybean Pulp." *Asian Journal of Animal and Veterinary Advances*, vol. 7, no. 6, 2012, pp. 502-11, doi:10.3923/ajava.2012.502.511. Accessed 25 Aug. 2018.

Hong, H. A., J. M. Huang, R. Khaneja, L. V. Hiep, and M. C. Urdaci. "The safety of *Bacillus subtilis* and *Bacillus indicus* as food probiotics." *Journal of Applied Microbiology*, vol. 105, 1 Aug. 2008, pp. 510-20. Accessed 3 Dec. 2017.

Huma, Nuzhat, Faqir M. Anjum, Samreen Sehar, Muhammad I. Khan, and Shahzad Hussain. "Effect of soaking and cooking on nutritional quality and safety of legumes." *Nutrition and Food Science*, vol. 38, no. 6, 2008, doi:10.1108/00346650810920187. Accessed 28 Dec. 2017.

Hurt, R D. *Food and Agriculture during the Civil War*. Santa Barbara, CA, Praeger, 2016, pp. 65-79. Accessed 23 Jan. 2018.

Ito, A, et al. "Effects of Soy Products in Reducing Risk of Spontaneous and Neutron-induced Liver-tumors in Mice." *International Journal of Oncology*, vol. 2, no. 5, 1993, pp. 773-76.

Iwamoto, Jun, et al. "Menatetrenone (Vitamin K2) and Bone Quality in the Treatment of Postmenopausal Osteoporosis." *Nutrition Reviews*, vol. 64, no. 12, 2006, pp. 509-17.

Jiménez-Martínez, Cristian, et al. "Diminution of quinolizidine alkaloids, oligosaccharides and phenolic compounds from two species of *Lupinus* and soybean seeds by the effect of *Rhizopus oligosporus*." *Journal of the Science of Food and Agriculture*, vol. 87, no. 7, 2007, pp. 1315-22.

Joshi, V K. *Indigenous Fermented Foods of South Asia*. Boca Raton, CRC Press, 2016.

Juckett, Gregory, et al. "The Microbiology of Salt Rising Bread." *West Virginia Medical Journal*, vol. 104, 2008, pp. 26-27.

Jukanti, A. K. "Nutritional Quality and Health Benefits of Chickpea (Cicer Arietinum L.): a Review." *British Journal of Nutrition*, vol. 108, no. S1, 2012, pp. S11-26.

Katz, Sandor E. *The Art of Fermentation: An In-Depth Exploration of Essential Concepts and Processes from Around the World*. White River Junction, Chelsea Green Publishing, 2012.

Keuth, S, and B Bisping. "Vitamin B12 by Citobacter freundii or Llebsiella pneumoniae during tempeh fermentation and proof of enterotoxin absence by PCR." *Applied and Environmental Microbiology*, vol. 60, no. 5, 1994, pp. 1495+.

Khalil, Abdul W., et al. "Comparison of sprout quality characteristics of design and kabuli type chickpea cultivars (Cicer arietinum L.)." *LWT - Food Science and Technology*, vol. 40, no. 6, 2007, pp. 937-45.

Khush, Gurdev S. "Origin, dispersal, cultivation and variation of rice." *Plant Molecular Biology*, vol. 35, 1997, pp. 25-34.

Kim, Jung, and Kwan-Woo Lee. "Fermented kimchi reduces body weight and improves metabolic parameters in overweight and obese patients." *Nutrition Research*, vol. 31, no. 6, 2011, pp. 436-43.

Kuo, Hsiao T., Po H. Wu, Hui H. Chung, Chi M. Jiang, and Ming C. Wu. "Koji Fermentation Improve the Protein Digestibility of Sorghum." *Journal of Food Processing and Preservation*, vol. 37, no. 5, 2013, pp. 419-23.

L'vov, V., and E. Sadykov. "Activity of angiotensi converting enzyme inhibitors in venoms of central Asia snakes." *Chemistry of Natural Compounds*, vol. 36, no. 3, 2000, pp. 314-17.

Lee, Byong H. *Fundamentals of Food Biotechnology*. 2nd ed., Chichester, UK, John Wiley & Sons, Ltd, 2015, pp. 275-78.

Lee, Gyoung-Ah, Gary W. Crawford, Li Liu, Yuka Sasaki, and Xuexiang Chen. "Archaeological Soybean (*Glycine max*) in East Asia: Does Size Matter?" *PLOS ONE*, vol. 6, no. 11, 4 Nov. 2011, journals.plos.org/plosone/article?id=10.1371/journal.pone.0026720. Accessed 18 Apr. 2017.

Lima, Fernando, Louise Kurozawa, and Elza Ida. "The effects of soybean soaking on grain properties and isoflavones loss." *LWT - Food Science and Technology*, vol. 59, no. 2, 2014, pp. 1274-82.

Luna, Ruth A., and Jane A. Foster. "Gut brain axis: diet microbiota interactions and implications for modulation on anxiety and depression." *Current Opinion in Biotechnology*, vol. 32, 2015, pp. 35-41.

Maangchi's Real Korean Cooking. New York, Houghton Mifflin Harcourt Publishing, 2015.

Maangchi, and Lauren Chattman. *Maangchi's Real Korean Cooking: Authentic Dishes for the Home Cook*. New York, Houghton Mifflin Harcourt Publishing, 2015.

Madode, Yann E., Martinus J. Nout, Evert-Jan Bakker, Anita R. Linnemann, and Djidjoho J. Hounhouigan. "Enhancing the digestibility of cowpea (*Vigna unguiculata*) by traditional processing and fermentation." *LWT - Food Science and Technology*, vol. 54, no. 1, 2013, pp. 186-93.

Marco, Maria L., et al. "Health benefits of fermented foods: microbiota and beyond." *Current Opinion in Biotechnology*, vol. 44, 2017, pp. 94-102.

Mouritsen, Ole G., and Klavs Styrbaek. *Umami: Unlocking the Secrets of the Fifth Taste*. New York, Columbia University Press, 2014.

Murooka, Yoshikatsu, and Mitsuo Yamshita. "Traditional healthful fermented products of Japan." *Journal of Industrial Microbiology Biotechnology*, vol. 35, no. 8, pp. 791-98.

Myneni, V D., and E Messy. "Immunomodulatory effect of vitamin K2: Implications for bone health." *Oral diseases*, vol. 24, no. 1-2, 2018, pp. 67-71.

National Research Council. *Toxicants occurring naturally in foods*. Washington, DC, National Academy of Sciences, 1973, pp. 363-71.

Nguyen, Andrea. "Soybean Buying Guide: Where and How to Find Good Beans." *Viet World Kitchen*, https://www.vietworldkitchen.com/blog/2012/10/soybean-buying-guide.html. Accessed 11 Oct. 2012.

Njoroge, Daniel M., Peter K. Kinyanjui, Claire M. Chigwedere, Stephanie Christiaens, and Anselimo O. Makokha. "Mechanistic insight into common bean pectic polysaccharide changes during storage, soaking and thermal treatment in relation to the hart-to-cook defect." *Food Research International*, vol. 81, 2016, pp. 39-49.

Nout, M J., and J L. Kiers. "Tempe fermentation, innovation and functionality: update into the third millenium." *Journal of Applied Microbiology*, vol. 98, no. 4, 2005, pp. 789-805.

"#1 Mold." *Surprisingly Awesome*, edited by Adam Davidson and Adam McKay, Gimlet Media, 3 Nov. 2015, https://www.gimletmedia.com/surprisingly-awesome/1-mold. Accessed 22 May 2018.

Ogasawara, Masashi, Yuki Yamada, and Makoto Egi. "Taste enhancer from the long-term ripening of miso (soybean paste)." *Food Chemistry*, vol. 99, no. 4, 2006, pp. 736-41.

Onda, T., F. Yanagida, M. Tsuji, T. Shinohara, and K. Yokotsuka. "Time series analysis of aerobic bacterial flora during Miso fermentation." *Letters in Applied Microbiology*, vol. 37, no. 2, 2003, pp. 162-68.

Owens, J D. *Indigenous Fermented Foods of Southeast Asia*. 1st ed., Boca Raton, CRC Press, 2015.

Paredes-Lopez, O, and G I. Harry. "Changes in Selected Chemical and Antinutritional Components during Tempeh Preparation Using Fresh and Hardened Common Beans." *Journal of Food Science*, vol. 54, no. 5, 1989, pp. 968-70.

Prasad, Saroj Kumar, and Manoj Kumar Singh. "Horse gram- an underutilized nutraceutical pulse crop: a review." *Journal of Food Science Technology*, vol. 52, no. 5, 25 Mar. 2014, doi:10.1007/s13197-014-1312-z, https://www.ncbi.nlm.nih.gov/pmc/articles/PMC4397296/. Accessed 3 Feb. 2017.

Roos, Baukje. "Dietary oats and modulation of atherogenic pathways." *Molecular Nutrition & Food Research*, vol. 56, no. 7, 1 July 2012, pp. 1003-10014, doi:10.1002/mnfr.201100706. Accessed 20 Dec. 2017.

Saberi, Helen. *Cured, Fermented and Smoked Foods: Proceedings of the Oxford Symposium on Food and Cookery 2010*. Totnes, England, Prospect Books, 2011, pp. 84–87.

Sano, Yoshihito, Rie Kuramitsu, and Shunsuke Muramatsu. "Antibacterial activity of pyridine derivatives produced by *Aspergillus oryzae* in soy sauce production." *Journal of Biotechnology*, vol. 136, 2008.

Sarkar, Prabir K., and M J. Nout. *Handbook of Indigenous Foods Involving Alkaline Fermentation*. Boca Raton, CRC Press, 2015.

Sasaki, Takuji. "Rice genome analysis to understand the rice plant as an assembly of genetic codes." *Photosynthesis Research*, vol. 70, no. 1, 3 Oct. 2004, pp. 119–27.

Sena, Cristina M., and Raquel M. Seica. "Soybean: Friend or Foe." *Soybean and Nutrition*, 2011.

Schinkel, Christiane, and P Gepts. "Allozyme Variability in the Tepary Bean, *Phaseolus acutifolius* A. Gray." *Plant Breeding*, vol. 102, no. 3, 1989.

"Sea Island Red Peas." *Ark of Taste*, Slow Food USA, www.slowfoodusa.org/ark-item/sea-island-red-peas. Accessed 3 Sept. 2018.

Seo, Sang-Hyun, and Seong-Jun Cho. "Changes in allergenic and antinutritional protein profiles of soybean meal during solid-state fermentation with *Bacillus subtilis*." *LWT — Food Science and Technology*, vol. 70, 2016, pp. 208–12.

Shechter, Ari, and Gary J. Schwartz. "Gut-brain nutrient sensing in food reward." *Appetite*, vol. 122, 2018, pp. 32-35.

Shen, Jian-Dong, Qiu-Feng Cai, Guang-Ming Liu, Ling-Jing Zhang, and Min-Jie Cao. "A comparison study of the impact of boiling and high pressure steaming on the stability of soybean trypsin inhibitor." *International Journal of Food Science & Technology*, vol. 48, no. 9, 2013, pp. 1877-83.

Shumoy, Habtu, Sara Lauwens, Molly Gabaza, Julie Vandevelde, and Frank Vanhaecke. "Traditional fermentation of tef injera: Impact on in vitro iron and zinc dialysability." *Food Research International*, vol. 102, 2017, pp. 93-100.

Shurtleff, William, and Akiko Aoyagi. *The Book of Miso: Savory, High-Protein Seasoning*. 2nd ed., vol. I, Berkeley, Ten Speed Press, 2001.

———— *The Book of Tempeh: The Delicious, Cholesterol-Free Protein*. 2nd ed., Harper Colophon Books, 1985.

———— *History of Koji— Grains and/or Soybeans Enrobed with a Mold Culture (300 BCE to 2012): Extensively Annotated Bibliography and Sourcebook*. Lafayette, CA, SoyInfo Center, 2012.

———— *History of Natto and Its Relatives (1440–2012): Extensively Annotated Bibliography and Sourcebook*. Lafayette, CA, SoyInfo Ceter, 2012. Accessed 4 July 2017.

Solomon, Karen. *Asian Pickles*. Berkeley, Ten Speed Press, 2014.

Staff, Munchies. "This Is What Egoless Cuisine Tastes Like." *Munchies*, Vice, 18 July 2017, munchies.vice.com/en_us/article/bjjvnw/this-is-what-egoless-cuisine-tastes-like. Accessed 31 July 2018.

Starzynska-Janiszewska, Anna, Robert Dulinski, Bozena Stodolak, Barbara Mickowska, and Agnieszka Wikiera. "Prolonged tempe-type fermentation in order to improve bioactive potential and nutritional parameters of quinoa seeds." *Journal of Cereal Science*, vol. 71, 2016, pp. 116-21.

Steinkraus, Keith H. *Handbook of Indigenous Fermented Foods, Second Edition, Revised and Expanded*. 2nd ed., Boca Raton, CRC Press, 1995, pp. 2–5.

Sugiura, Keiji, and Mariko Sugiura. *Soybean and Nutrition*. 2011.

Sumi, Hiroyuki, et al. "A novel fibrinolytic enzyme (Nattokinase) in the vegetable cheese Natto: a typical and popular soybean food in the Japanese diet." *Experientia*, vol. 43, 1987, pp. 1110-11.

Székács, András, and Béla Darvas. "Forty Years with Glyphosate, Herbicides - Properties, Synthesis and Control of Weeds." *InTech*, doi:10.5772/32491, https://www.intechopen.com/books/herbicides-properties-synthesis-and-control-of-weeds/forty-years-with-glyphosate. Accessed 16 Aug. 2017.

Takeuchi, Yoshiko. *Japanese Superfoods*. London, New Holland Publishers, 2016.

Tinay, A. H., A. M. El, Abdel GaDIR, and M EL Hiday. "Sorghum fermented kisra bread. I-nutritive value if kisra." *Journal of the Science of Food and Agriculture*, vol. 30, no. 9, 1 Sept. 1979, pp. 859-63.

Toriyama, K, K L. Heong, and B Hardy. *Rice is Life: Scientific Perspectives for the 21st Century*. Manila, International Rice Research Institute (IRRI), 2005, pp. 294-315.

Tsumura, Yuke, Aki Ohyane, Yamashita Kuniko, and Yoshiaki Sone. "Which characteristic of Natto: appearance, odor, or taste most affects preference for Natto." *Journal of Physiological Anthropology*, vol. 31, no. 1, p. 13.

Turner, Nancy D., and Joanne R. Lupton. "Dietary Fiber." *Advances in Nutrition*, vol. 2, no. 2, 2011, pp. 151–52.

Urbano, G., M. Lopez-Jurado, P. Aranda, C. Vidal-Valverde, and E. Tenorio. "The role of phytic acid in legumes: antinutrient or beneficial function?" *Journal of Physiology and Biochemistry*, vol. 56, no. 3, 2010, pp. 283–94.

Wacher, Carmen, et al. "Microbiology of Indian and Mestizo pozol fermentations." *Food Microbiology*, vol. 17, no. 3, 2000, pp. 251-56.

Wakin, Daniel J. *The Man with the Sawed-Off Leg and Other Tales of a New York City Block*. Arcade Publishing, 2018. Accessed 25 Feb. 2018.

Wood, B J. *The Microbiology of Fermented Foods*. Springer Science & Business Media, 2012, pp. 658-78.

Yanagisawa, Yashhide, and Hiroyuki Sumi. "Natto Bacillus contains a large amount of water-soluble vitamin K (Melaquinone-7)." *Journal of Food Biochemistry*, vol. 29, no. 3, 2005, pp. 267-77.

Yang, Rachel, and Jess Thomson. *My Rice Bowl: Korean Cooking Outside the Lines*. Seattle, Sasquatch Books, 2017.

Zhao, Xin, et al. "Comparisons of Shuidouchi, Natto, and Cheonggukjang in their physicochemical properties, and antimutagenic and anticancer effects." *Food Science and Biotechnology*, vol. 22, no. 4, Aug. 2013, pp. 1077-84.

Zhong, Ling, Megan Blaxland, and Ting Zuo. "Without Rice, Even the Cleverest Housewife Cannot Cook: Sustainable Livelihoods Research in a Poor Chinese Village." *Asian Social Work and Policy Review*, vol. 9, no. 1, Feb. 2015, pp. 3-17, doi:10.1111/aswp.12046. Accessed 7 Jan. 2018.

METRIC CONVERSIONS

Ounces to grams	multiply ounces by 28.35
Grams to ounces	multiply grams by 0.035
Pounds to grams	multiply pounds by 453.5
Pounds to kilograms	multiply pounds by 0.45
Cups to liters	multiply cups by 0.24
Fahrenheit to Celsius	subtract 32 from Fahrenheit temperature, multiply by 5, then divide by 9
Celsius to Fahrenheit	multiply Celsius temperature by 9, divide by 5, then add 32

SOURCE GUIDE

Beans and Grains

Anson Mills
http://ansonmills.com
They offer U.S.-grown heirloom rice, hominy corn (and lime), as well as peas.

Laura Soybeans
www.laurasoybeans.com
They sell GMO-free soybeans from family farms, including natto beans.

Organic Grocery USA
https://organicgroceryusa.com
This is an online market based in South India. The company is Beyonce Infotech. They ship organic Indian dals and other beans like horse gram straight from India.

Patel Brothers
http://patelbrothersusa.com/store
An Indian grocery store chain since 1974, there are 50 brick-and-mortar locations as well as an online market where you can source many of the dals and beans from this book.

Ramona Farms
www.ramonafarms.com
Located along the Gila River in Arizona, they grow tepary beans and other heirloom grains and beans.

Rancho Gordo
www.ranchogordo.com
A great source for unusual and rare beans and a few grains.

Signature Soy
http://signaturesoy.com
Started by sisters, this company offers patented, non-GMO soybeans from family farms, including natto.

True Leaf Market
www.trueleafmarket.com
They sell organic black soy beans as sprouting seeds, but they work well for tempeh and natto.

Zürsun Idaho Heirloom Beans
www.zursunbeans.com
They offer an amazing selection of beans and grains grown on small farms in the United States.

Koji

Aedan Fermented Foods
https://aedansf.com/
This website by the San Francisco artisan miso maker is lovely. Not only can you order products such as fresh rice koji and many products from amazake to miso, but you will also find inspiring recipes to use these ingredients.

Cultures for Health
www.culturesforhealth.com/
Their selection of bean starter cultures includes tempeh and natto, and they sell inoculated barley koji and organic brown rice koji.

GEM Cultures
www.gemcultures.com
Carries the biggest variety of koji spores; they also carry inoculated rice and barley. You will find natto spores as well as other milk and grain cultures, including sourdough.

Higucki Matsunosuke Shoten Co. Ltd
www.higuchi-m.co.jp/english
Seventh-generation koji company in Japan.

Natural Import Company
www.naturalimport.com
Carries many Japanese imports including organic brown rice koji and mitoku barley koji. Also carries Nattomoto natto spores.

Rhapsody Natural Foods
https://rhapsodynaturalfoods.com
This Vermont company carries the most varieties of inoculated koji grains that we have seen. They have a red barley miso koji, amazake koji, long-term rice miso koji, and a short-term rice miso. They also have a selection of tempeh, natto, amazake, and rice bran for making a nukadoko pot.

South River Miso Company
www.southrivermiso.com
Carries a wonderful selection of miso and is a good resource for U.S.-made organic brown rice koji.

Natto

For spores, see Cultures for Health, GEM Cultures, the Tempeh Starter Store, Organic-cultures.com, and Natural Import Company. For fresh natto, see the companies below as well as Rhapsody Natural Foods.

Megumi Natto

www.meguminatto.com
Sells both large- and small-bean fresh natto.

NYrture New York Natto

www.nyrture.com
Carries a variety of small-bean fresh natto, including turmeric and black soybean.

Wanpaku Natto

www.wanpakunatto.com
Sells fresh natto.

Tempeh

Cultures for Health

www.culturesforhealth.com
They offer two types of tempeh starter: *Rhizopus oryzae* for soy tempeh and *R. oligosporus* for non-soy tempeh. They also sell inoculated koji rice and koji barley as well as natto starter.

Organic-Cultures.com

http://store.organic-cultures.com/info.html
They offer a huge selection of cultures — everything in this book and more, including a nuka pot starter. They sell the same black oncom starter as The Tempeh Starter Store (*Rhizapus Oligasporus*), though the photos on both sites show red oncom (*Neurospora intermedia* var. *oncomensis*) — a culture we have not been able to find stateside — not black oncom.

Smiling Hara

www.smilingharatempeh.com
They sell tempeh made from hemp seeds, available in soy-free varieties.

Squirrel and Crow

https://eattempeh.com
They make exclusively soy-free tempeh, including tempeh made from lentils, chickpeas, and peanuts.

The Tempeh Starter Store

www.tempehstarter.com
This site sends starter from Indonesia. We used their tempeh starter (*Rhizopus oligosporus*), tape, oncom, and other Indonesian ferments. They also sell natto starter.

TempehSure

https://tempehsure.com
They sell tempeh culture (*Rhizopus Oligosporus*) in single-use packs as well as a kilo-size bulk starter. They also are the group working on bringing Gunter Pfaff's dry tempeh incubation solution to the public.

Top Cultures

www.topcultures.com
This Belgium company sells *Rhizopus oryzae* spores in three different sizes.

More Information

Betsy's Tempeh

www.makethebesttempeh.org
This website provides recipe and instructions for how to make tempeh in a tempeh incubator designed by Gunter Pfaff.

Contraband Ferments

www.contrabandferments.com
This organization is based in Brooklyn, New York, and was founded by Cheryl Paswater. It offers a fermentation CSA, workshops, and private coaching.

Cultures Group

https://culturesgroup.net
This blog, founded by Chef Ken Fornataro, explores the science of fermentation and the preparation of fermented foods.

Fermentation on Wheels

www.fermentationonwheels.com
A traveling educational hub, founded by Tara Whitsitt, with a mission to bridge communities, inspire sustainability, and teach fermentation.

Our Cook Quest

http://ourcookquest.com
Started by Rich Shih, this website provides an exchange of ideas for cooks of all levels. Workshops and private lessons are also offered.

ACKNOWLEDGMENTS

First we would like to thank the microbes — *Aspergillus, B. subtillis, Rhizopus, L.bacillus,* and all the other varieties — for showing up, teaching, and humbling us as we wrote. This book would not exist without them or the fermentation community that supported this work. We are honored by all the friends we've made along the way.

We are indebted to Deborah Balmuth, who had the idea for this book and chose us to do the work. This project stretched us as eaters, fermenters, and humans on this planet, in good albeit sometimes uncomfortable ways.

Jeremy Umansky, thank you and your family not only for the work you do, but for welcoming Kirsten into your home, sharing your knowledge, and becoming our *Aspergillus* mentor. Your support of our koji journey has been profound. Rich Shih, thank you for reaching out and then joining this project — #kojibuildscommunity, and we are happy you are part of ours. Ken Fornataro, you are a wealth of knowledge and experience; we thank you for the hours you spent answering random questions.

We are grateful for the rest of our miso support team that started with Sandor Katz, who connected us with the right people, first with Betty Stechmeyer. Thank you, Betty, for all the enjoyable conversations and your longtime knowledge of the microbes. Thank you, Christian and Gaella Elwell. Gaella, the delightful afternoon we spent together in your home was one of the highlights of writing this book. We thank Cheryl Paswater for her enthusiasm of beans and miso. We thank Bill Shurtleff for a lovely afternoon of good conversation, good food, and a better understanding of the lineage of fermentation in the United States.

Natto! Ann Yonetani, thank you for helping us gain a deeper understanding of *Bacillus subtilis,* reading our words, and giving us support. Heidi Nestler, you taught us how to make our first batch and, more importantly, to love the flavor and the texture. Claudia Lucero, you tasted our many early "hide-the-natto" experiments and let us use your kitchen to do so.

Team tempeh! Thank you to Sarah Yancy and Chad Oliphant for everything you gave freely — time, support, and knowledge. We are indebted to Betsy Shipley for sharing her and Gunter Pfaff's adventures and allowing us to share a tiny bit of his work with the world. Tara Whitsitt, thank you for the fire you kindled with that first fresh tempeh we tasted in Eugene and for creating a recipe to contribute to this book. Jon Westdahl and Julia Bisnett, we thank you for your knowledge and the beautiful tempeh on the cover.

Other people who talked to us or cheered us on: This means you, Vicki Hames, despite the fact you weren't so sure . . . Steve Sando, thank you for sharing your story, a lot of laughs, and a mountain of beans. Thank you to Josh Fratoni, the tempeh makers in Indonesia, Karen Diggs, Eric Edgin, and Jerry Trottmann, and a special shout-out and thank-you to Colin Dyck @Mudslide_Stoneware whose love of beauty, function, and fermentation makes him a joy to talk to when our paths cross. He gifted us the miso crock pictured in the step-by-step photos.

Thank you to our amazing editor, Sarah Guare, who made ferments alongside the manuscript. You had so much patience as this book kept expanding with every round and again when we (grumbling) had to pare it down to fit within the pages. Dina Avila and Arthur Hitchcock are the phototography dream team — need we say more? To everyone at Storey (those we know had a hand in the book — Deanna, Jennifer, Michaela, Emily, and Sarah A. — and those we don't), it has been a joy to work with each of you. We cannot thank you enough for always making our books better. It takes a village . . .

And finally, to our children and their much-loved partners. It hasn't always been easy to come home to ferment experiments and our enthusiasm for them, but you clearly love us unconditionally. What else could parents ask for.

INDEX

Page numbers in *italics* indicate photos; numbers in **bold** indicate charts.

INDEX